Essential Semiconductor Physics

New Era Electronics: A Lecture Notes Series

Series Editors: Vijay Raghunathan *(Purdue University, USA)*
Muhammad Ashraf Alam *(Purdue University, USA)*
Mark S Lundstrom *(Purdue University, USA)*

Published

Vol. 4 *Essential Semiconductor Physics*
by Mark S Lundstrom

Vol. 3 *p-Bits and q-Bits:*
Probabilistic and Quantum Computing
by Supriyo Datta

Vol. 2 *Quantum Communication:*
The Physical Layer of Future Optical Networks
by Mahdi Hosseini

Vol. 1 *Transistors!*
by Mark S Lundstrom

New Era Electronics: a Lecture Notes Series ▪ Volume 4

Essential Semiconductor Physics

Mark Lundstrom
Purdue University, USA

World Scientific

NEW JERSEY • LONDON • SINGAPORE • BEIJING • SHANGHAI • HONG KONG • TAIPEI • CHENNAI • TOKYO

Published by

World Scientific Publishing Co. Pte. Ltd.
5 Toh Tuck Link, Singapore 596224
USA office: 27 Warren Street, Suite 401-402, Hackensack, NJ 07601
UK office: 57 Shelton Street, Covent Garden, London WC2H 9HE

Library of Congress Control Number: 2025031924

British Library Cataloguing-in-Publication Data
A catalogue record for this book is available from the British Library.

New Era Electronics: A Lecture Notes Series — Vol. 4
ESSENTIAL SEMICONDUCTOR PHYSICS

ISBN 978-981-98-1836-5 (hardcover)
ISBN 978-981-98-1837-2 (ebook for institutions)
ISBN 978-981-98-1838-9 (ebook for individuals)

For any available supplementary material, please visit
https://www.worldscientific.com/worldscibooks/10.1142/14454#t=suppl

Desk Editors: Soundararajan Raghuraman/Joseph Ang

Typeset by Stallion Press
Email: enquiries@stallionpress.com

Preface

Since the middle of the 20th century, transistors, integrated circuit chips, and other semiconductor devices have transformed our world. Today, semiconductor technology is a pervasive, foundational technology for modern society. It is an example of a science-driven technology, one that was enabled by development of quantum mechanics, solid state physics and chemistry, and materials science in the first half of the 20th century. To understand this technology, a basic understanding of the essentials of semiconductor physics is needed. That is the subject of these lecture notes.

The scope of semiconductor physics is extensive. One could spend a lifetime studying it, and many of us have done just that, but the goal here is a broad, descriptive introduction to the topic. This "book" is a set of lecture slides and notes on the essential semiconductor physics that everyone working in the field should understand. The lectures are not about semiconductor devices; instead, they are about the underlying physics that makes semiconductor devices possible. The lectures are designed to be broadly accessible to students in science or engineering and to working engineers. The treatment is physical and intuitive – descriptive, not heavily mathematical.

These lectures present an electrical engineering perspective, but those in other fields may find it a useful introduction to the approach that has guided the development of semiconductor technology for more than 75 years. Those who use semiconductor devices will gain an understanding of the physics that underlies their operation. Technology developers should find these lectures a useful starting point for diving deeper into the physics, chemistry, and materials science relevant to semiconductors. Those who have taken advanced courses should see how specific topics fit into a broader framework.

A comment about the format of this "book" may be helpful to the reader. It is a collection of lectures, slides with notes, from a five-week course on semiconductor fundamentals. Each lecture begins with a quick review of the key points of the previous lecture. Reading these lecture notes should feel more like taking a class than reading a traditional textbook. The lectures are grouped into five parts.

Part 1: Materials properties and doping
Part 2: Rudiments of quantum mechanics
Part 3: Equilibrium carrier concentrations
Part 4: Carrier transport, generation, and recombination
Part 5: The semiconductor equations

A great many books have been written about each of these topics, and readers are encouraged to dive deeper into specific topics as needed. References to get you started are included, but you will find that a basic understanding of the essentials goes a long way. I hope that you find these lectures an effective introduction to an interesting and important topic.

Mark Lundstrom
Purdue University
April 2025

Acknowledgments

For more than 50 years now, I've been been "along for the ride" as semiconductor technology advanced from small scale integration to the powerful and pervasive technology that we have today. For more than 45 years, I have also had the privilege of teaching many students, some on campus and many more online, about semiconductor science and technology. I've been fortunate to do this work with some wonderful colleagues, one of whom was the late Professor Robert F. Pierret. Bob was a role model from whom I learned much about how to teach and write effectively. I learned a good deal of semiconductor physics from Bob too. His books are recommended references for those who wish to dive deeper into the topics discuss in these lecture notes.

These lecture notes evolved over a period of many years as I taught graduate and undergraduate students and working engineers. I owe much to those many students who helped me shape, refine, and correct the evolving drafts of these lecture notes. I hope that those reading these lectures will find them an effective introduction to a fascinating field. In particular, I thank my colleague, Professor M. Ashraf Alam and his students, Jabir Bin Jahangir, Marco Fratus, Chun-An Shih, Mir Md Fahimul Islam, In Huh, and Yi-Chieh Chen for their careful proofreading. I'm sure that some errors remain, but far fewer than there would have been without their help. Special thanks to Jabir Bin Jahangir for producing the plots in Secs. 18.4 and 18.6. Thanks also to Joey Lopez, not only for proofreading the text, but also for suggestions that improved the clarity and flow. I won't say, as one of my colleagues did, that the errors that remain "were put there to see if the reader is paying attention", but hopefully, there aren't too many.

I was able to finally finish this project because of some additional support provided by my university through Vice President for Special Projects

Cristina Farmus and her team. That extra help pushed this project over the finish line.

Special thanks also to Zvi Ruder, my editor at World Scientific Publishing Company, for his support and encouragement. I've had the pleasure of working with Zvi on two lecture notes series and on four World Scientific books of my own. In this case, I was fortunate that Zvi was willing to do an experiment with a format that is somewhere between that of a traditional book and a set of lecture slides. If this experiment is successful for authors and for readers, then there may be more volumes like this one to come in our lecture notes series.

Research universities with their interwoven missions of discovery, learning, and engagement are very special places. Land grant research universities, with their charge to prepare as many students as possible with a high-quality, affordable education are, in my view, even more special. I've been very fortunate to have spent the better part of my career at a great land grant university, Purdue University, and I'm deeply grateful that I had that opportunity.

Mark Lundstrom
Purdue University
April 2025

Contents

Preface v

Acknowledgments vii

Part 1: Materials Properties and Doping **1**

1. Energy Levels to Energy Bands 3

1.1 Introduction 3
1.2 From energy levels to energy bands 5
1.3 Band gap and intrinsic carrier concentration 10
1.4 Energy band diagrams 12
1.5 Hot electrons and hot holes 14
1.6 Insulators, metals, and semiconductors 15
1.7 Summary 16
1.8 References 16

2. Crystalline, Polycrystalline, and Amorphous Semiconductors 17

2.1 Introduction 17
2.2 Crystal lattices 18
2.3 Silicon crystal structure 21
2.4 Polycrystalline materials 25
2.5 Amorphous materials 26
2.6 Summary 27
2.7 Reference 27

3. Miller Indices 29

3.1 Introduction 29
3.2 Describing planes with Miller indices 30
3.3 Describing directions with Miller indices 34
3.4 Direction normal to a plane 35
3.5 Angle between two planes 37
3.6 Summary 39
3.7 Reference 39

4. Properties of Common Semiconductors 41

4.1 Introduction 41
4.2 Elemental and compound semiconductors 42
4.3 Band gap and intrinsic carrier concentration 43
4.4 Silicon and silicon dioxide 44
4.5 Carrier recombination 45
4.6 Summary 46
4.7 Reference 46

5. Free Carriers in Semiconductors 47

5.1 Introduction 47
5.2 Effective mass 48
5.3 Energy vs. momentum 50
5.4 Direct and indirect gap semiconductors 53
5.5 Summary 56
5.6 References 56

6. Doping 57

6.1 Introduction 57
6.2 Substitutional doping 58
6.3 N-type doping 60
6.4 Binding energy of a dopant 62
6.5 P-type doping 65
6.6 Deep dopants 68
6.7 Zinc Blende crystals and amphoteric dopants 69
6.8 Carrier concentration vs. temperature 71
6.9 Summary 74
6.10 Reference 75

Part 2: Rudiments of Quantum Mechanics **77**

7. The Wave Equation 79

7.1 Introduction 79
7.2 The need for quantum mechanics 80
7.3 The wave equation 86
7.4 Particles and wave packets 96
7.5 Summary 99
7.6 Reference 99

8. Quantum Confinement 101

8.1 Introduction 101
8.2 Quantum mechanical particle in a box 102
8.3 Quantum confinement in semiconductors 109
8.4 2D electrons and subbands 113
8.5 Summary 116
8.6 References 116

9. Quantum Tunneling and Reflection 117

9.1 Introduction 117
9.2 Quantum tunneling and reflection 118
9.3 Quantum mechanical tunneling 119
9.4 Quantum mechanical reflection 128
9.5 Summary 129
9.6 Reference 129

10. Electron Waves in Crystals 131

10.1 Introduction 131
10.2 Free electrons in the crystal potential 133
10.3 Energy vs. momentum 135
10.4 Bandstructure 137
10.5 Constant energy surfaces 140
10.6 Model bandstructures 142
10.7 Summary 145
10.8 References 145

11. Density of States 147

11.1 Introduction 147

11.2 Density of states in k-space 148
11.3 Density of states vs. energy 152
11.4 Density of states discussion 160
11.5 Summary 163
11.6 References 164

Part 3: Equilibrium Carrier Concentrations 165

12. The Fermi Function 167

12.1 Introduction 167
12.2 The Fermi function 168
12.3 The effect of temperature 171
12.4 Nondegenerate semiconductors 172
12.5 Fermi levels and energy band diagrams 175
12.6 Summary 177
12.7 References 177

13. Fermi-Dirac Integrals 179

13.1 Introduction 179
13.2 Carrier concentration calculation in k-space 180
13.3 Carrier concentration calculation in energy space 182
13.4 The Fermi-Dirac integral 186
13.5 Electron and hole concentrations 188
13.6 Summary 190
13.7 References 190

14. Carrier Concentration vs. Fermi Level 191

14.1 Introduction 191
14.2 Fermi levels and carrier concentrations 192
14.3 The equilibrium np product 194
14.4 Summary 203
14.5 Reference 203

15. Carrier Concentration vs. Doping Density 205

15.1 Introduction 205
15.2 Space charge neutrality 206
15.3 General expressions for carrier concentrations 208
15.4 The extrinsic region 212

15.5 Summary 213
15.6 Reference 213

16. Carrier Concentration vs. Temperature 215

16.1 Introduction 215
16.2 The intrinsic region 217
16.3 The freeze-out region 220
16.4 The metal-insulator transistion 226
16.5 The Fermi level vs. temperature 227
16.6 Summary 228
16.7 Reference 228

Part 4: Carrier Transport, Recombination, and Generation **229**

17. Current Equation 231

17.1 Introduction 231
17.2 A hydrodynamic analogy 232
17.3 Current in a nano device 234
17.4 Transmission 237
17.5 Channels 240
17.6 Fermi window 241
17.7 Small bias, $T > 0$ K 243
17.8 Current in a bulk semiconductor 246
17.9 Quasi-Fermi levels 249
17.10 Summary 250
17.11 Reference 250

18. Drift Current 251

18.1 Introduction 251
18.2 Near-equilibrium, diffusive transport 252
18.3 Drift current and drift velocities 254
18.4 Mobility 259
18.5 High-field transport 263
18.6 Current, conductivity, and resistivity 265
18.7 Relation to the general current equation 268
18.8 Summary 270
18.9 Reference 270

19. Diffusion Current 271

19.1 Introduction . . . 271
19.2 Diffusion fluxes and diffusion currents . . . 272
19.3 The physics of diffusion . . . 274
19.4 Relation to the general current equation . . . 278
19.5 Summary . . . 280
19.6 Reference . . . 280

20. Drift-Diffusion Equation 281

20.1 Introduction . . . 281
20.2 Review of the general current equation . . . 282
20.3 The drift-diffusion equation . . . 283
20.4 Built-in electric fields . . . 286
20.5 Summary . . . 287
20.6 Reference . . . 288

21. Carrier Recombination 289

21.1 Introduction . . . 289
21.2 Band-to-band radiative recombination . . . 291
21.3 Auger recombination . . . 295
21.4 Defect-assisted (SRH) recombination . . . 297
21.5 Multiple recombination processes . . . 300
21.6 Band structure and recombination . . . 302
21.7 Recombination and generation . . . 304
21.8 Summary . . . 307
21.9 References . . . 308

22. Carrier Generation 309

22.1 Introduction . . . 309
22.2 Optical absorption . . . 310
22.3 Field and impact ionization . . . 318
22.4 Summary . . . 322
22.5 Reference . . . 323

Part 5: The Semiconductor Equations 325

23. The Semiconductor Equations 327

23.1 Introduction . . . 327

23.2 The unknowns in a semiconductor problem 328
23.3 The continuity equation 331
23.4 Gauss's law 334
23.5 The semiconductor equations 336
23.6 The semiconductor equations in equilibrium 338
23.7 Summary 341
23.8 References 342

24. Energy Band Diagrams 343

24.1 Introduction 343
24.2 The Fermi level is constant in equilibrium 344
24.3 Band bending 345
24.4 Energy band diagrams 349
24.5 Energy band diagram of an NP junction 351
24.6 A graded doping example 356
24.7 Non-equilibrium energy band diagrams 358
24.8 Summary 360
24.9 References 361

25. Quasi-Fermi Levels 363

25.1 Introduction 363
25.2 Equilibrium vs. non-equilibrium 364
25.3 Quasi-Fermi levels 367
25.4 Quasi-Fermi level discussion 371
25.5 Summary 374
25.6 References 374

26. Minority Carrier Diffusion Equation 375

26.1 Introduction 375
26.2 Minority carrier diffusion equation (MCDE) 376
26.3 Ex. 1: Steady-state, uniform illumination 380
26.4 Ex. 2: Transient decay to equilibrium 382
26.5 Ex. 3: Steady-state diffusion in a long semiconductor 385
26.6 Ex. 4: Steady-state diffusion in a short semiconductor 388
26.7 Ex. 5: Steady-state diffusion when $L \approx L_n$ 390
26.8 Ex. 6: Steady-state diffusion with surface recombination 392

26.9 Summary 395
26.10 References 396

Epilogue 397

Index 401

PART 1
Materials Properties and Doping

Lecture 1

Energy Levels to Energy Bands

1.1 Introduction

The goal of these lectures is to provide the reader with the basic knowledge of semiconductor physics sufficient to understand the operation of almost any type of semiconductor device. In Part 1, we will rely on some knowledge from undergraduate chemistry and physics courses and show how those concepts are extended and applied to semiconducting materials. This part is a 50,000 foot view of the "forest"; we won't be examining the "trees" in detail. Those who wish to dive deeper should refer to the references. Most of the topics introduced in Part 1 will be revisited later in the course (e.g. band structure, energy band diagrams, effective mass, etc.) Other topics just introduce some terminology that you should be aware of (e.g. crystalline, polycrystalline, and amorphous semiconductors, Miller indices, etc.) Energy bands, energy band diagrams, band structure, and doping are some of the most important concepts discussed in Part 1. Finally, **boldface text** indicates that the term is important and should be remembered.

Upon completion of Part 1, you will be able to:

- Explain what is meant by "conduction band," "valence band," "forbidden gap" (or band gap), "electron," "hole," and "intrinsic carrier concentration."
- Explain the difference between a metal, an insulator, and a semiconductor in terms of their energy bands.

- Explain the difference between crystalline, polycrystalline, and amorphous semiconductors and be familiar with the crystal structure of Si and GaAs.
- Given the Miller indices, draw the corresponding plane or vector.
- Be able to draw the energy vs. momentum (or energy vs. wave vector, k) diagram for direct and indirect semiconductors.
- Be able to explain how N- and P-type doping is accomplished and the carrier concentration vs. temperature characteristics of semiconductors.

Before we continue, a brief word about units, which sometimes causes confusion. When we say the thermal energy is $3k_BT/2$, we often quote the numerical value in electron volts (eV), but remember that electron volt is not a legitimate system international (SI) unit. It's a convenient unit to express energies in, but it's not a proper unit for doing calculations.

When computing the average energy, we should take 3/2 times Boltzmann's constant, 1.38×10^{-23} J/K times the temperature, T, in Kelvin, 300 K. The answer is in Joules. The average thermal energy is 6.21×10^{-21} Joules. That's not a convenient number to remember. It is more convenient to express these energies in electron volts. To convert an energy in Joules to electron volts, divide the energy in Joules by the magnitude of the charge on an electron, $q = 1.6 \times 10^{-19}$ C, and get 0.039 eV. To use this energy in a formula with SI units, we need to convert back into Joules and insert that into the formula. An electron volt is not a proper SI unit, so be careful about that when you are working out semiconductor problems.

The remainder of Lecture 1 consists of a set of slides with brief notes accompanying each slide.

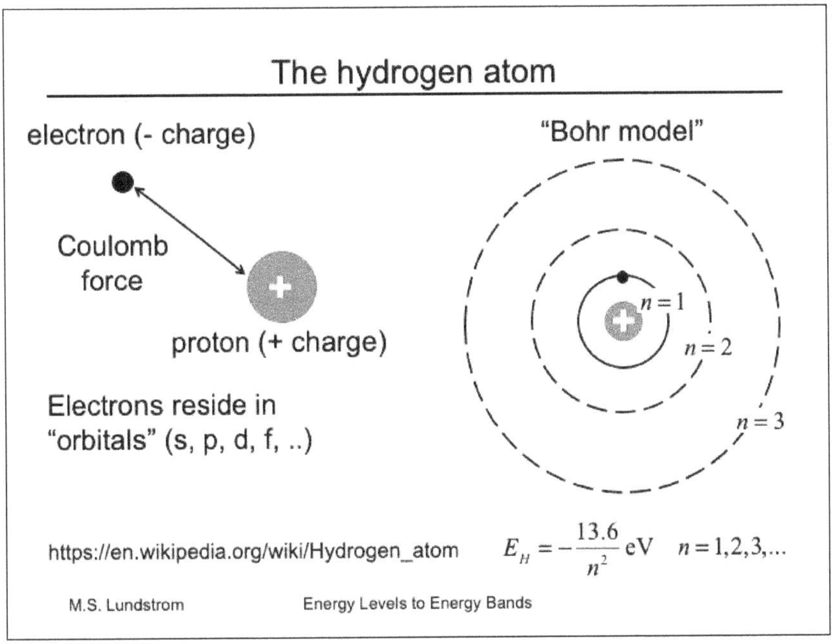

1.2 From energy levels to energy bands

Let's begin with the simplest atom — the hydrogen atom. This is probably where you started in first year chemistry. As shown in the slide above on the right, the hydrogen atom has a positively charged proton and negatively charged electron. The two are held together by the Coulomb force — opposite charges attract. You'll recall that the electrons reside in **states**.

As shown above on the right, the states are labeled with a **principal quantum number**, n. The lowest energy state is $n = 1$; $n = 2$ is a higher energy, and the energy increases as the principal quantum number increases.

Associated with each of these principal quantum numbers is a set of **orbitals**, the s, p, d, and f orbitals. You may recall that these orbitals have interesting shapes. The circular orbits shown here are from the first model of the hydrogen atom, the **Bohr model**, which was developed before orbitals were understood.

You may also remember the formula shown above, the energy of the electron is minus 13.6 electron volts divided by the principal quantum number squared. These are some basic fundamentals about hydrogen atoms.

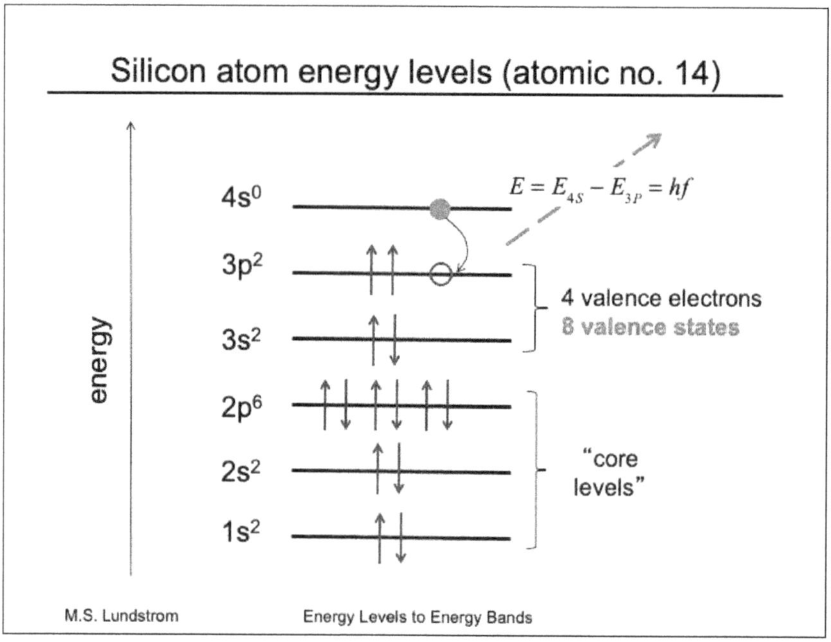

In the early 20th century, scientists probed atoms by exciting electrons to higher energy orbitals and letting them drop down and release energy in the form of light. Planck's constant, h, times the frequency, f, gave the energy of the emitted photon. Understanding the discrete colors that were emitted led to the development of quantum mechanics.

Silicon has 14 electrons that reside in a set of states. Associated with each principal quantum number, n, is a set of orbitals. The $n = 1$ s orbital holds two electrons — one with spin up, one with spin down. Two more electrons occupy the 2s orbital. There are p orbitals associated with the x, y, and z directions; each one can hold two electrons, so six electrons occupy the 2p orbitals. The $n = 3$ state holds two electrons in the 3s orbital. There are six states available in the 3p orbitals, so there are two electrons in the 3p orbitals, and four 3p states are empty.

Electrons in the lowest energy states are bound very tightly to the core and do not interact much with the outside world. The electrons that we're interested in are in the highest energy states. These **valence** electrons are responsible for chemistry and chemical bonds. There are four of them for each Si atom, but you will recall that there are also four empty 3p orbitals. So we have four valence electrons and four empty valence states.

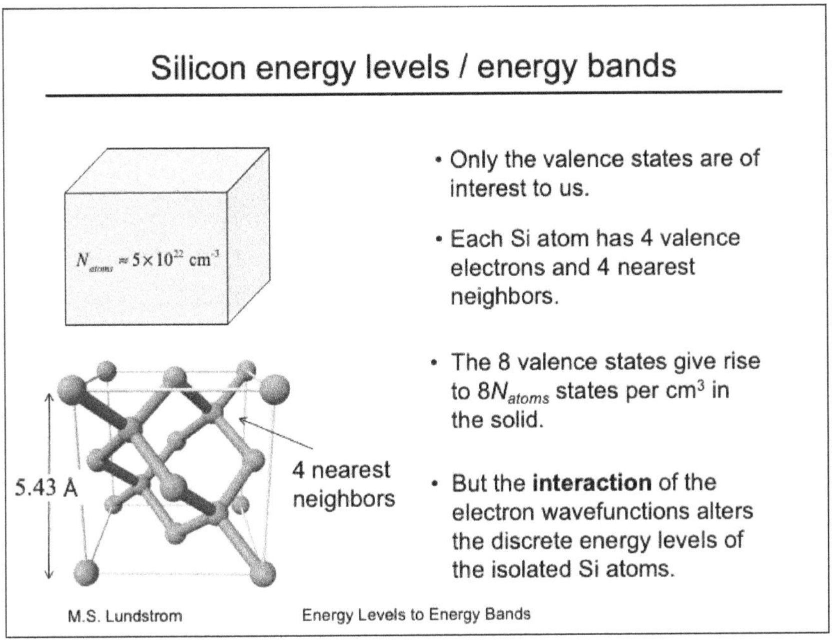

What we are trying to understand is not an isolated silicon atom, but a silicon crystal in which there is a large number of atoms per cubic centimeter — about 5×10^{22}. These atoms are arranged in a crystal lattice that has a bit of a complicated structure as shown above. This particular arrangement of atoms is known as the **diamond lattice**. Common semiconductors like silicon and germanium crystalize in a diamond lattice.

As you can see, there is a silicon atom at each corner of every cube. There are also atoms in the center of each of the 6 faces of the cube, and then there are four more inside — along body diagonals. If you look closely at the figure above, you can see that each silicon atom is surrounded by four nearest neighbors. That is important. The eight valence states for each atom give rise to a large number of valence states in the crystal, but only half of them are filled. Sharing these electrons with four nearest neighbors effectively fills all 8 valence states.

Something very important happens when the silicon atoms are put in a lattice a short distance away from each other, the orbitals of adjacent atoms interact, and the discrete energy levels of the isolated silicon atoms change. In fact, we no longer have discrete energy levels.

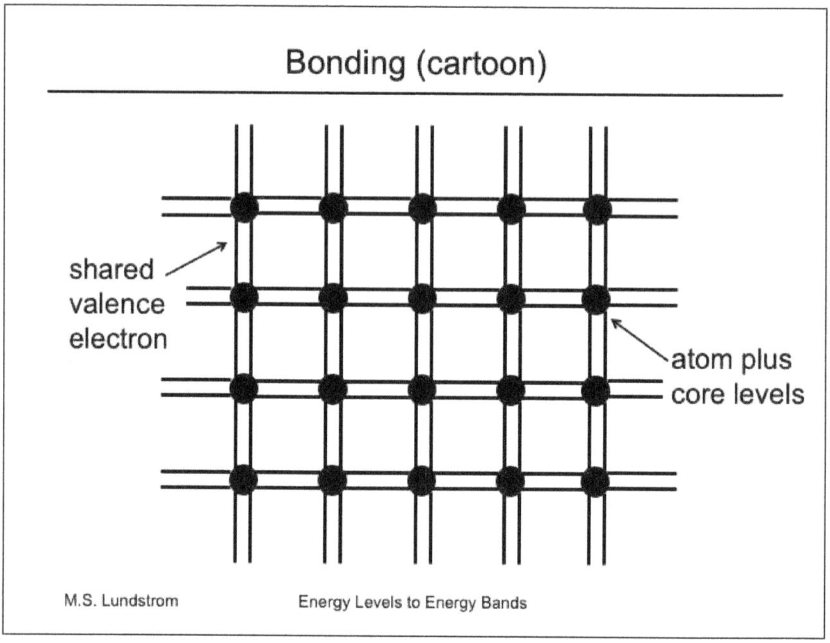

Before we continue, let me present a much simpler picture of the silicon lattice that we will use frequently in these lectures. Rather than drawing the complicated 3D structure, we will draw the little cartoon shown above. The important point about the silicon or other common semiconductor lattices is that each atom (represented by the black dots here) is surrounded by four nearest neighbors. The lines in this figure represent the valence electrons. There are four associated with each atom. Every atom shares its valence electrons with its four nearest neighbors, then we effectively have the eight states per atom filled.

That's the basis of a **covalent chemical bond**. It's what holds the silicon crystal together.

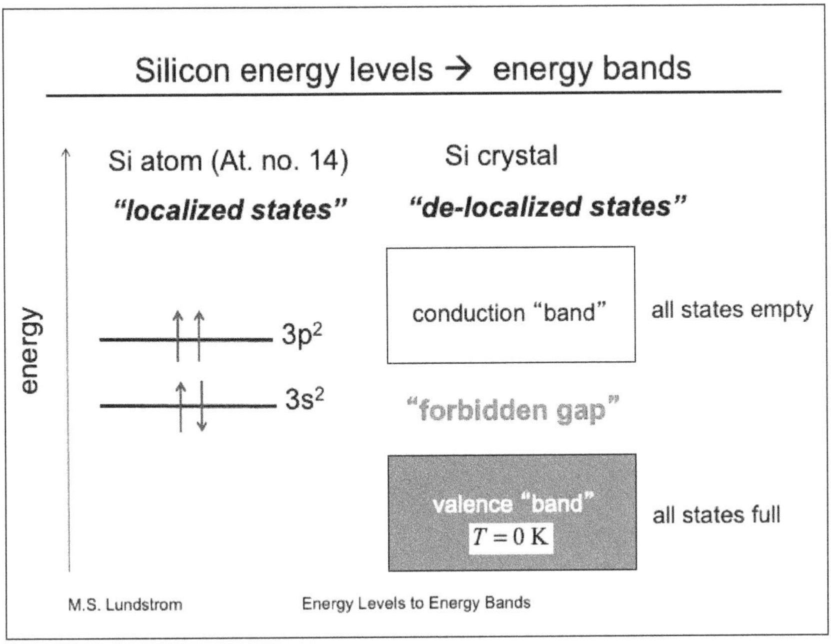

There are 5×10^{22} atoms per cubic centimeter in silicon, so there are 5×10^{22} times 8 valence states, but the states change because of the interaction between orbitals when they overlap in a covalent bond. Instead of a large number of discrete energy levels, the states spread out in energy. Some of the states move lower in energy. These are the bonding states, and they are filled with electrons, the **valence electrons** that form the covalent bonds and hold the atoms together.

There are also an equal number of empty states (the anti-bonding states). These empty states get pushed up in energy, and they are available to hold electrons, but there are no electrons to hold. There is a gap between the states that are filled and the states that are empty. Inside of that gap there are no states for electrons. This very important feature of semiconductors is called the **forbidden gap**. The situation we have described is for $T = 0$ K. All of the valence band states are filled, all of the conduction band states are empty, and there are no states between them. In an isolated silicon atom, the states are **localized**; the electrons sit near specific silicon atoms. In a silicon crystal lattice, the states become **de-localized**, and they spread throughout the entire crystal, which means that electrons can move throughout the entire crystal.

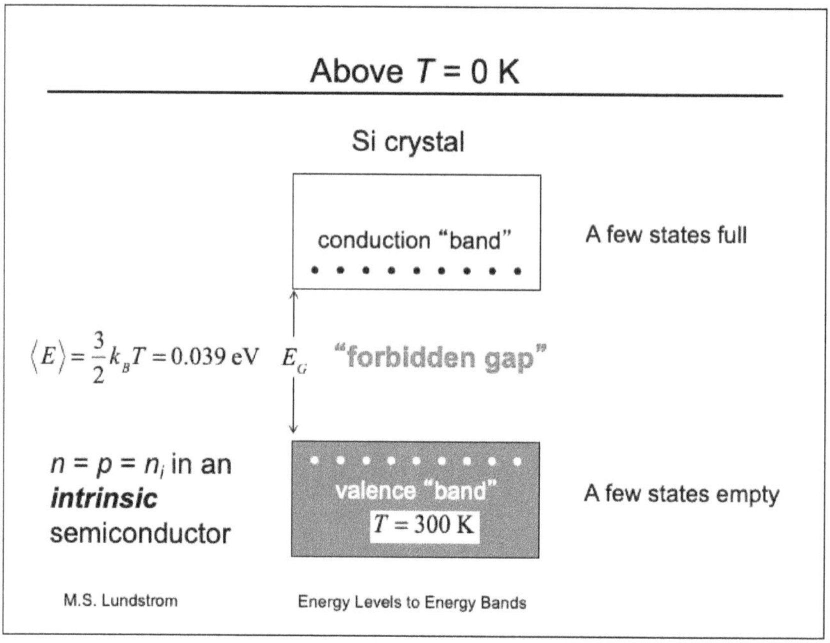

1.3 Band gap and intrinsic carrier concentration

What happens if we warm the silicon lattice up? At 300 K, there is thermal energy, which is proportional to Boltzmann's constant times temperature. The silicon atoms jiggle around in random thermal motion with an average kinetic energy of $3k_BT/2$.

Since there is **thermal energy**, there is a small probability that some valence bonds will be broken. When that happens, there are empty states in the valence band, just a few. They go to the conduction band. For a typical semiconductor at room temperature, only a tiny fraction of the large number of valence band states are empty. There is a large number of empty states in the conduction band, and a tiny fraction of them are filled with electrons. The valence band and conduction bands, are separated by the forbidden gap, or **band gap**, E_G. A broken bond produces a **hole**, an empty state in the valence band, and an electron, a filled state in the conduction band, so the number of holes and electrons is equal. This important number is the **intrinsic carrier concentration**. Electrons in the conduction band are mobile and conduct electrical currents. Holes in the valence band are also mobile and conduct electrical currents.

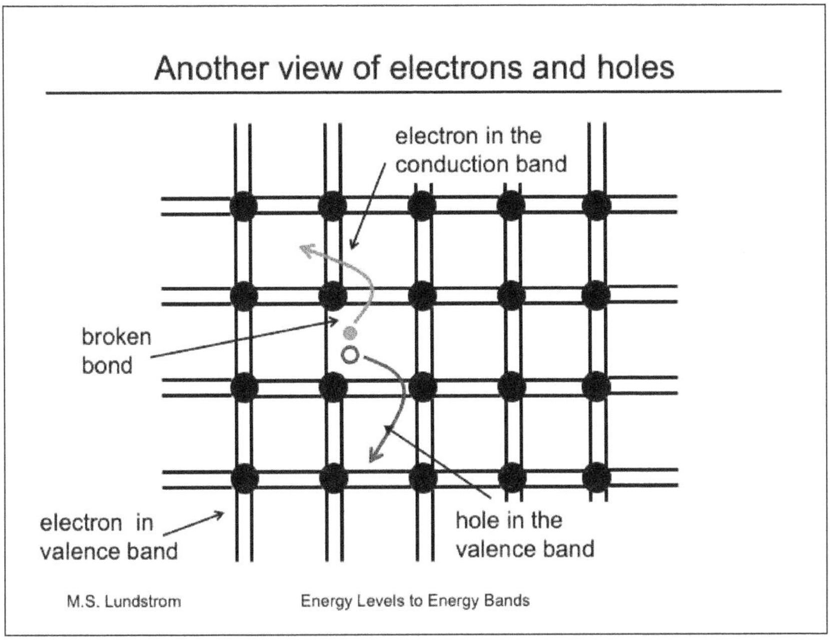

Another way to visualize what happens when we break bonds is shown above in our cartoon Si lattice with each atom surrounded by four nearest neighbors. The lines are the covalent bonds. If there is some thermal energy, there is a small probability that some bonds will be broken. When a bond breaks, it produces an empty state in the valence band. What happened to the electron? That electron is now free to wander around the crystal. That is an **electron in the conduction band**.

Broken bonds produce **holes in the valence band**, and holes are free to move too. For example, an electron in the valence band could hop over and fill up the hole shown above. The hole has moved. Then another electron in a nearby bond could hop over and fill up that hole, so the hole moves. A hole is an absence of a negative charge; it behaves like a positively charged mobile particle. We will think of electrons and holes as being two types of mobile charge carriers that carry the currents in semiconductors.

Finally, an electron in the conduction band might dropdown and fill a hole in the valence band. As discussed in Lecture 21, these **recombination** processes occur in semiconductors and can be important.

The figure shown above and the energy band picture discussed previously are two ways to think about broken bonds and intrinsic carriers.

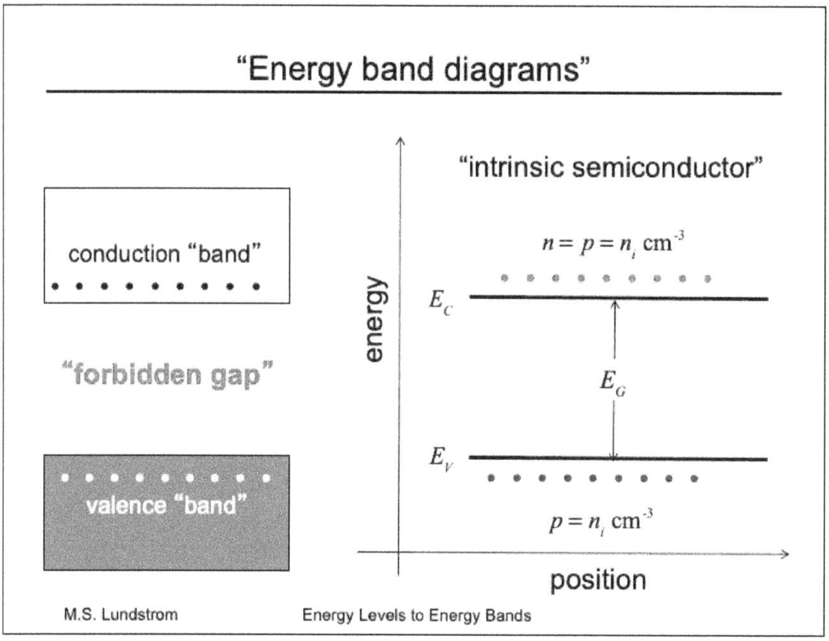

1.4 Energy band diagrams

An **energy band diagram** is a plot of electron energy vs. position. When we draw these energy band diagrams, and we will draw them over and over and over again in these lectures, we will simplify them. We'll find that the empty states in the valence band are all very near the top of the valence band. The filled states in the conduction band are all very near the bottom of the conduction band, so there is no point in showing the top of the conduction band and the bottom of the valence band.

As shown above on right, when we draw energy band diagrams, we simply draw the top of the valence band, E_V, and the bottom of the conduction band, E_C. They are separated by the band gap E_G. An energy band diagrams is a plot of E_V and E_C vs. position. We will discuss energy band diagrams in Lecture 24.

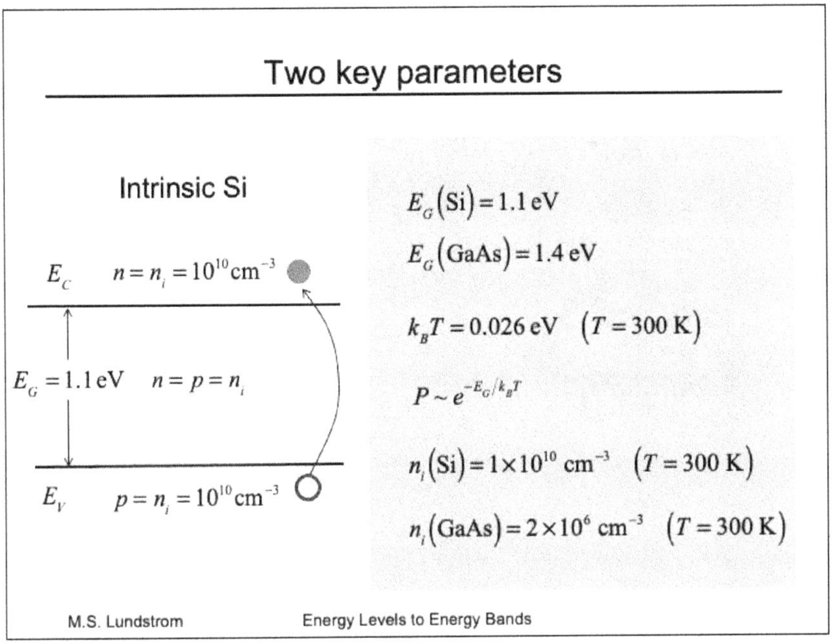

A pure semiconductor is called **intrinsic**. At room temperature, the thermal energy is sufficient to break a few bonds and produce a small number of holes in the valence band, about 10^{10} per cubic centimeter in Si. There is an equal number of electrons in the conduction band. This number, the **intrinsic carrier concentration** is a key parameter. Another one is the **band gap** - a measure of how difficult it is to break bonds. For silicon at $T = 300$ K, it is 1.125 eV; other semiconductors have different band gaps. For example, gallium arsenide (GaAs), has a band gap of 1.422 eV.

The average thermal energy at room temperature, $3k_BT/2$, is 0.039 eV. The probability of breaking a covalent bonds decreases exponentially as the band gap increases. The bigger the band gap, the harder it is to break bonds, so the number of electrons and holes at a given temperature is extremely sensitive to the band gap. For GaAs, the band gap is larger than for Si, so it's harder to break bonds, and the intrinsic carrier concentration is much less than that of Si.

Finally, 10^{10} might sound like a large number, but remember that there are 5×10^{22} atoms per cm^3 in a Si crystal, so 10^{10} is a tiny fraction of the covalent bonds that have been broken.

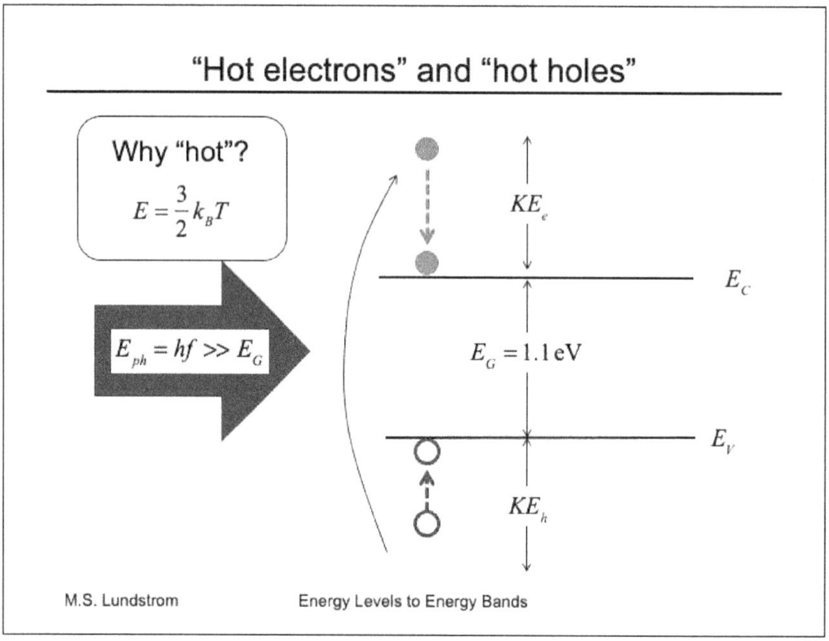

1.5 Hot electrons and hot holes

Light can be thought of as a wave or as a particle. The particle is called a **photon**. The energy of a photon is $E = hf$, Planck's constant, h, times the frequency of the light. If the frequency is high, or the wavelength short (e.g. blue or violet or ultraviolet) then the photons will have a lot of energy, $E \gg E_G$, and can lift an electron deep in the valence band way up into the conduction band creating very energetic electrons in the conduction band, and energetic holes deep in the valence band. We call these carriers **hot carriers** (hot electrons and hot holes) because temperature is a way of measuring the kinetic energy of a carrier.

Electrons are typically near the bottom of the conduction band, and holes are near the top of the valence band. This happens because very quickly (within femtoseconds) hot electrons lose energy and drops down to the bottom of the conduction band. In the valence band, electrons above the hot hole drop down and fill the hole, so the hole moves up in energy, and within a few femtoseconds it is near the top of the valence band. We will assume in these lectures that electrons are always near E_C and holes are near E_V and will ignore processes that occur on very short time scales.

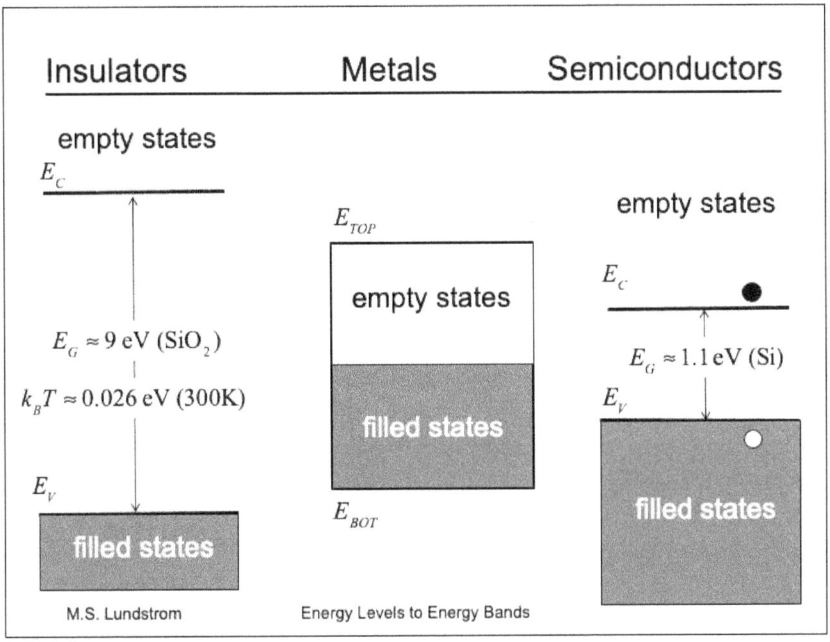

1.6 Insulators, metals, and semiconductors

We can now understand the difference between **metals**, **insulators**, and **semiconductors**. Metals conduct electricity well, insulators don't, and semiconductors are in between, but their properties can be varied from nearly insulating to nearly metallic. As discussed in Lecture 6, this is done by **doping**.

On an energy band diagram, an insulator is a material with a very large band gap. It is very difficult to break the bonds and create any carriers that can conduct current. For example, silicon dioxide, which is glass, has a band gap of about 9 electron volts.

For metals, there is no band gap at all. We fill up the states to accommodate the electrons, and right in the middle of a band all of the electrons have been accommodated; the states above this point are empty. So the electrons are free to move, if they need more energy they just move up a little bit and find another state, there is no forbidden gap.

Semiconductors have moderate band gaps, small enough that at room temperature, a few of the bonds are broken and there are a few electrons in the conduction band and a few holes in the valence band.

1.7 Summary

We have covered a lot of ground, so let's summarize. The most important concept introduced in this lecture is that for isolated silicon atoms, there is a set of discrete **energy levels**, but in semiconductor crystals, the electrons interact with each other, and the energy levels broaden and become bands of energies, **energy bands**.

Because everything typically happens in near the top of the valence band and near the bottom of the conduction band, those are the two regions of energy that we will focus on in these lectures. The electrical current in a semiconductor device is carried by the electrons that are in the conduction band, or by the holes, which we think of as positive charge carriers in the valence band.

The **band gap** and **intrinsic carrier concentration** are the two most important parameters for a semiconductor. They are tightly related – the wider the band gap, the smaller the intrinsic carrier concentration, which is the concentration of broken covalent bonds.

A lot of material has been introduced in this lecture, but do not be concerned, we will be revisiting these concepts over and over again in these lectures, and you will have a chance to develop some familiarity with them. Next, we continue the discussion in Lecture 2, which is about how the atoms are arranged in a semiconductor material.

1.8 References

Those who wish to dive deeper into the concepts introduced in this lecture may wish to consult the following two texts:

[1] R.F. Pierret, *Advanced Semiconductor Fundamentals*, (2nd edition), Pearson Education, Inc., 2003 (see Ch. 2, Sec. 2.1, and Ch. 3).

[2] S. Tiwari, *Semiconductor Physics: Principles, Theory, and Nanoscale*, Oxford Electroscience Series, Vol. 3, 2020 (see Ch. 3).

Lecture 2

Crystalline, Polycrystalline, and Amorphous Semiconductors

2.1 Introduction

In this lecture, we'll continue our discussion of the material properties of semiconductors by describing three kinds of semiconductors, **crystalline**, **polycrystalline**, and **amorphous**. Let's begin by defining some terms.

(1) The structure of a semiconductor crystal is specified by a **crystal lattice**.
(2) Polycrystalline semiconductors consist of many crystalline **grains** with different orientations.
(3) In amorphous semiconductors, the atoms are more or less randomly distributed throughout the solid, but these materials tend not to be completely random. We saw in the previous lecture, that each silicon atom in a crystal is surrounded by four nearest neighbors. In amorphous silicon, each silicon atom is surrounded on average by about four nearest neighbors.

The slides in the next sections explain these terms in more detail.

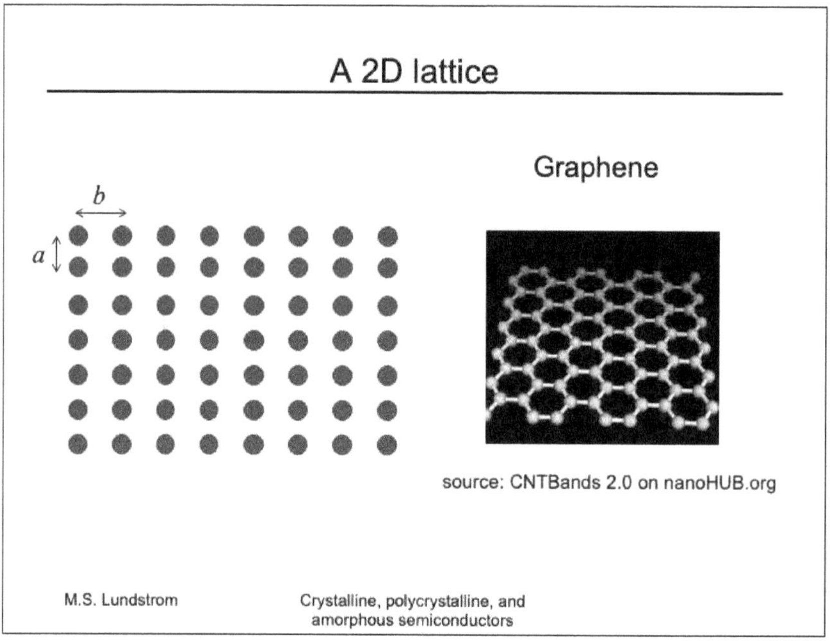

2.2 Crystal lattices

The first topic is what we mean by a crystal lattice. To keep things simple, let's work in two dimensions. In the slide above, the figure on the left is a 2D crystal lattice. The lattice points are separated a distance "a" in one direction and "b" in the other direction.

This 2D example is not completely artificial because there are 2D electronic materials. A notable one is graphene, which was subject of a 2010 Nobel Prize. As shown above on the right, graphene is a one atom thick, planar sheet of carbon in a honeycomb lattice. Graphene is just one example of a 2D material. Researchers are exploring ways to exploit the interesting electronic properties of 2D materials like graphene.

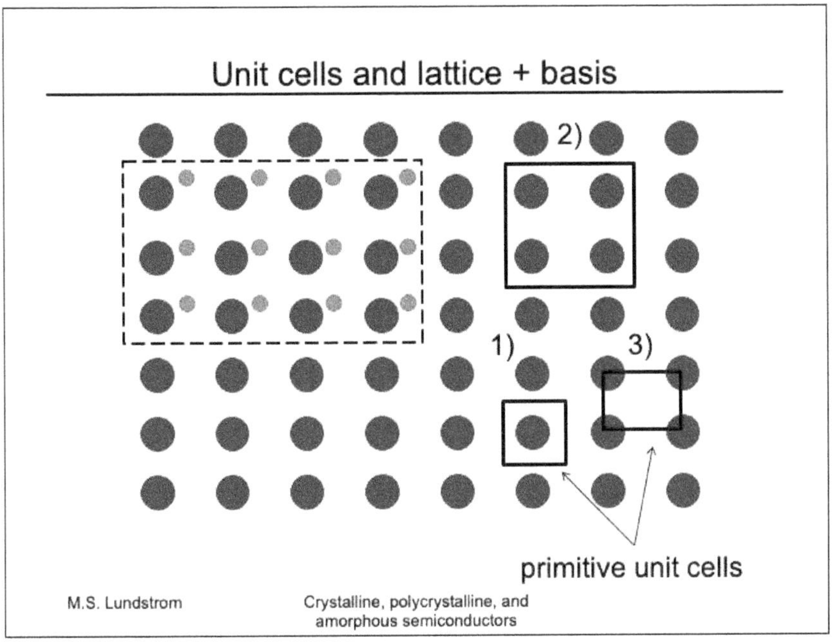

In these lectures, we will focus on 3D, bulk semiconductor materials and will only use 2D pictures to illustrate concepts that apply to both 2D materials and to 3D as well. One important concept is the **unit cell**. A unit cell can be used to reproduce the entire crystal simply by repetitively translating it. Three different unit cells are shown in the figure above. We could build up the entire crystal from these unit cells. As discussed below, unit cells are not unique.

If the unit cell consists of only one atom, then we call it a primitive unit cell. You can also see in the figure above that a primitive unit cell is not unique. There are four lattice points in unit cell #3, but we can only count one quarter of each of atom in this unit cell, because they're shared with other rectangles beside it, so there is one atom in unit cell #3 and one in unit cell #1.

We'll often talk about a lattice that specifies the structure of the crystal. Sometimes in a semiconductor material, there will also be a **basis**. As shown in the dashed rectangle on the upper left, at each lattice point, there can be additional atoms that are associated with it. So we talk about a lattice plus a basis.

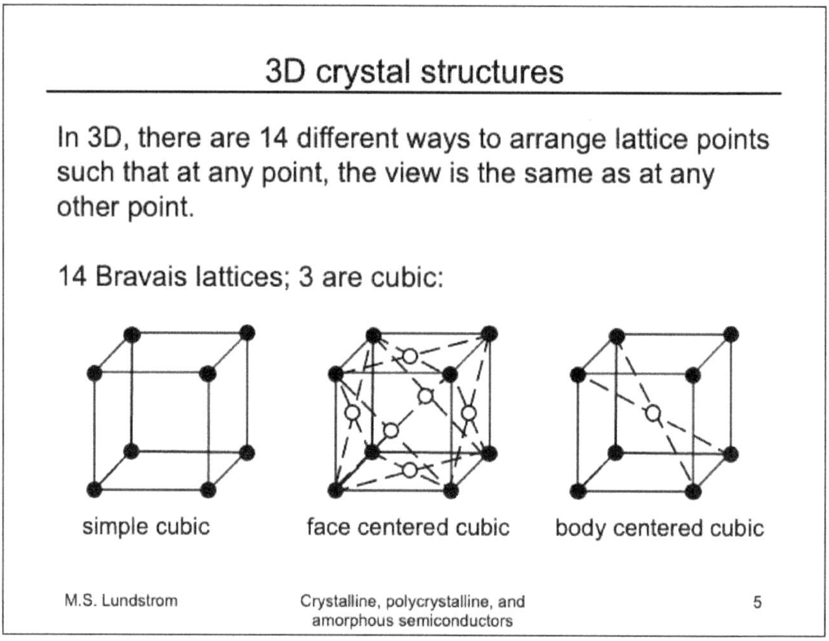

In three dimensions, if we ask: "How many ways are there that we can arrange a set of points, such that if we select any point and look around, the environment will be identical to what we would see from any other point." It turns out there are only 14 different ways to do that. These are called the 14 **Bravais lattices**.

Three of these Bravais lattices are very simple and are shown above. These three are cubic lattices. There's a **simple cubic** lattice, which has a lattice point at each of the eight corners of the cube. There is a **face centered cubic** lattice, which in addition has an atom at each of the six faces, and there is a **body centered cubic** lattice in which there's a lattice point in the middle of the cube. Some materials crystallize in these simple cubic lattices, but these are only 3 of the 14 possible lattices that materials can crystallize in.

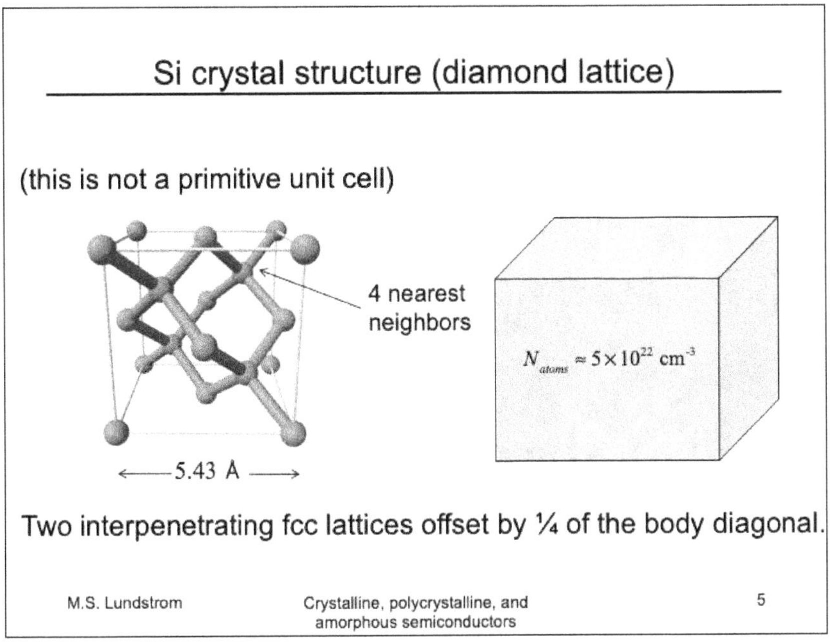

2.3 Silicon crystal structure

The figure on the left above is the way we've been drawing the silicon lattice. This is not a primitive cell because you can see that there's more than one atom in this unit cell, but it's convenient and easy to understand and draw the cubic cell. The **lattice constant**, a, the edge of the cube, is 5.43 Angstroms. This value is known to a high degree of precision from x-ray crystallography. We're not showing all of the Si atoms in the surrounding cubes, but you'll see that every Si atom has four nearest neighbors. The angle between the bonds is about 109 degrees. This is one of the early successes of quantum mechanics, being able to predict these bonding structures and what the angles should be.

If you're good at visualizing things in three dimensions, you might be able to see that this is actually two interpenetrating, face centered cubic lattices that are offset by one quarter of the diagonal from the body of this cube. This is an important lattice, called the **diamond lattice**; silicon, germanium, and diamond all crystalize in this structure.

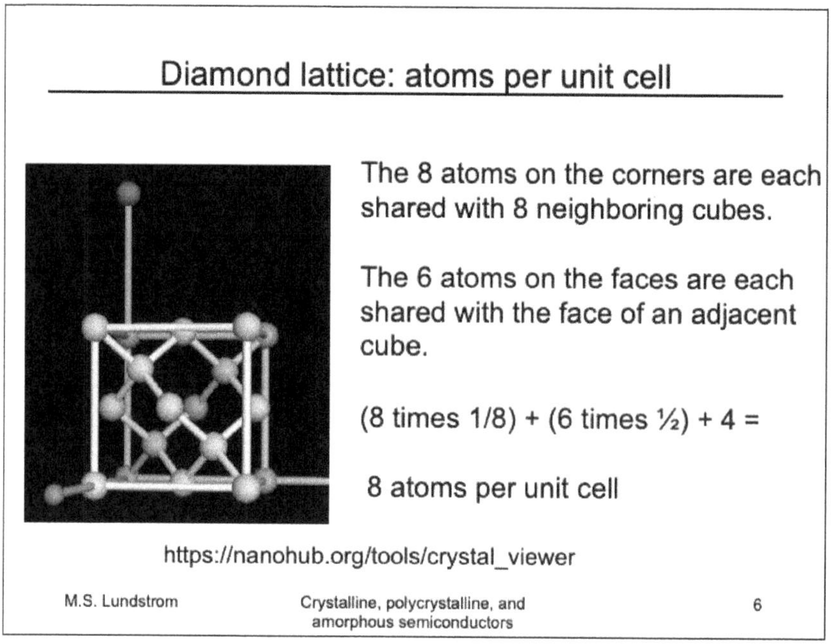

Let's take a more careful look at the diamond lattice and count the atoms. You can see that there is an atom on each of the 8 corners of the cube, but we can't count those eight atoms totally in this cube because they're shared. If we stack all of these cubes together in a 3D structure, each one of those eight atoms is shared with eight adjacent cubes, so we can only count 1/8 of each of them for this particular cube.

As you can see, there are also atoms on each of the 6 faces. In a 3D structure, we have to share the atom on each face with an adjacent cube. So if we count the number of atoms in this unit cell, there are 8 on the corners, but we can only count 1/8 of them because they're shared with 8 neighbors. There are 6 on the faces, but we can only count 1/2 of them, because they're shared with an adjacent cube on each face.

If you look carefully at this structure, you can also see that there are also four atoms inside the cube. So we have $8 \times (1/8) + 6 \times (1/2) + 4 = 8$ atoms in this important unit cell for semiconductors.

Silicon: # atoms / cc

Lattice constant: a = 5.4307 Å

8 atoms per unit cell.

$$N_{atoms} = \frac{8}{\left(5.4307 \times 10^{-10}\right)^3} \ /\mathrm{m}^3$$
$$= 4.99 \times 10^{28} \ /\mathrm{m}^3$$
$$= 4.99 \times 10^{22} /\mathrm{cm}^3$$

https://nanohub.org/tools/crystal_viewer

M.S. Lundstrom Crystalline, polycrystalline, and amorphous semiconductors

Knowing how many atoms there are in the unit cell, we can do some calculations. The lattice constant of silicon is known to be 5.4307 angstroms, and we have 8 atoms per unit cell. I mentioned earlier that the number of atoms per cubic centimeter in a silicon lattice is about 5×10^{22}, so we can see now how that's computed. We have 8 atoms. The volume of this cell is a^3. We plug numbers in, and the answer comes out as 4.99×10^{28} per cubic meter.

Now, per cubic meter is the proper SI unit to do calculations in, but semiconductor people prefer to state concentrations per cubic centimeter (cc), so we have to convert per cubic meter to per cubic centimeter, and quote the density of atoms as about 5×10^{22} per cubic centimeter.

Remember the intrinsic carrier concentration, the number of electrons and holes that result from the breaking of covalent bonds, is about $n_i = 1 \times 10^{10}$ per cc at room temperature in silicon, which is more than 12 orders of magnitude smaller than the density of atoms. The intrinsic carrier concentration is a very small number.

Silicon: density and NN spacing

Lattice constant: a = 5.4307 Å

Density = total mass/vol. of unit cell.

Atomic mass of Si: 28.0855 amu

1 amu = 1.6605 x 10^{-27} kg

$$\rho = \frac{8 \times 28.0855 \times 1.6605 \times 10^{-27}}{\left(5.4307 \times 10^{-10}\right)^3} \quad \text{kg/m}^3$$

$$\rho = 2.329 \text{ g/cm}^3$$

Body diagonal = sqrt(3) a.

NN spacing = sqrt(3)a/4

NN = 2.35 Å

M.S. Lundstrom

We can also compute the density of Si in grams per cubic centimeter. The density is the total mass in this unit cell divided by the volume of the unit cell. The atomic mass of silicon is 28. Remember the atomic mass unit (amu) is about the mass of a proton, about 1.6×10^{-27} kilograms. If we take the total mass in the unit cell, divide it by the volume of the unit cell, we get the density in proper SI units, kilograms per cubic meter. We then convert to grams per cubic centimeter, and find that the density is about 2.329 grams per cubic centimeter.

We can ask other questions, like "What is the **nearest neighbor spacing**?" That calculation is shown in the shaded area above. Remember that the edge of the cube is $a = 5.4307$ Angstroms. The diagonal of the cube is the square root of $\sqrt{3}\,a$. If you look carefully at this lattice, you can see there's an atom in the upper left corner, and there is an atom one quarter of the body diagonal away from it. That's its nearest neighbor. The nearest neighbor spacing is 1/4 of the body diagonal ($\sqrt{3}\,a/4$, which is 2.35 Angstroms. The lattice constant is $a = 5.4307$ Angstroms, but the nearest neighbor spacing is 2.35 Angstroms.

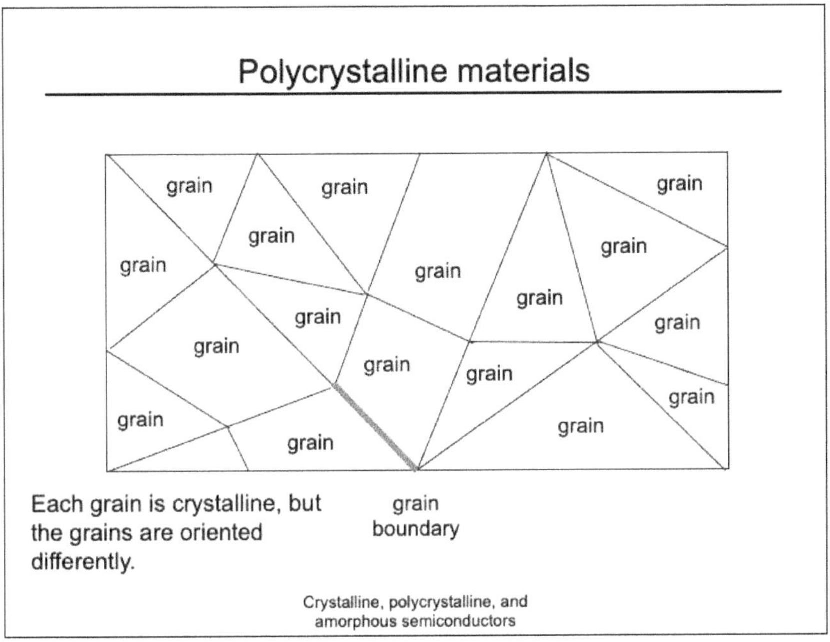

2.4 Polycrystalline materials

We've been talking about crystal lattices. What is a **polycrystalline** material? A polycrystalline material is just a material that consists of many small crystals, but crystals that are oriented in different ways, so that things don't line up at the boundaries. The boundaries between the different orientations, are called **grain boundaries**. At a grain boundary the atoms don't match up well, so a grain boundary is the source of defects, and these defects can affect the electronic performance of a material. They usually degrade the performance of an electronic device.

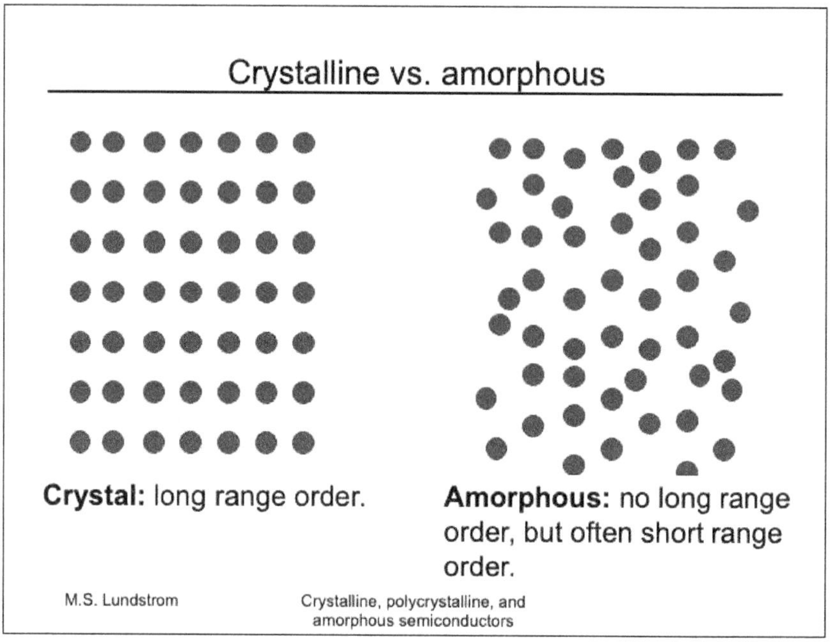

2.5 Amorphous materials

There are crystalline and polycrystalline semiconductors, and there are also **amorphous** semiconductors, which don't have long range order. In an amorphous semiconductor, atoms are not in a precise location specified by a crystal lattice and basis. Although they don't have long range order, amorphous semiconductors tend to have short range order. For example, in a silicon lattice, each atom is surrounded by four nearest neighbors. In amorphous silicon, each atom is surrounded on average by about four silicon atoms.

The electronic properties of amorphous silicon tend to be somewhat similar to the electronic properties of crystalline silicon because there are four nearest neighbors in crystalline Si and four nearest neighbors on average in amorphous Si.

2.6 Summary

We can summarize the three main categories of semiconductors. Crystalline semiconductors are the most expensive, because it's difficult to grow them without defects to ensure that every atom is where it should be, but they have the highest electronic quality because they have the fewest defects.

Polycrystalline semiconductors are generally less expensive to produce, but the grain boundaries can impede current flow or cause carriers to recombine. They're used when the highest quality isn't needed, but where cost is very important — for example, in solar cells, where you need large areas of semiconductors, polycrystalline materials are often used.

Finally, amorphous semiconductors are the least expensive. Their electronic properties are the poorest, but they're adequate for many purposes, and when cost is critical — for example, in the thin film transistors in flat panel displays, amorphous silicon might be used.

We will be concentrating in this course on crystalline materials, because they're the easiest to understand, but the general concepts that we will discuss apply broadly to polycrystalline and to amorphous semiconductors as well.

In the next lecture, we'll talk about how we specify directions and planes in crystals, because directions and planes in a crystal can have an important effect on the operation of a device.

2.7 Reference

Those who wish to dive deeper into the concepts introduced in this lecture, may wish to consult the following text:

[1] R.F. Pierret, *Advanced Semiconductor Fundamentals*, (2nd edition), Pearson Education, Inc., 2003 (see Ch. 1, Secs. 1.2.1–1.2.4).

Lecture 3

Miller Indices

3.1 Introduction

In these lectures, we are focusing on crystalline semiconductors, but before we continue, there's one technical detail to discuss; how do we specify planes and directions in crystals? It turns out that there's a very convenient and easy way to do this by specifying **Miller indices**. This is especially easy to do for cubic systems like silicon. The goal in this lecture is to understand what Miller indices are all about.

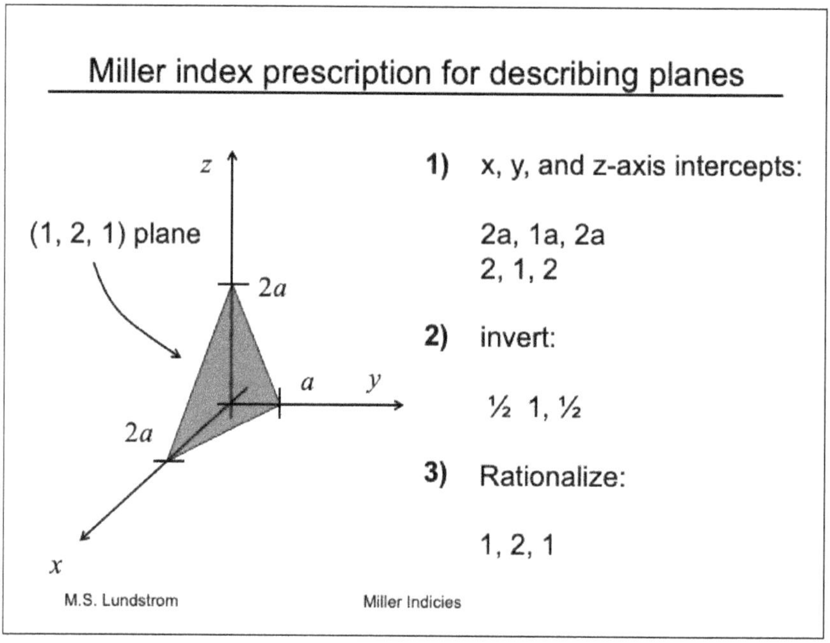

3.2 Describing planes with Miller indices

Shown above is a plane with the x, y, and z axes shown. We're going to do everything in terms of the lattice constant, a, in this cubic crystal. In this example, you can see that the x, y, and z intercepts are $2a$ in the x direction, $1a$ in the y direction, and $2a$ in the z direction. There's no need to carry the "a" around, because we understand that everything is measured in units of a, so we express these intercepts as 2, 1, 2. That's the first step of our prescription.

The next step of the prescription is to invert these intercepts, so 2 becomes one-half, 1 becomes 1, and 2 becomes one-half. Now we have a set of inverted indices: 1/2, 1, 1/2.

The final step in our prescription is to rationalize these numbers and make each one of these numbers a whole number. If we multiply everything by 2 then we have a set of numbers, 1, 2, 1. That's how we'll define this plane. We will put those numbers in parentheses; the parentheses mean we're talking about a plane. The plane shown above is a (1 2 1) plane.

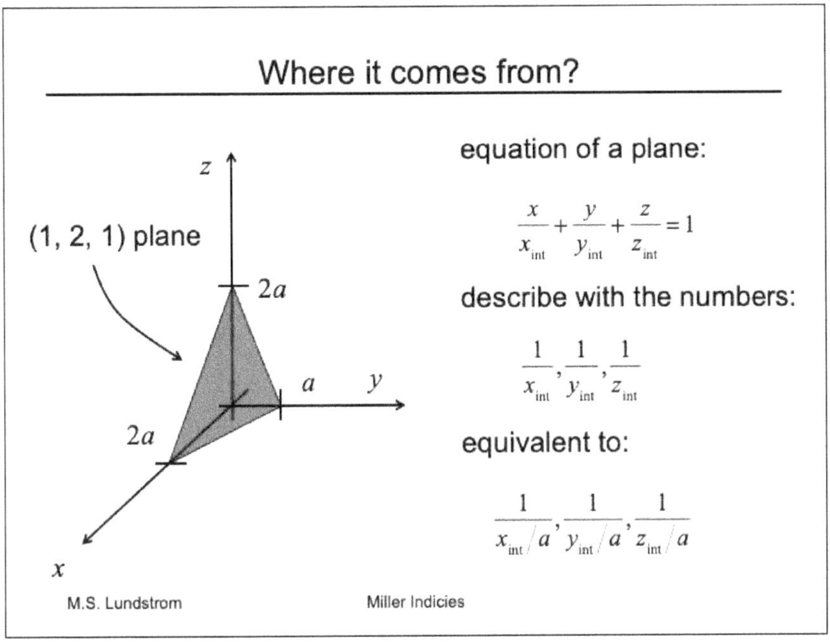

Now you might say, "okay that's a simple enough prescription, but where does it come from?" It comes from the equation for a plane.

If you recall your high school algebra, you will remember that one way to describe planes is with an equation of the form shown above on the upper right. A plane is described by x over the x intercept of the plane, y over the y intercept, plus z over the z intercept equals 1. The important elements of this equation for a plane are the 3 numbers, 1 over x intercept, 1 over y intercept, 1 over z intercept. Since we're doing everything in units of lattice spacing, we can express each of those intercepts in units of the lattice spacing, and then we can rationalize them to make a set of whole numbers. That's where the prescription for Miller indices planes comes from.

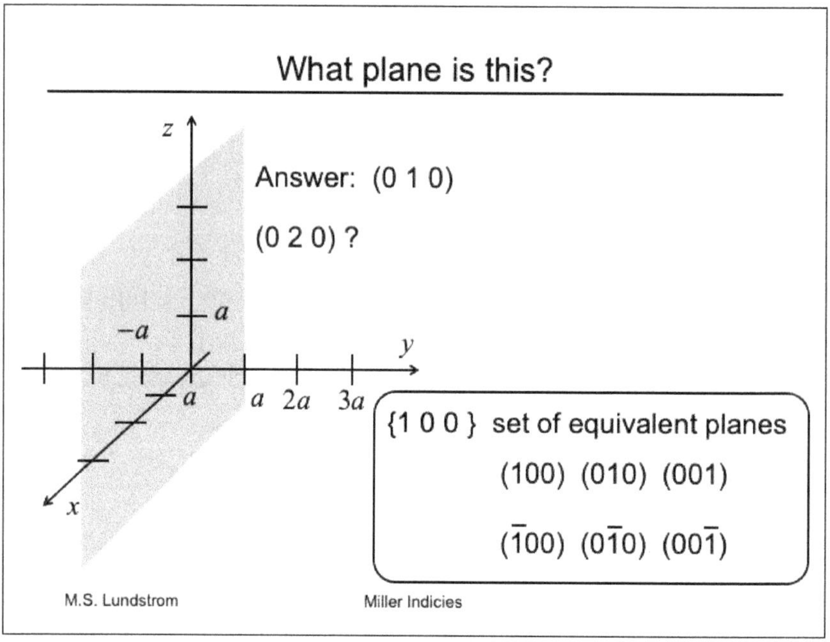

Applying our prescription to the plane shown above, we see that it is a (0 1 0) plane. (The x and z intercepts are infinity and one over infinity is zero). If we move the y intercept over from a to $2a$, our prescription shows it's also a (0 1 0) plane. They're really the same plane because in a crystal lattice, every point is the same as every other point, so I could move my origin a distance a in the y-direction and that second plane would look just like the one above.

If the y-intercept is at $-a$, our prescription shows that the plane is $(0\ -1\ 0)$. In Miller index notation, instead of writing -1, we put a bar over the 1, so this would be a $(0\bar{1}0)$ plane, but, again, all we have done is to translate the origin, so these planes are really all the same. In fact, there is a whole set of planes that are equivalent because the crystal lattice is symmetric.

If we rotate a cubic lattice by 90 degrees, or 180 degrees, or 270 degrees, everything looks the same, so all of these specific planes are really the same. We put these numbers in squiggly parentheses to indicate that we're talking about all of the planes that are crystallographically equivalent; (h k l) is a specific plane and {h k l} is a set of **equivalent planes.**

Silicon: atoms / cm^2 on a {100} plane

Lattice constant: 5.4307 Å

Atoms on face = (4 times ¼) +1 = 2

$N_S = 2/a^2$

$N_S = 6.81\text{x } 10^{14} /cm^2$

https://nanohub.org/tools/crystal_viewer

M.S. Lundstrom Miller Indicies

If we're building an electronic device on the surface of a silicon wafer, we might need to know how many atoms there are per square centimeter on the surface. Let's say that the front face of the cube shown above, a (100) plane, is the surface of my electronic device. What is the number of atoms per square centimeter on this surface?

On the face of the cube, there are four atoms — one on each corner, but they're shared with adjacent cube faces on the surface, so we can only count one quarter of each of those atoms for this particular cube face. There's also an atom right in the middle of the cube face that is not shared with anything because this is assumed to be the surface of my semiconductor, so there's no cube above it.

We conclude that the number of atoms on this cube face is 4 x 1/4 for the ones on the corners plus the 1 in the middle. There are 2 atoms on this (1 0 0) face of the cube. The density of atoms per square centimeter is 2 divided by the area of the face, a^2. Plugging in numbers, we find 6.8×10^{14} atoms per square centimeter. Do this calculation on other planes, {1 1 0} planes, {1 1 1} planes, and you'll find different numbers. It's sometimes important to pick a surface that has a low density of atoms.

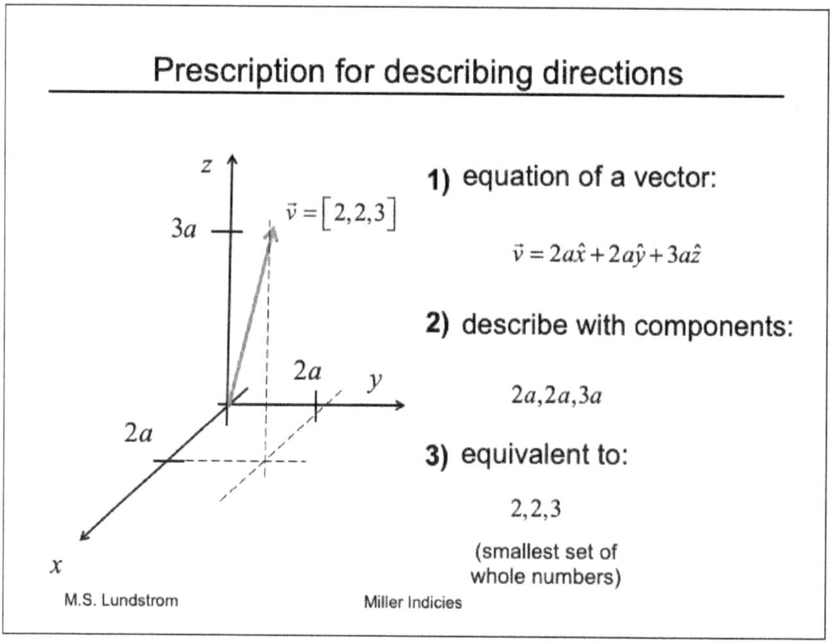

3.3 Describing directions with Miller indices

To describe a direction in a crystal lattice, we specify a vector in the crystal lattice. The way to specify a vector is to give its x component, y component, and z component. The particular vector shown above has an x component of $2a$, a y component of $2a$ and a z component of $3a$. We list those three components as $2a, 2a$ and $3a$, but again we are doing everything in terms of lattice constant so there's no need to carry the "a" around with us, and we can just describe this vector by the numbers, 2, 2 and 3.

To summarize, the first step was to remember how we specify a vector. The second step was to identify its components, and the third step was to drop the lattice constant a, which is the same in all directions. We then enclose those numbers in square brackets. The square brackets indicate that we are talking about a vector and not a plane.

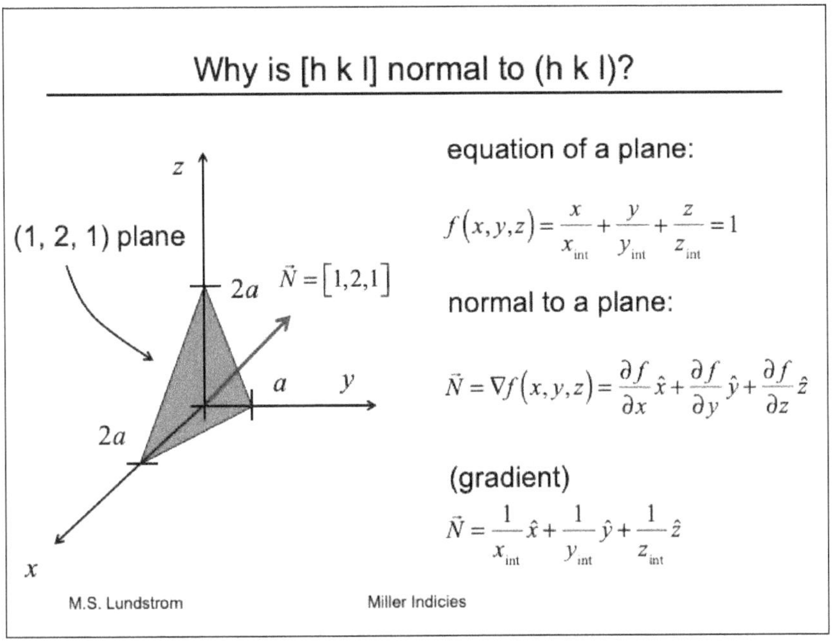

3.4 Direction normal to a plane

Another question is: Given a plane, how can we determine the direction normal to the plane? What direction is perpendicular to the surface of this (1 2 1) plane? The answer is very simple, the vector [1 2 1] is perpendicular to the plane (1 2 1). That's nice and easy to remember and simple. You might ask, why is that so?

We can understand where the fact that [h k l] is normal to (h k l) comes from if we go back to the equation of our plane. You may recall that the way we deduce the vector that is normal to a plane is to take the gradient of the equation that describes the plane. The gradient is the derivative with respect to x in the x direction plus the derivative with respect to y in the y direction plus the derivative with respect to z in the z direction. If we perform that gradient on this function, we'll find that the normal is 1 over x intercept in the x direction and 1 over y intercept in the y direction, 1 over z intercept in the z direction. That is the vector, [1 2 1]. The three intercepts that describe the plane are the three components of the vector normal to the plane.

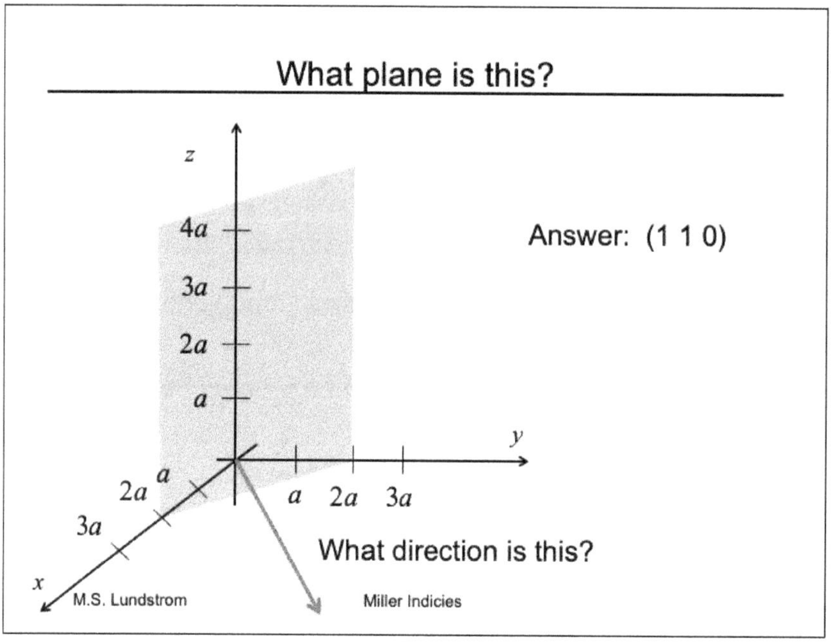

Shown above is another plane. Using our prescription, we find that this is a (1 1 0) plane. The direction normal to the plane is a [1 1 0] direction. In general, the direction [a b c] is normal to the plane (a b c).

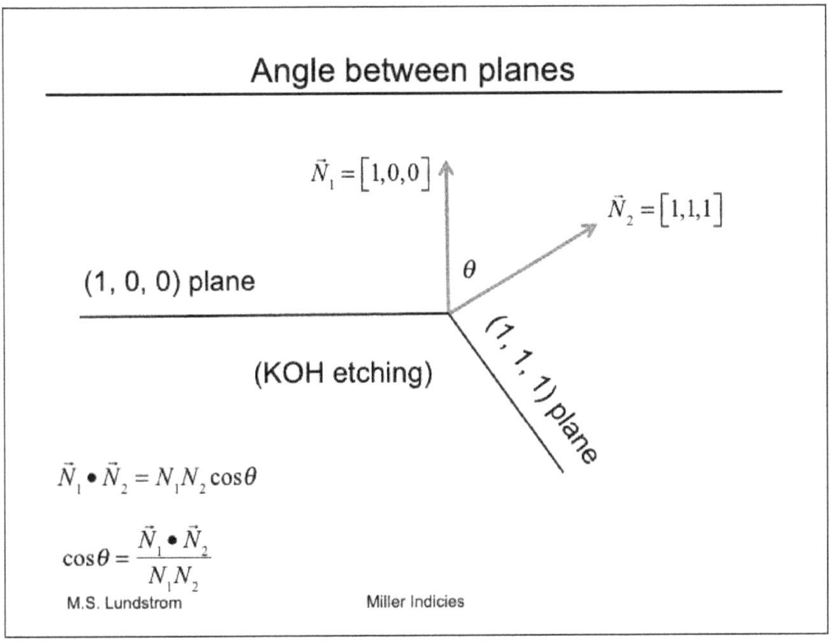

3.5 Angle between two planes

There is more that we can do with Miller notation. We might ask, for example, if we have a (1 0 0) plane of silicon and a (1 1 1) plane of silicon, what is the angle between these two planes? (By the way, there are etchants like KOH that preferentially etch certain crystal directions, so we can actually produce structures like this, and they are used for some semiconductor applications). To find the angle between these two planes, remember that the normal to the (1 0 0) plane is the [1 0 0] direction. The normal to the (1 1 1) plane is the [1 1 1] direction. The angle between the planes is the angle between these two vectors. You'll remember that the dot product of two vectors is the magnitude of the first vector times the magnitude of the second vector times the cosine of the angle between them, $\vec{a} \cdot \vec{b} = |a||b| \cos(\theta)$ From the dot product we can solve for the cosine of the angle between these two vectors or the cosine of the angle between the two planes.

Angle between (1 0 0) and (1 1 1) planes

$$\cos\theta = \frac{\vec{N}_1 \bullet \vec{N}_2}{N_1 N_2}$$

$$\vec{N}_1 = [h_1, k_1, l_1]$$

$$\vec{N}_2 = [h_2, k_2, l_2]$$

$$\cos\theta = \frac{h_1 h_2 + k_1 k_2 + l_1 l_2}{\sqrt{h_1^2 + k_1^2 + l_1^2}\sqrt{h_2^2 + k_2^2 + l_2^2}}$$

$$\vec{N}_1 = [1,0,0]$$

$$\vec{N}_2 = [1,1,1]$$

$$\cos\theta = \frac{1+0+0}{\sqrt{1^2+0^2+0^2}\sqrt{1_2^2+1_2^2+1_2^2}}$$

$$\cos\theta = \frac{1}{\sqrt{3}}$$

$$\theta = 54.7^\circ$$

M.S. Lundstrom Miller Indicies

Let's work out an example. We have $\cos\theta = \vec{N_1} \cdot \vec{N_2}$ over the magnitude of N_1 times the magnitude of N_2. We describe $\vec{N_1}$ by three indices, $(h_1\, k_1\, l_1)$, and we describe $\vec{N_2}$ by $(h_2\, k_2\, l_2)$. To take the dot product,we multiply the x components, multiply the y components, multiply the z components, and then add them up. We find the magnitude of a vector by taking the square root of the x component squared plus the y component squared plus the z component squared. So we have the formula shown above at the lower left that gives us the cosine of the angle between these two planes. We take the arccosine (the inverse cosine), and we have the angle.

As shown above on the right, we can now work out the example. We have a (1 0 0) plane, vector $\vec{N_1}$ normal to that and a (1 1 1) plane and a vector $\vec{N_2}$ normal to that. Using our prescription, we take the dot product of the two vectors, find the magnitude of the two vectors, and find that $\cos\theta = 1/\sqrt{3}$. We take the arccosine of $1/\sqrt{3}$ and we find that the angle between those two planes is 54.7 degrees. As mentioned earlier, this structure is one that can be produced by selectively etching crystal planes with chemicals that etch (1 0 0) planes rapidly and (1 1 1) planes slowly, which exposes the (1 1 1) planes.

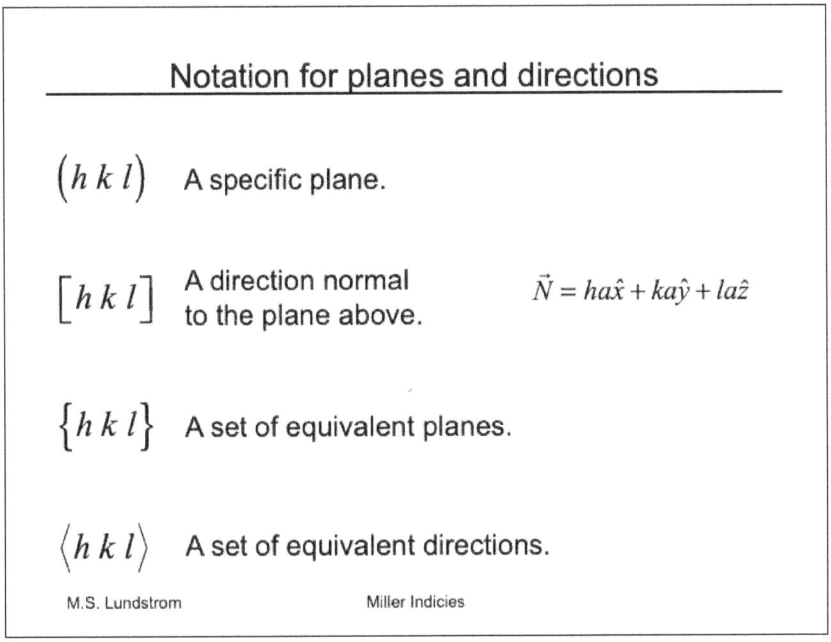

3.6 Summary

Miller indices provide a simple way to describe both planes and directions in crystal lattices. For cubic systems, the prescription is easy to understand, and it's easy to remember if you work through several examples.

- The notation (h k l) in parentheses refers to a specific plane.
- The notation [h k l] in square brackets refers to a vector. In fact, it's the vector normal to the (h k l) plane.
- The notation {h k l} refers to a set of equivalent planes.
- The notation $\langle hkl\rangle$ refers to a set of equivalent directions. Remember, the x, y, and z directions are equivalent because if we rotate a cubic crystal by 90 degrees, everything looks the same.

3.7 Reference

For more on Miller indices, consult the following text:

[1] R.F. Pierret, *Advanced Semiconductor Fundamentals*, (2nd edition), Pearson Education, Inc., 2003 (see Ch. 1, Secs. 1.2.5, 1.2.6).

Lecture 4

Properties of Common Semiconductors

4.1 Introduction

In this short lecture, we discuss common elemental and compound semiconductor and explain, for example, what we mean by a column IV semiconductor, a III-V semiconductor, and a II-VI semiconductor. We also discuss briefly the two most important properties of any semiconductor, its **band gap** and its **intrinsic carrier concentration**.

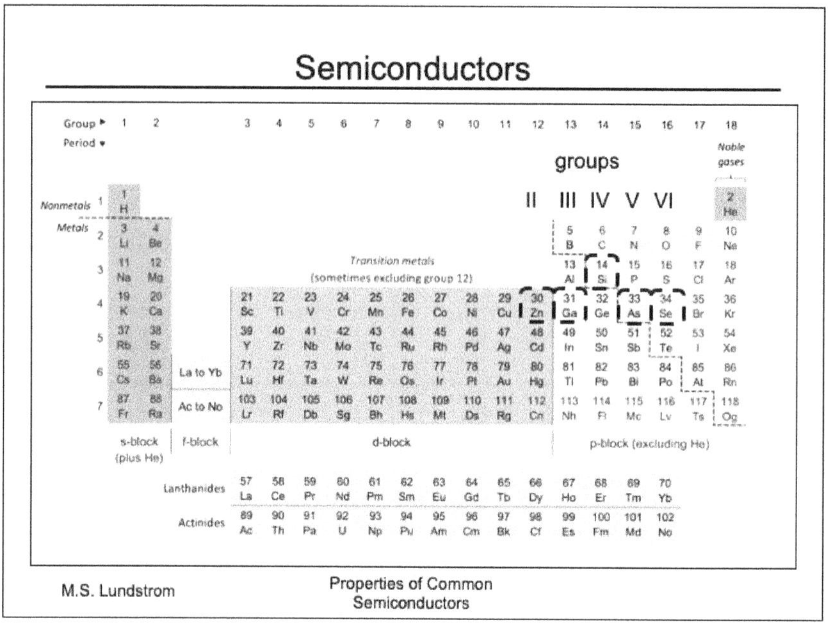

4.2 Elemental and compound semiconductors

Elements in the same column (group) of the periodic table have the same number of valence electrons. The most common semiconductors come from group IV with four valence electrons per atom. (In modern notation, this is group 14, but it is universally known called group 4 by the semiconductor community.) Silicon and germanium are **group IV semiconductors**. The crystalline form of carbon, diamond, which is also in group IV, is usually considered to be an insulator because its band gap is so wide.

Elements in group III have three valence electrons; elements in group V have five. Compounds of elements from these two groups have, on average, four valence electrons per atom. These are the **III-V (three-five) semiconductors**, such as gallium arsenide (GaAs).

Elements in group II have two valence electrons; elements in group VI have six. Zinc is from group II, selenium is from group VI, and ZnSe is a **II-VI (two-six) semiconductor**. On average, there are four valence electrons per atom. Other materials can be semiconductors too, but the most widely used semiconductors come from group IV or groups nearby.

Two key numbers

Intrinsic Si

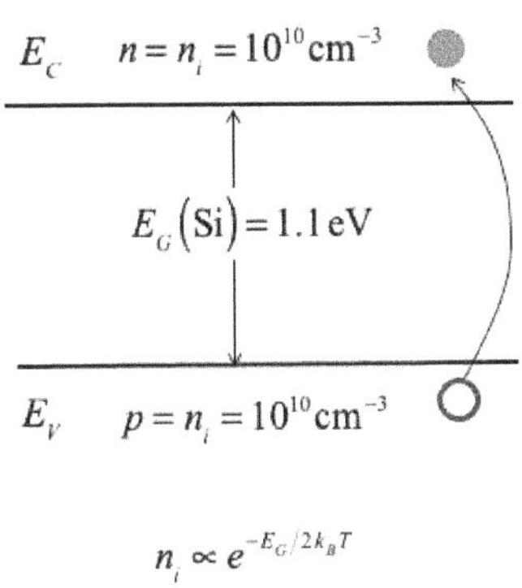

	E_G (eV)	n_i
Ge	0.663	2.0×10^{13} cm^{-3}
Si	1.125	1.0×10^{10} cm^{-3}
InP	1.344	
GaAs	1.422	2.3×10^{6} cm^{-3}
CdTe	1.5	
6H SiC	2.99	
4H SiC	3.26	
GaN	3.4	

M.S. Lundstrom   Properties of Common Semiconductors

4.3 Band gap and intrinsic carrier concentration

Two key numbers are very important, the first is the **band gap**, the separation between the top of the valence band and the bottom of the conduction band. In silicon, that number is 1.125 eV at $T = 300$ K. The other important number is n_i, the **intrinsic carrier concentration**, which comes from breaking covalent bonds and creating holes in the valence band and electrons in the conduction band. In Si, $n_i \approx 1 \times 10^{10}$ cm^{-3} at 300 K.

There is a very close relation between E_G and n_i; n_i depends exponentially on E_G. As the band gap gets wider (or the temperature lower), it gets exponentially harder to break covalent bonds, and n_i gets exponentially smaller. As the band gap gets smaller (or the temperature higher), it gets easier to create electron-hole pairs, and n_i increases.

As shown above on the right, germanium has a band gap about 0.5 eV smaller than Si, and $n_i \approx 2 \times 10^{13}$ cm^{-3}, which is 1,000 times more than the n_i of Si. Gallium arsenide has a band gap that's about 0.3 eV larger than Si, and GaAs has an n_i that is almost four orders of magnitude smaller than Si. We will see later how to actually calculate n_i given the properties of the semiconductor and the temperature.

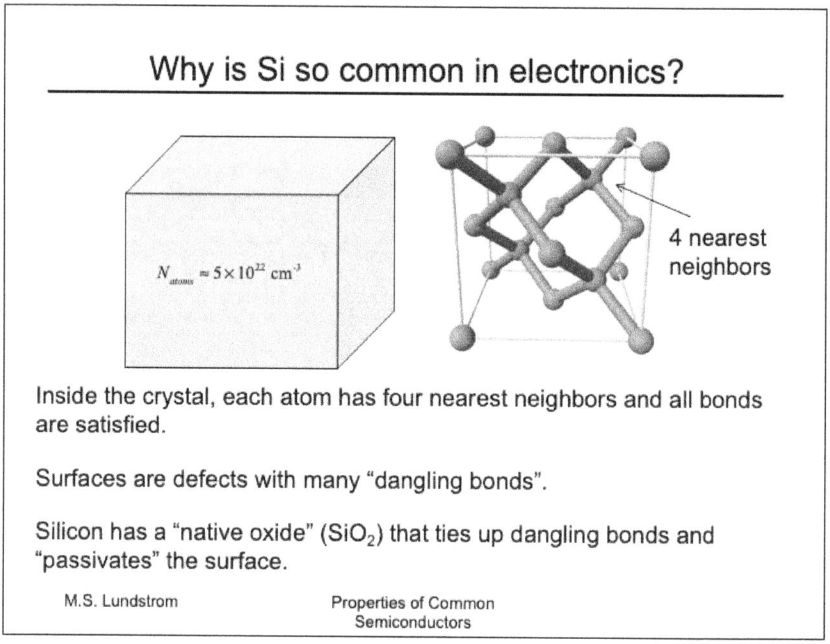

4.4 Silicon and silicon dioxide

Let's pause briefly and ask: "There are many different types of semiconductors, why is silicon the one that is most often used?" It's not because it has the best electronic properties, there's another reason; Si has a remarkable property that makes it easy to manufacture high performance devices from Si.

Each Si atom has four nearest neighbors, but we have to terminate the crystal. There are many defects at a surface because the atoms on the surface don't have all of the nearby atoms to bond with, which leads to **dangling bonds**. Dangling bonds are defects that are usually electronically active and degrade the performance of a device.

Silicon has a native oxide, silicon dioxide (SiO_2), which is a hard, high quality insulator that has the ability to tie up and **passivate** dangling bonds. Passivating these defects renders them electrically inactive, so that we can build electronic devices on the surfaces of silicon. Other semiconductors don't have this nice property, and this is what has led to the widespread use of silicon.

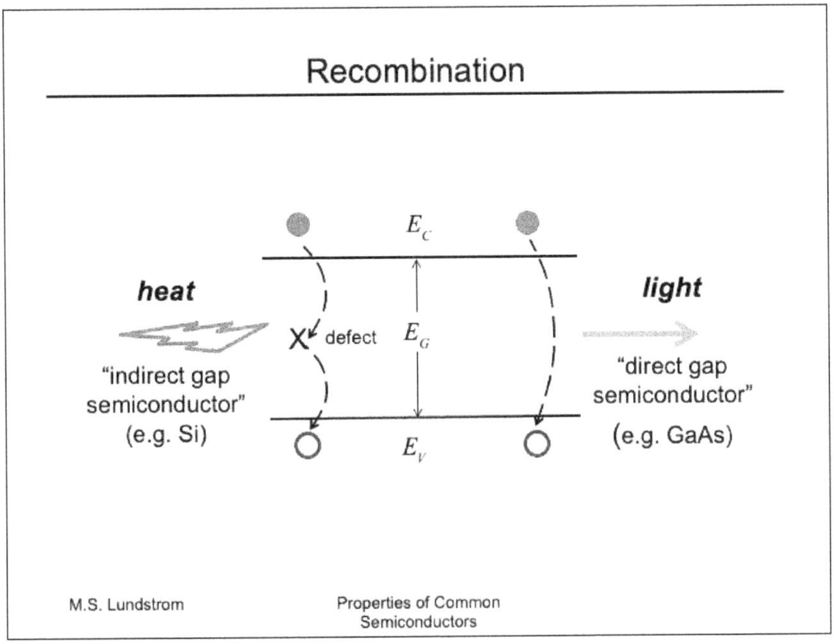

4.5 Carrier recombination

We mentioned earlier that it's possible for an electron in the conduction band to drop down into a hole in the valence band. This is a **recombination** event, after which there is one fewer electrons in the conduction band and one fewer holes in the valence band. When this happens, the energy has to go somewhere.

One possibility is that the energy goes into light; photons are emitted. This happens, for example, in gallium arsenide, a III-V semiconductor. Semiconductors like GaAs are called **direct gap** semiconductors; we'll talk about what that means in the next lecture, but essentially, electrons drop directly from the conduction band to the valence band. Direct gap semiconductors are used to make light emitting diodes and lasers.

Another possibility is that the excess energy just heats up the lattice. This happens in silicon and other **indirect gap** semiconductors. Instead of dropping directly from the conduction band to valence band, electrons first drop into a defect state in the band gap, then drop into a hole in the valence. The semiconductor gets hotter, which usually degrades the performance of electronic devices.

4.6 Summary

The **band gap** and the **intrinsic carrier concentration**, which is related to the band gap, are two important parameters of a semiconductor. Silicon is the most commonly used semiconductor, and this is not because it has the best electronic properties. The primary reason it is used so extensively is that it has a native oxide (silicon dioxide, SiO_2) that passivates and renders defects on the surface electrically inactive. Other semiconductors are used when silicon isn't suitable. One example, silicon doesn't emit light efficiently when electrons and holes recombine. so silicon is not suitable for light-emitting diodes. For light-emitting diodes, III-IV semiconductors are often used.

There are several other important properties of semiconductors – we've just mentioned a few of them. In Lecture 17, we'll discuss how fast electrons and holes in semiconductors move in a semiconductor. The faster they move, the more current they can carry. The **mobility** and **diffusion coefficient** are two important parameters for describing electron and hole transport. Another parameter, **effective mass**, is also important for electron and hole transport; we'll discuss effective mass in the next lecture.

It is also important to know the **thermal conductivity** of the semiconductor, because we don't want semiconductor devices to over heat. If we apply a large voltage to a semiconductor, the electric field may be so strong that it can rip electrons out of the covalent bonds. This **breakdown field** is another important parameter. There are several parameters that describe electron-hole recombination processes, which we will discuss in Lecture 21. These are just a few of the semiconductor parameters that we'll discuss as the lectures proceed.

In the next lecture we will continue by discussing the properties of the charge carriers (the electrons and holes) that move throughout a semiconductor device and that produce the electrical currents that make semiconductor devices operate.

4.7 Reference

Numerical values of some common semiconductor parameters are listed in the book by Pierret.

[1] R.F. Pierret, *Advanced Semiconductor Fundamentals*, (2nd Ed.), Pearson Education, Inc., 2003.

Lecture 5

Free Carriers in Semiconductors

5.1 Introduction

Let's continue our discussion of semiconductors by examining the behavior of the free carriers — electrons in the conduction band and holes in the valence band. What do we mean by "free carriers"? Holes in the valence band and electrons in the conduction band reside in extended states and are free to move about in the semiconductor lattice, and they carry their charge with them; these "charge carriers" cause electrical currents to flow.

In semiconductor work, q is the **magnitude of the charge** on an electron, $q = 1.6 \times 10^{-19}$ Coulombs. Remember also that the actual charge on an electron is negative, so we write the charge on an electron as $-q$. Different conventions are used by different communities; this is the convention that is typically used by the semiconductor community. (In physics, q has a sign — negative for electrons, and "e" $= |q|$ is the magnitude of the charge.) Holes in the valence band of a semiconductor carry a charge of $+q$, and electrons in the conduction band carry a charge of $-q$.

The question for this lecture is: "What are the properties of the free charge carriers?"

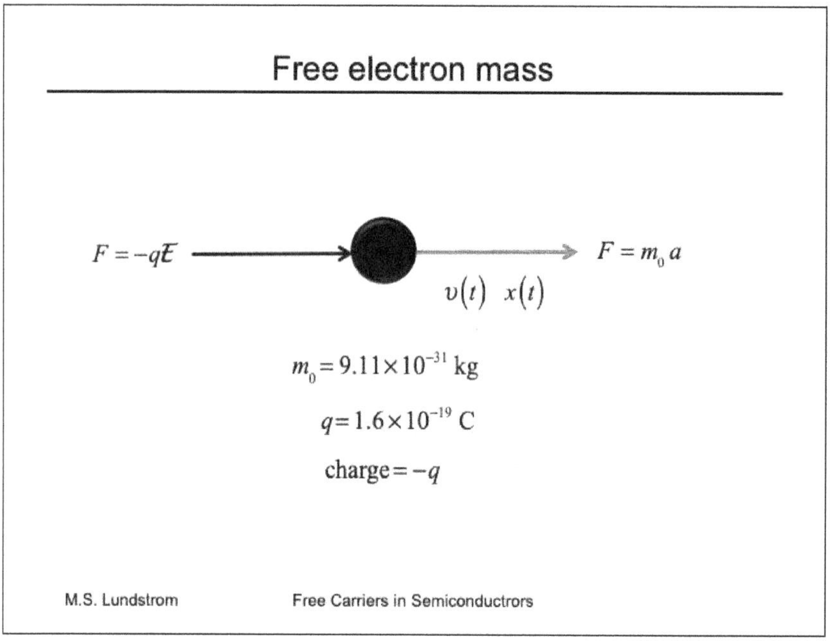

5.2 Effective mass

Because the motion of free electrons in semiconductors is complicated, let's begin with a simpler problem – a free electron in vacuum. A free electron has a rest mass of $m_0 = 9.11 \times 10^{-31}$ kilograms. Rest mass refers to the fact that we will be dealing only with electrons that are moving with velocities that are much less than the speed of light, c.

You may recall from first-year physics that electric fields exert forces on charged particles. The force on an electron is minus q times the electric field, $F_e = -q\mathcal{E}$. The force accelerates the electron according to Newton's law; force equals mass times acceleration. The electron moves, and we can compute its velocity as a function of time, its position as a function of time, and its energy as a function of time, all with Newton's laws of motion.

You may know, however, that an electron is a quantum mechanical particle that behaves like both a particle and a wave. For most semiconductor work, we can treat electrons as particles, as long as we are careful about some things like the appropriate mass to use. This lecture in a short introduction to these issues, which will be discussed much more in Part 2.

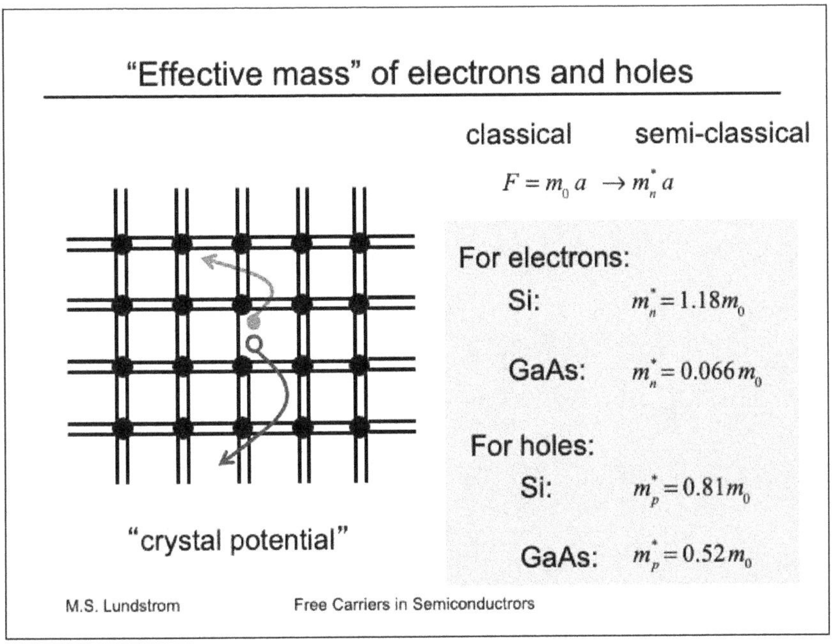

As electrons in the conduction band move within the crystal lattice, they experience Coulomb forces due to the bonds (the **crystal potential**) and due to the charge of other electrons and the holes too. One of the triumphs of 20th century condensed matter physics was the understanding of how this complicated problem could be described simply.

It turns out that there is a simple way to described the motion of electrons in the conduction band that preserves the use of Newton's laws. For a classical particle, $F = m_0 a$, where F is the force, m_0 is the mass, and a is the acceleration. For electrons in a crystal lattice, the complicated forces that act upon the electron in the crystal lattice, can be embedded into an **effective mass**. For electrons in semiconductors, $F = m_n^* a$, where m_n^* is the electron effective mass. There are well-developed techniques to compute and measure effective masses.

It's similar for holes in the valence band, which can be thought of as positively charged particles that respond to forces with an effective mass, m_p^*. In general, the hole effective mass is different from the electron effective mass. We will assume that the effective masses are known to us and will look them up relying on measurements or calculations that have been done for the specific numerical values.

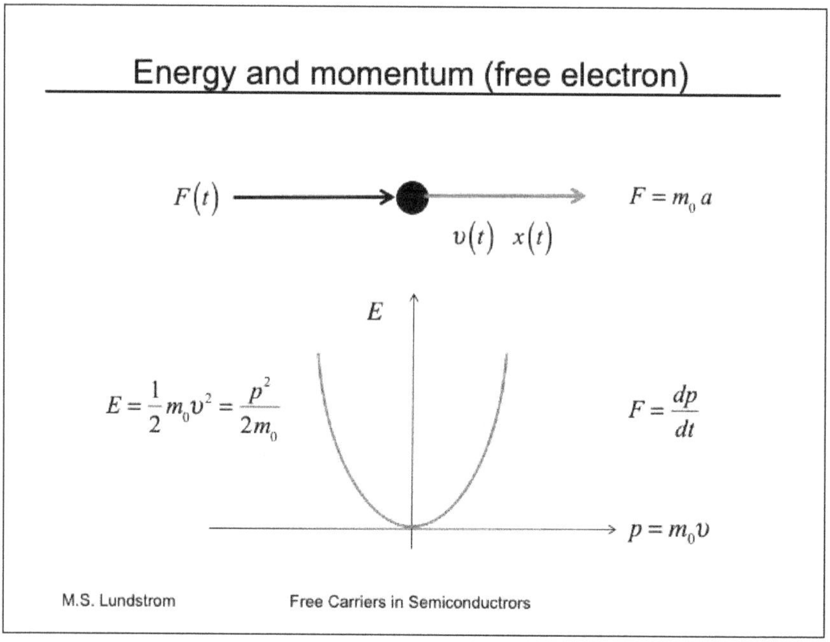

5.3 Energy vs. momentum

Another important concept that we will revisit in Part 2 is the relation between energy and momentum.

You'll recall for a free particle in vacuum, the kinetic energy is $E = (m_0 \upsilon^2)/2$. Momentum is mass times velocity, $p = m_0 \upsilon$, so we can write the energy as $E = p^2/(2m_0)$. If we plot the energy vs. momentum, $E(p)$, it will be a parabola as shown above.

Note also that we can write $F = m_0 a$ equivalently as $F = dp/dt$. The force is the time rate of change of the momentum. This is for a classical, free particle. How do we describe the motion of electrons in semiconducting crystals?

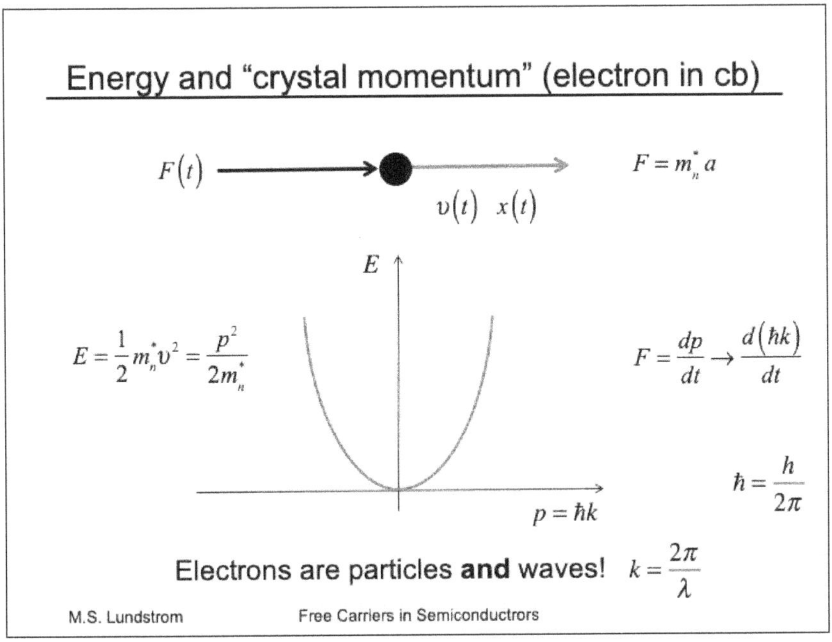

For free electrons in semiconductors, the energy vs. momentum relation can often be described as a parabola, just as for a free particle, but the “momentum” or the **crystal momentum** is not actually the momentum of the electron.

Recall that electrons behave as both particles and as waves. As waves, they have a **wavelength**, λ, and a **wavevector**, $k = 2\pi/\lambda$. As we will discuss in Part 2, for a free electron in a vacuum, the momentum is $p = \hbar k$, where $\hbar = h/2\pi$ with h being Planck's constant.. For an electron in a semiconductor, the crystal momentum is $p = \hbar k$. The crystal momentum is a quantity that behaves like the momentum. For example, momentum is conserved when electrons in vacuum collide; crystal momentum is conserved when electrons in semiconductors collide. When we plot $E(p)$ vs. p for electrons in semiconductors, the p means crystal momentum. Instead, we'll often plot $E(k)$, the electron's energy vs. its wavevector, k.

For a classical particle, $F = dp/dt$ where F is the actual force on the particle and p is its momentum. For electrons in a semiconductor, $F = d(\hbar k)/dt$. Here, F is not the total force on the electron, forces due to the crystal potential are embedded in the effective mass. This additional force, F, is the “external” force due to applied electric and magnetic fields.

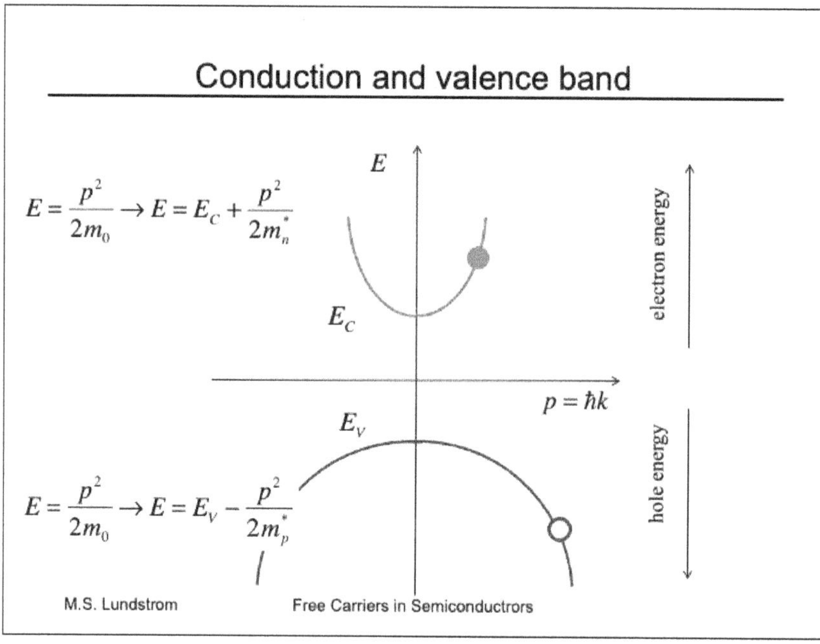

There are electrons in the conduction band and holes in the valence band. We've seen that we can write the energy vs. momentum for electrons in the conduction band as the bottom of the conduction band, E_C, plus its kinetic energy, $p^2/2m_n^* = \hbar^2 k^2/2m_n^*$. What about holes in the valence band? For the valence band, $E(k)$ bends down in energy. It has to bend down with increasing momentum because the states in the valence band are all below the top of the valence band, E_V, so we can write $E(p)$ as the top of the valence band, E_V, minus $p^2/2m_p^*$.

To understand this a little more, think about creating a hot hole. (Recall from Lecture 1, Sec. 5 that this can be done by absorbing light with a very high energy.) Hot holes lie deep in the valence band at a very low energy, but the higher energy electrons in the valence band will drop down in energy and fill the holes. So the holes will move up, and within a femtoseconds the holes will be near the top of the valence band.

Our $E(k)$ vs. k plots are plots of **electron** energy vs. k. The hot hole has a lot of kinetic energy, as it sheds its energy, it relaxes up in electron energy. In $E(k)$ plots, the hole energy increases downward while electron energy increases upwards.

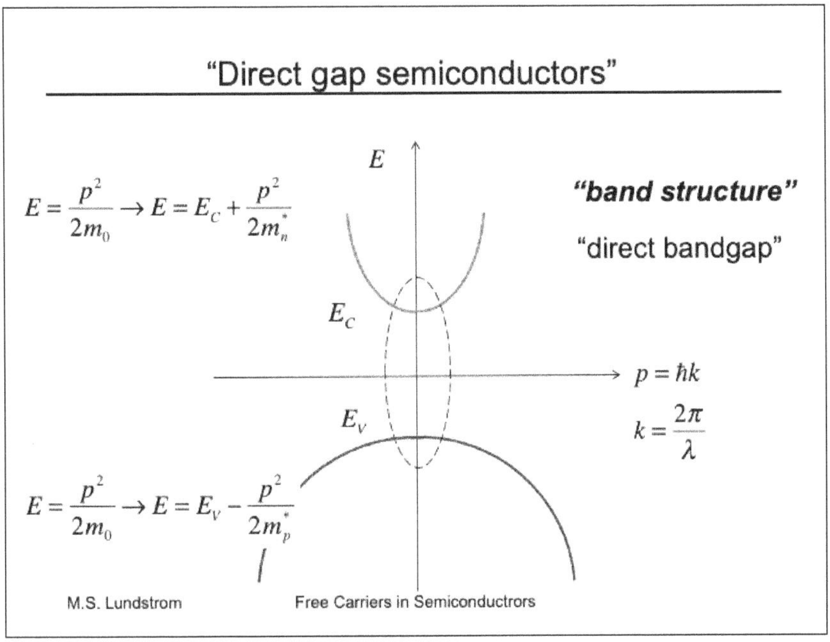

5.4 Direct and indirect gap semiconductors

We call $E(k)$ or $E(p)$ plots like the one above **band structure plots**. I've mentioned earlier that there are two different types of semiconductors. There are semiconductors that emit light when electrons and holes recombine – those are called **direct gap semiconductors**. We can now explain what a direct gap semiconductor is. The band structure of a direct gap semiconductor is shown above. In a direct gap semiconductor, the minimum of the conduction band and the maximum of the valence band occur at the same crystal momentum, so when an electron and a hole recombine, it's an electron and a hole with the same crystal momentum. These direct gap semiconductors like gallium arsenide have a very useful property, they emit light when carriers recombine. This is the direct recombination process discussed in Lecture 4, Sec. 5. Finally, it's also important to know that direct gap semiconductors also absorb light very effectively.

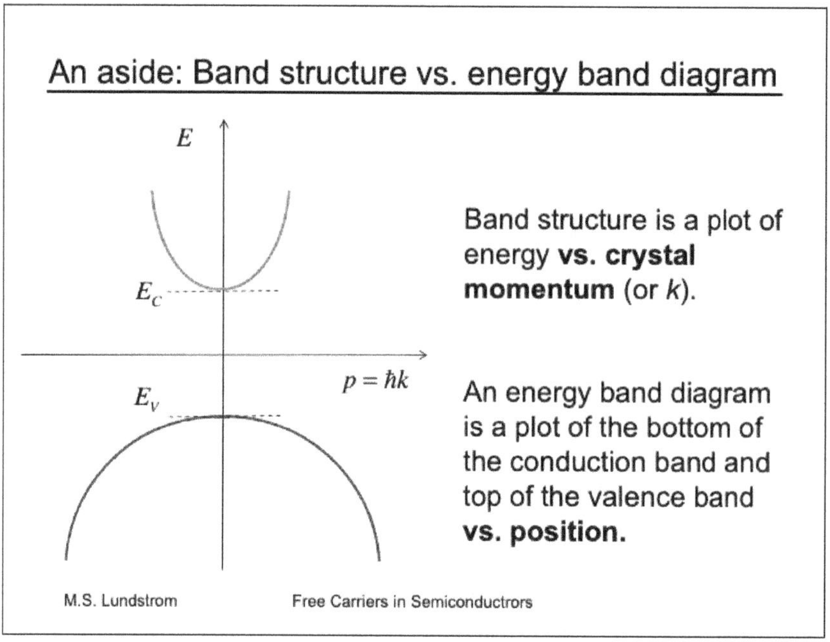

Before we move on, let's be clear what we mean by a **band structure plot** and an **energy band diagram.**

When we plot $E(k)$ vs. k, we are producing a band structure plot. A **band structure plot** is plot of energy vs. crystal momentum or energy vs. k. The quantity $\hbar k$, where $k = 2\pi/\lambda$, behaves like the momentum of an electron in a crystal.

An **energy band diagram** is another type of plot. It is simply a plot of the bottom of the conduction band, E_c and the top of the valence band, E_v vs. position.

In this course, we will make use of both types of plots — band structure plots, $E(k)$ vs. k, and energy band diagrams, $E_c(x)$ and $E_v(x)$ vs. position, x. In Part 2 we'll discuss band structure plots more, but as the course progresses, energy band diagrams will become more and more useful to us.

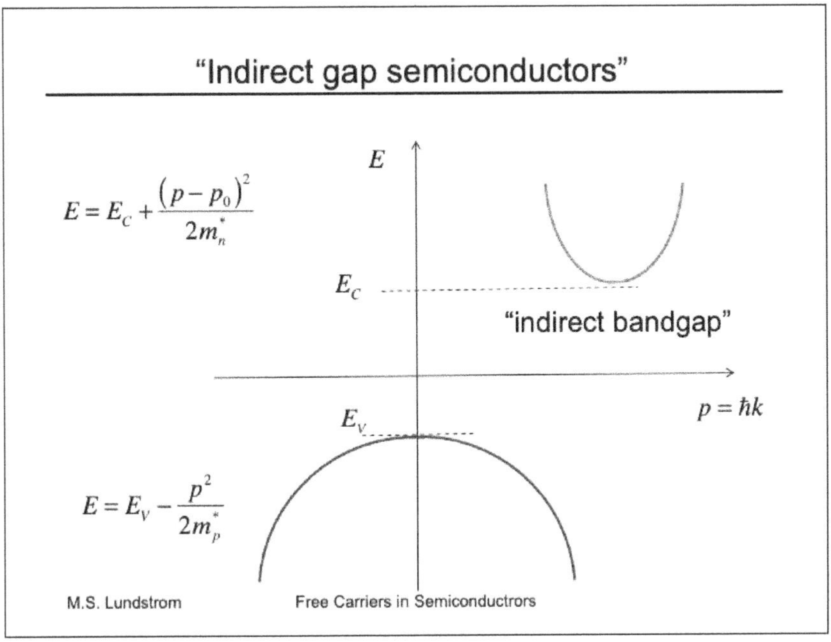

Semiconductors like silicon are **indirect gap semiconductors**. In these materials, recombination processes don't tend to emit light. How do we draw an indirect gap semiconductor on an energy band diagram? Well, it looks something like the figure above. This is a band structure plot where the bottom of the conduction band and the top of valence band occur at different crystal momentum. Because of the mismatch in momentum, it is difficult in these kinds of semiconductors for electrons and holes to recombine in a way that gives off light. Instead, recombination tends to produce heat rather than light. Indirect recombination processes are discussed in Lecture 4, Sec. 5.

Finally, it's important to also know that indirect gap semiconductors don't absorb light as effectively as direct gap semiconductors.

5.5 Summary

Conduction band electrons are free to move about the crystal and carry currents. They can often be treated as Newtonian particles with an effective mass that accounts for the complicated interactions with the crystal potential. Valence band holes can be thought of positively charged particles that are free to move about within the crystal and also conduct current. They can often be treated as Newtonian particles with a different effective mass. We will regard effective masses as known quantities that can be looked up.

In this lecture, we introduced the concept of band structure, the energy of an electron in the conduction band vs. its crystal momentum and the energy of a hole in the valence band vs. its crystal momentum. While band structures can get complicated, in these lectures, we'll assume a simple parabolic energy band structure, energy varies as crystal momentum squared. Part 2 will provide a little more insight into band structure.

There are two kinds of semiconductor band structures, direct gap and indirect gap. For direct gap semiconductors, the minimum of E_c and the maximum of E_v occur at the same crystal momentum. They absorb light efficiently, and produce light efficiently when free carriers recombine. For indirect gap semiconductors, the minimum E_c and the maximum E_v occur at different crystal momentum. They absorb light less efficiently, and produce more heat than light when free carriers recombine.

We'll revisit these concepts and have a chance to get more and more familiar with them as the lectures proceed. We have not yet, however, talked about the most important property of semiconductors, which is the ability to dope them. Doping, which changes the electronic properties of semiconductors in ways that enable the operation of electronic devices, is the subject of the next lecture.

5.6 References

Those who wish to dive deeper into the concepts introduced in this lecture may wish to consult the following two texts:

[1] R.F. Pierret, *Advanced Semiconductor Fundamentals*, (2nd Ed.), Pearson Education, Inc., 2003 (see Ch. 3).

[2] Sandip Tiwari, *Semiconductor Physics: Principles, Theory, and Nanoscale*, Oxford Electroscience Series, Vol. 3, 2020 (see Ch. 4).

Lecture 6

Doping

6.1 Introduction

Doping is what makes semiconductor devices possible. As you know, metals are good electrical conductors, and insulators are poor electrical conductors. Semiconductors are in between, but they have a remarkable property. We can control their electronic properties such that they can become reasonably metallic or reasonably insulating or anything in-between; that is what makes semiconductors so useful.

Because early work with semiconductors was with materials that had unknown concentrations of defects in them, they produced all kinds of effects that made it very difficult to understand what was going on and very difficult to reproduce experiments. In 1931, the physicist, Wolfgang Pauli, said: "One shouldn't work on semiconductors, that is a filthy mess; who knows whether any semiconductors exist."

Things have changed, and we are now able to grow very high purity semiconductors and reproducibly introduce dopants that are intentionally put into the semiconductor lattice to alter the electronic properties in specific ways. How this occurs is the subject of this lecture.

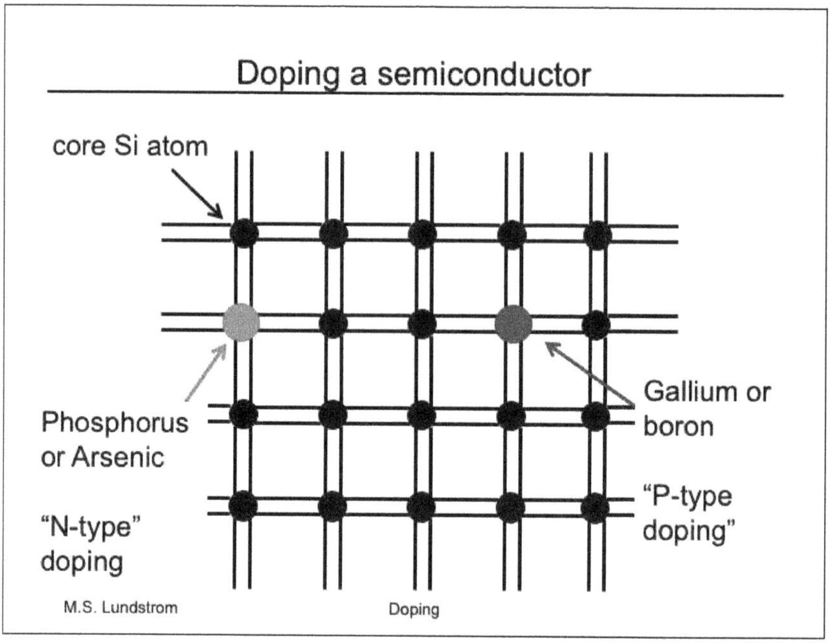

6.2 Substitutional doping

Shown above is our cartoon of the silicon or a group IV semiconductor lattice. Each filled circle represents an atom, each line represents a covalent bond, and each atom has four nearest neighbors. We see above the shared valence electrons and the core atoms.

Doping a semiconductor means that we introduce other atoms into the crystalline lattice. In **substitutional doping**, these atoms reside on a lattice site replacing a few Si atoms. They are sometimes called "impurities," but they are intentionally introduced. For example, we might introduce phosphorus or arsenic into silicon or germanium, and this would produce **N-type** or negative doping. We'll explain next what we mean by that. Or we might introduce an atom like gallium or boron and would call that **P-type**, or positive doping. We'll explain that too.

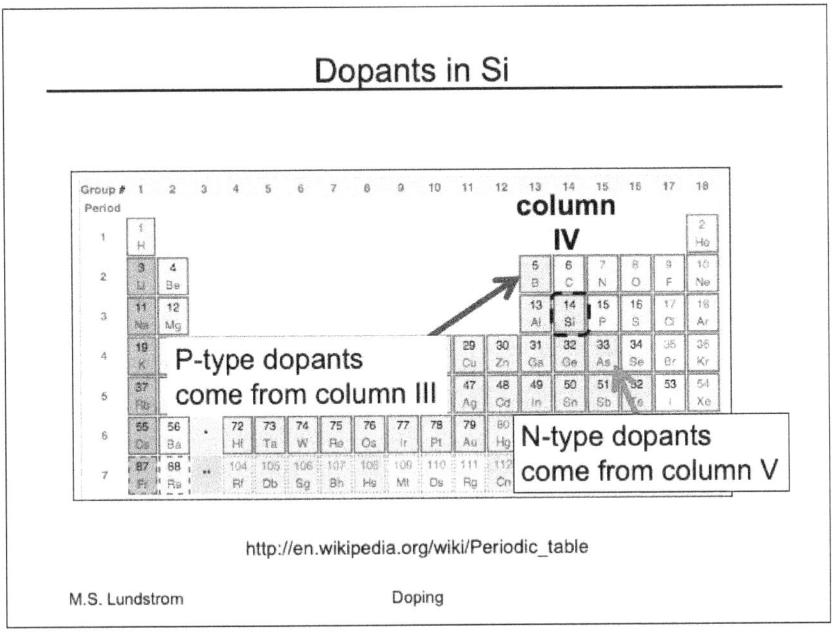

To understand how doping works, let's go back to the periodic table and consider a common semiconductor like silicon. Silicon has four valence electrons. We find that N-type dopants come from column V; elements in column V have five valence electrons. The P-type dopants come from column III; elements in column III have three valence electrons. In column V, we see atoms like arsenic and phosphorus. These are N-type dopants in silicon. In column III, we see boron, aluminum, and gallium — these are P-type dopants.

Just a reminder about how groups (columns) are labeled. In the modern, IUPAC notation, groups are labeled from 1 to 18. In an older notation, the carbon group (IUPAC 14) was labeled IV because there are 4 valence electrons in this elements in this column. In semiconductor physics, the older notation is universally used.

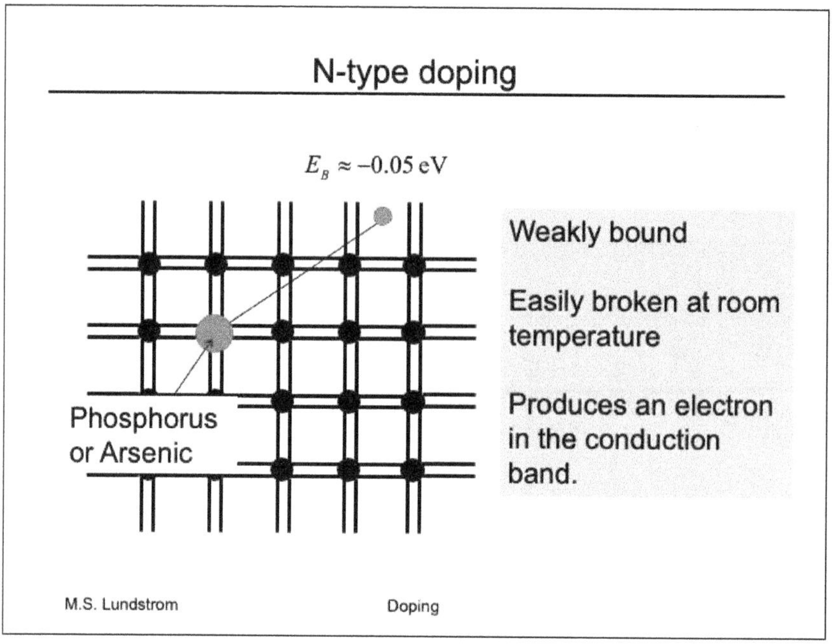

6.3 N-type doping

We'll discuss N-type doping first.

There are ways (e.g. diffusion, ion implantation) to replace a few of the silicon atoms in the lattice with atoms such as phosphorus or arsenic from column V. In **substitutional doping** a column V element replaces a column IV element in the crystal lattice. Column V atoms have five valence electrons, so four of those five can be shared with the four nearest neighbors thereby satisfying those covalent bonds, but there is one electron left over, and that one electron turns out to be very weakly bound to the phosphorous or arsenic atom. The binding energy is much less than a tenth of an electron volt. It's easy at room temperature for the thermal energy to jiggle the lattice and break that bond, which produces an electron that can wander around in the crystal lattice.

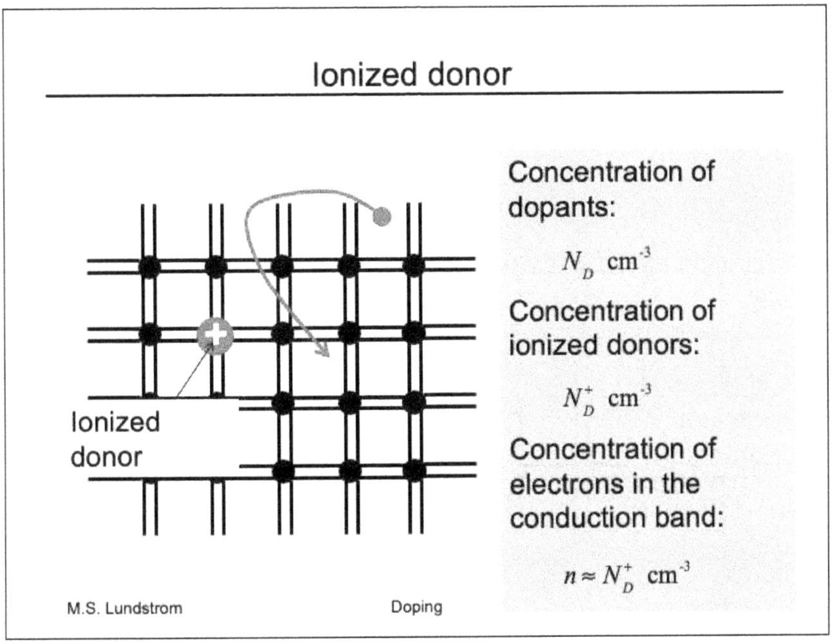

Remember that the fifth electron made the phosphorous atom neutral. We had an equal number of electrons and positively charged protons in the nucleus, so the net charge adds up to zero. When the fifth electron's bond is broken and the electron is free to wander away throughout the silicon lattice, the phosphorus atom is ionized. It is left with a net positive charge. We call this an **ionized donor**. The term, **donor** indicates that the impurity has donated an electron to the conduction band.

The concentration of phosphorus atoms is $N_D\,\mathrm{cm}^{-3}$, What's important though is the concentration of phosphorus atoms (or, more generally donor atoms) that are ionized. We'll label that as N_D^+, because an ionized donor has a positive charge. Ionized donors have donated an electron to the conduction band to produce the doping that we're after.

What is the concentration of electrons in the conduction band after introducing N-type dopants? There were intrinsic carriers there before we introduced the donors, about $1 \times 10^{10}\,\mathrm{cm}^{-3}$, but typical doping concentrations overwhelm the intrinsic carriers, so that the concentration of electrons in the conduction band is very nearly the concentration of ionized donors — the concentration of donor atoms that have donated their fifth electron to the conduction band. We conclude that $n \approx N_D^+$.

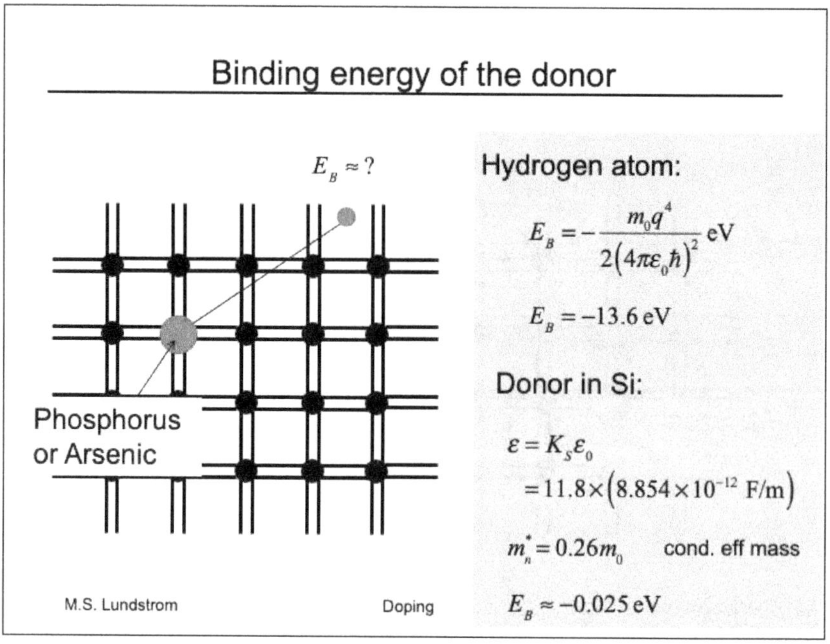

6.4 Binding energy of a dopant

Let's examine the binding energy of the fifth electron, the one that's left over after the other four electrons form covalent bonds. A hydrogen atom has only one electron and one proton. That's similar to what we have here. Four of the valence electrons are in covalent bonds, and we have one left over. There is a net positive charge of one on the donor atom and one electron not in a covalent bond balances that charge, much like a hydrogen atom. In first-year chemistry, you may have seen the formula above for the binding energy of a hydrogen atom. When we put in numbers, the result is 13.6 eV.

Substitutional dopants in silicon behave like hydrogen atoms, but one important differences is that the dielectric constant, ϵ, is much higher than in vacuum. The relative dielectric constant of silicon, K_s, is about 12. That quantity is squared, so that's over 100, so the binding energy is much less than 13.6 eV. Putting in numbers, we find that it's less than 0.1 eV. A more detailed calculation, shows that the binding energy is even less than that — more like 0.05 electron volts which is a very weakly bound electron. The bond is easily broken at room temperature.

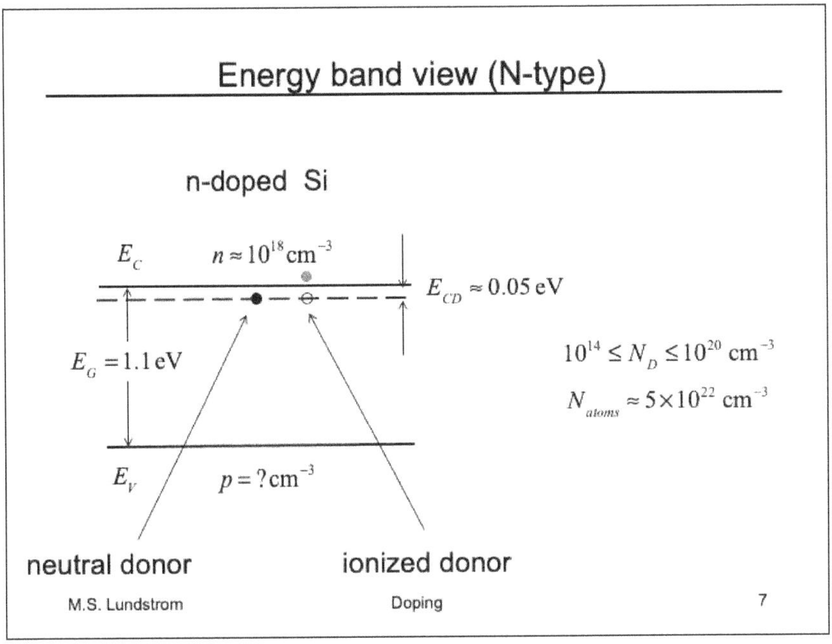

We saw energy band diagrams in Lecture 1; what changes if we have doped the semiconductor N-type? What changes is that dopants introduce states in the forbidden region where there were no states.

A donor atom introduces a **localized states** because the fifth electron sits on some specific donor atom; it is located a small energy below the conduction band (the binding energy of the 5th electron). The donor is neutral, but if we break that weak bond and release the electron, we now have an empty state on the donor, so it acquires a positive charge; it is now ionized, and the electron is now in the conduction band, in an **extended state**, free to move away from the specific dopant that produced the free electron.

Shown above is an energy band diagram for an N-type semiconductor. Moderate doping densities are in the range, $10^{16} - 10^{18}$ donors per cm^3. A concentration of 10^{14} would be considered a light doping. A concentration of 10^{20} would be considered a very heavy doping, about as heavy as we're able to dope it because there's a solid solubility limit for phosphorous or arsenic in silicon. Remember, however, that the number of atoms per cm^3 in Si is about 5×10^{22}, so even when we dope silicon as heavily as we can, fewer than 1% of the atoms are donor atoms.

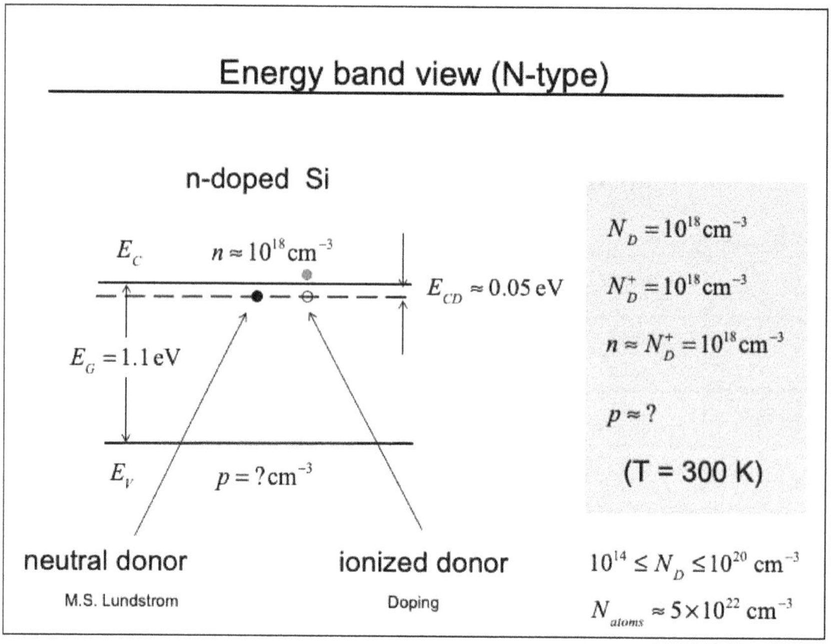

To continue our discussion...

For phosphorus and arsenic in silicon, the binding energy of the fifth electron is very weak, easily broken at room temperature, so the concentration of ionized dopants at room temperature is virtually identical to the concentration of dopant atoms. That means that the concentration of electrons in the conduction band is also very nearly equal to the number of ionized dopants. Remember, the intrinsic carrier density is $10^{10}\,\text{cm}^{-3}$, and we've overwhelmed those **intrinsic carriers** with **extrinsic carriers**.

Next, we might ask what is the hole concentration? We haven't changed anything in the valence band. Before we introduced the N-type dopants, the concentration of holes was $p = n_i = 10^{10}\,\text{cm}^{-3}$. Is p still $10^{10}\,\text{cm}^{-3}$? Well, interestingly enough the answer is no, it's different, in fact it's even lower. We'll talk about how to compute the hole concentration in an N-type semiconductor that later on in these lectures.

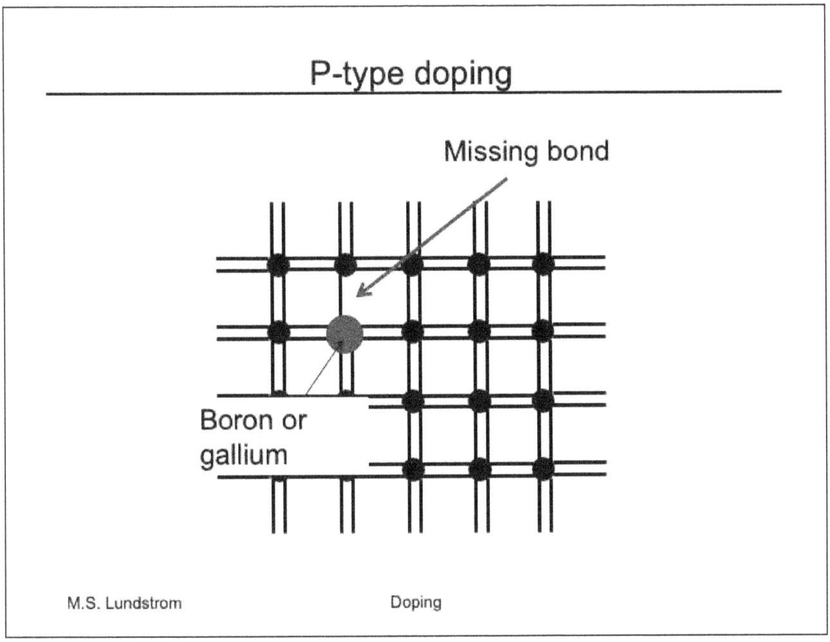

6.5 P-type doping

Let's discuss P-type doping. If we replace one of the silicon atoms with an atom from column III, then there aren't enough electrons to share with the four nearest neighbors to make covalent bonds. One of the bonds is missing. The P-type dopant, the **acceptor**, is neutral with its three valence electrons.

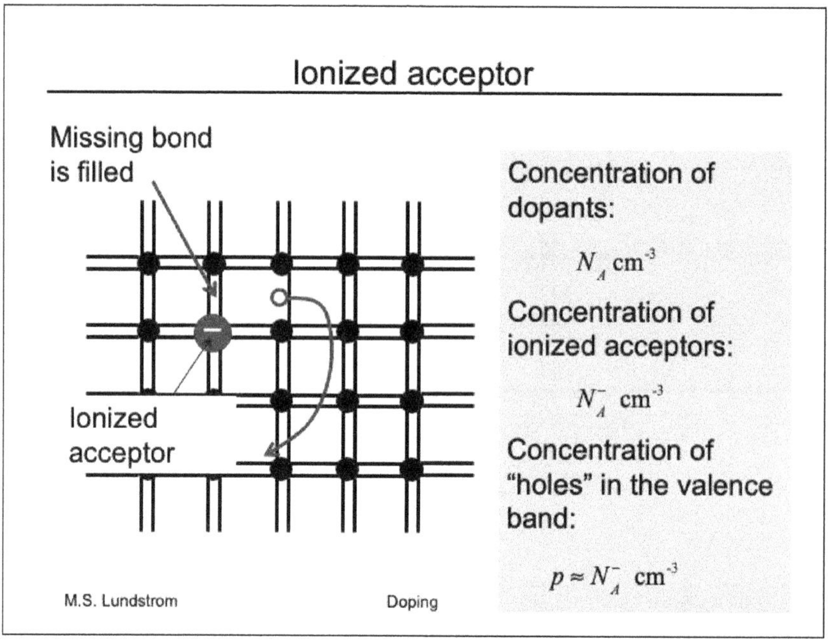

When an adjacent electron hops in and fills the missing covalent bond, the boron atom acquires a negative charge. With the extra electron, the boron atom is ionized. It was neutral with its three valence electrons, but now it has four electrons, so it acquires a negative charge. The hole produced in this process is now is free to move throughout the silicon lattice. We might think of the hole as analogous to a bubble in a carbonated beverage. The bubble is an absence of liquid that is free to move within the liquid.

Ionized acceptors produce holes in the valence band. If N_A is the concentration of acceptor atoms, boron or gallium or whatever, then N_A^- is the concentration of ionized acceptors that have acquired an extra electron and have created a hole that is free to move about the silicon lattice. The number of holes will be very nearly equal to the number of ionized acceptors because even a moderate P-type doping will overwhelm the intrinsic carriers.

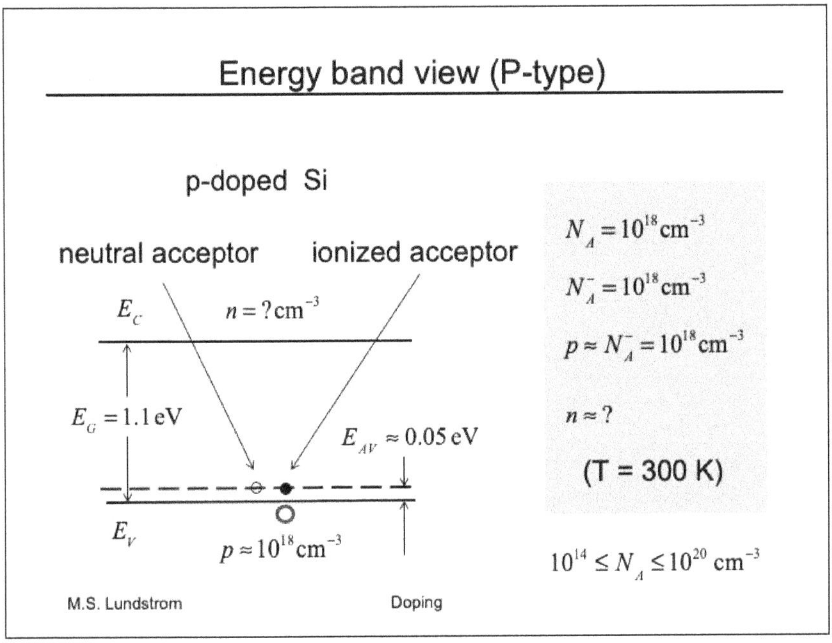

Shown above is an energy band diagram for a semiconductor doped P-type. It takes energy for an electron to hop from a state in the valence band to an empty state on an acceptor. We show the acceptor level energy a little bit above the top of the valence band. When the P-type dopant has its three valence electrons, it is electrically neutral. When an electron from the valence band hops up and occupies an empty state on the acceptor, it fills the missing covalent bond and leaves behind a hole in the valence band. Because there is one additional electron on the acceptor, it acquires a negative charge. Ionized acceptors are negatively charged.

A light P-type doping would be about $10^{14}\,\text{cm}^{-3}$. For silicon, the heaviest possible P-type doping is about $10^{20}\,\text{cm}^{-3}$, which is similar to the maximum N-type doping. For higher concentrations, dopants tend to cluster or they end up in interstitial sites between silicon atoms. In the doping concentration is not too high, then we have **substitutional doping** — a dopant takes the place of a silicon atom in the lattice.

Finally, we can ask: "What happens to the electron concentration in the conduction band?" It turns out that we've actually lowered the electron concentration from the $n_i = 10^{10}\,\text{cm}^{-3}$ before we introduced the acceptors. We'll discuss why this happens in Part 3.

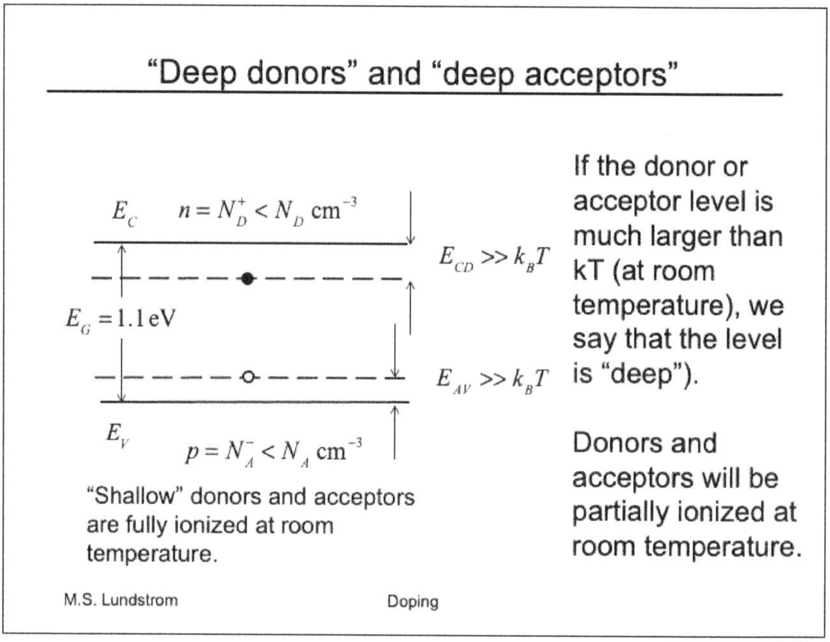

6.6 Deep dopants

One of the things that makes Si so nice to work with is that we have "good dopants," dopants that are fully ionized at room temperature. If the binding energy of the fifth electron on a donor is significantly larger than the thermal energy, $k_B T$, then it will be difficult to break the bond and move the electron into the conduction band. Similarly, if the acceptor level energy is too high, then it is difficult to take a valence band electron and move it onto the acceptor and create a hole in the valence band. We call these **deep donors** or **deep acceptors** because they're energetically deep. For some semiconductors, the dopants are deeper inside the forbidden gap than are the dopants in silicon. For example, the binding energy of the fifth valence electron in a donor may be large. Such dopants are not fully ionized at room temperature. For the rest of these lectures we'll be focusing on the simple case of **shallow dopants** that applies to silicon. In these lectures, we will assume that all dopants are fully ionized at room temperature.

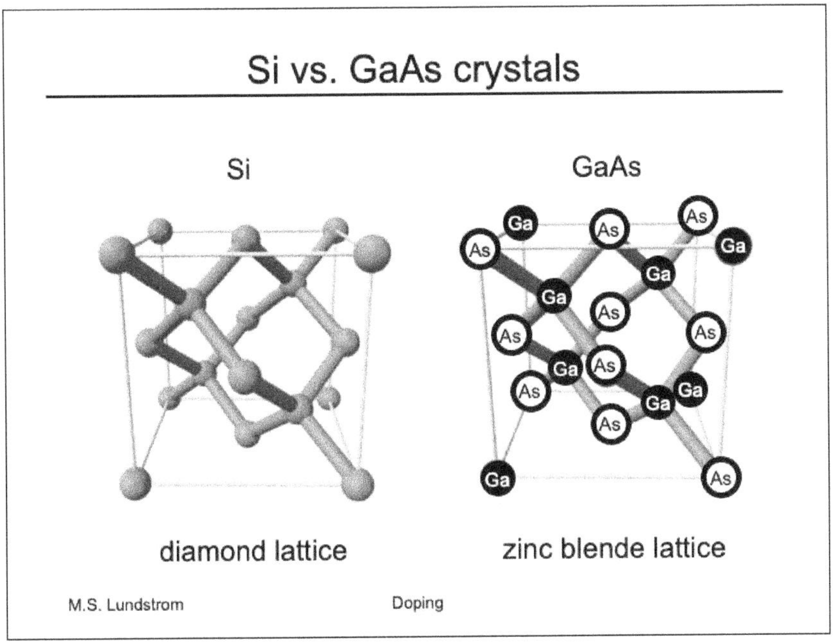

6.7 Zinc Blende crystals and amphoteric dopants

Now, let's make things a little more complicated. Consider a semiconductor like gallium arsenide doped with silicon. Remember, gallium is in column III, and arsenic is in column V. On average, GaAs has four valence electrons per atom. Silicon is in column IV, it has four valence electrons per atom. So, we can ask: "Is Si in GaAs an N-type dopant or is it a P-type dopant? Let's see if we can figure this out.

Shown above on the left is the silicon lattice, and the gallium arsenide lattice on the right is similar. We don't call the GaAs structure a diamond lattice, we call it a **zinc blende lattice**, but the structure is the same. The only difference is that every gallium atom is surrounded by four arsenic nearest neighbors, and every arsenic atom is surrounded by four gallium nearest neighbors. The crystal structure of gallium arsenide is very similar to the diamond structure of silicon, it just has two different types of atoms.

The diamond lattice got its name because it is the lattice that describes crystalline diamond. Similarly, zinc blende got its name because it is the lattice that describes crystalline zinc sulphide.

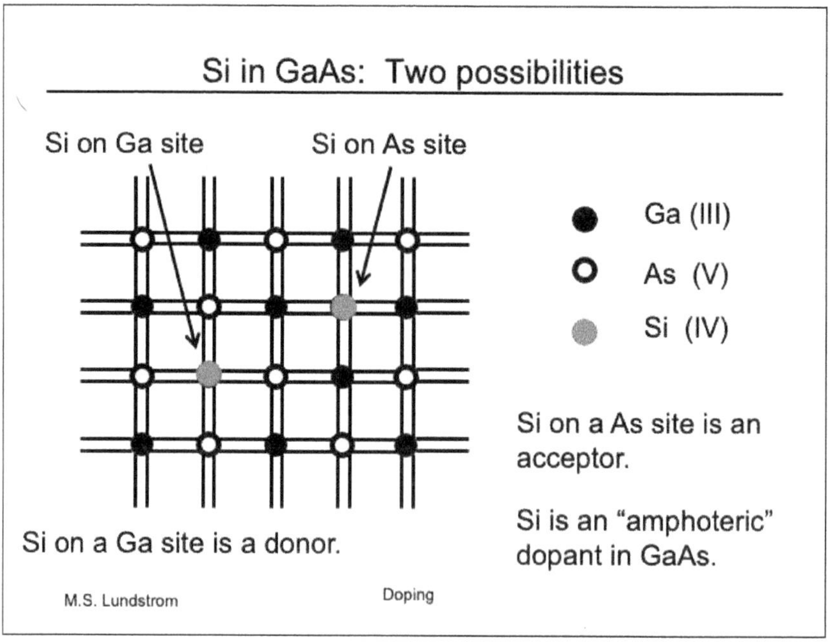

In our cartoon of the GaAs lattice, we draw Si doped GaAs as shown above. Every gallium atom is surrounded by four arsenic atoms. Every arsenic atom is surrounded by four gallium atoms.

Now let's ask what happens when we dope GaAs with Si. First, let's put a silicon atom on a gallium site. Gallium is from column III, it has three valence electrons. Silicon is from column IV, it has four valence electrons, so we have one extra electron. That means that silicon, if it happens to be sitting on a gallium site, is a donor.

Next, put silicon on an arsenic site. Arsenic is from column V, silicon is from column IV. Silicon has one fewer electrons than the arsenic atom it replaced, so silicon on an arsenic site is an acceptor.

We call silicon an **amphoteric dopant**, which means it can be either N-type, or P-type depending on which atom it replaces in the lattice. Depending on exactly how we introduce these dopants, and exactly how we do the annealing, we can favor one site vs. the other, so we can make N-type dopants or P-type dopants from the same element.

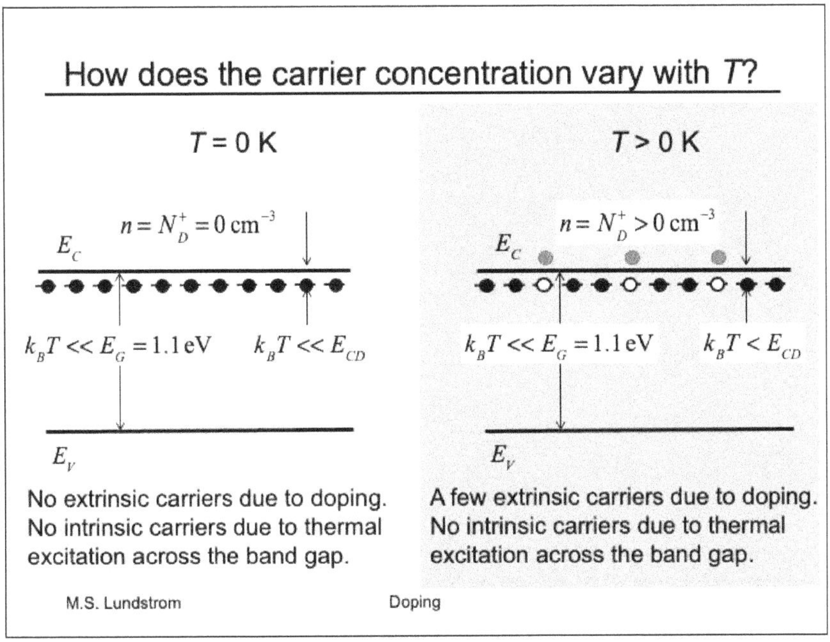

6.8 Carrier concentration vs. temperature

With this introduction to doping, we can now understand how the carrier concentration varies with temperature. The slide above shows the situation for a good, shallow N-type dopant. The energy needed to ionize the donor is small, so it is easily ionized at room temperature, but if we go down to $T = 0$ Kelvin, there is essentially no thermal energy. As shown above on the left, all of the fifth electrons would be sitting on their donor sites. There would be no electrons in the conduction band — no ionized donors and no extrinsic carriers due to doping. We would also have no intrinsic carriers due to thermal excitation across the band gap, because it takes an even larger thermal energy to break the covalent bonds. We conclude that Si is an insulator at $T = 0$ K because there are no mobile charge carriers.

If we raise the temperature at bit, the thermal energy will start breaking the weak bonds of the donor electron and a few electrons in the conduction band will be created. As shown on the right, we now we have some electrons in the conduction band, but the concentration of intrinsic carriers is still negligible.

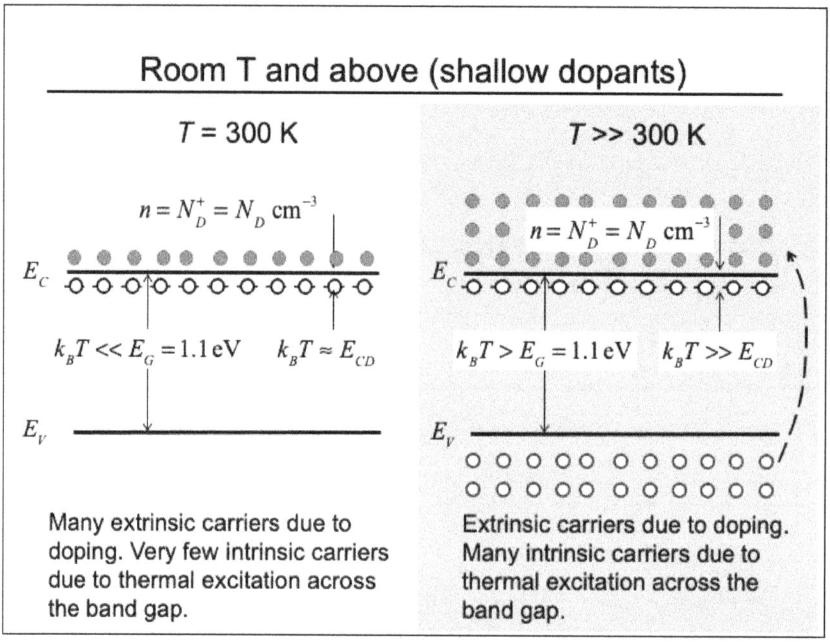

If we heat the semiconductor up to 300 K, then good dopants like phosphorus or arsenic in silicon will be fully ionized at 300 K. As shown above on the left, we will have one electron in the conduction band for every donor atom. There will be some intrinsic carriers too, but for typical doping densities in Si, their concentration at room temperature is negligible.

If we continue to heat the lattice, the increasing thermal energy will start breaking covalent bonds and creating electron-hole pairs by thermal excitation across the band gap. As shown on the right, the intrinsic carriers will begin to overwhelm the doping. When that happens, the semiconductor has become intrinsic. Even though we have dopants in the lattice, their concentration is now small compared with the intrinsic carrier density.

We build semiconductor devices by selectively doping different regions of a semiconductor. If we operate these devices at high temperatures, we can lose all of the intended effects of the dopants because they are overwhelmed by intrinsic careers. This is one of the reasons that when you heat devices up too much, semiconductor devices can cease to function.

For operation at very high temperature, semiconductor devices should be made with semiconductors such as SiC or GaN, which have a large band gap.

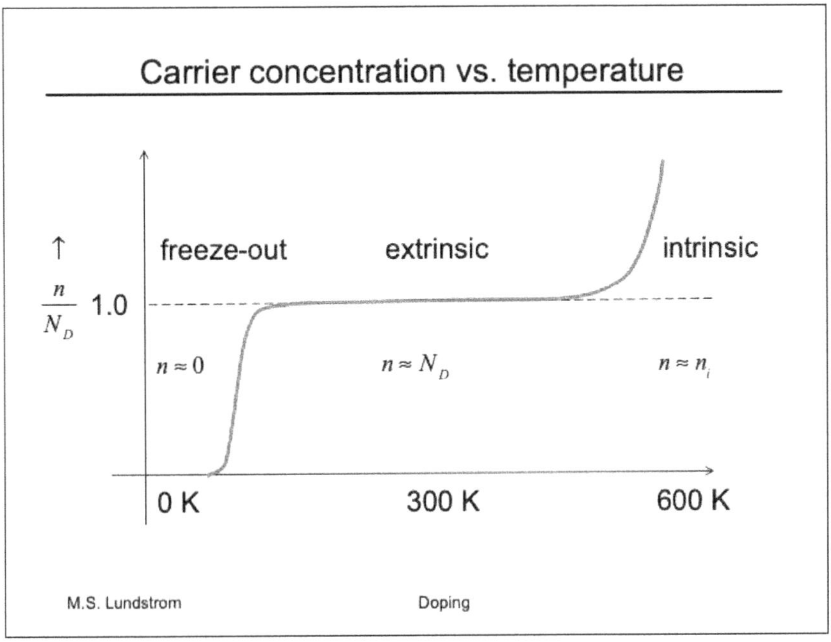

The electron concentration vs. temperature is shown above. Near $T = 0$ K, there won't be any electrons in the conduction band because there's too little thermal energy to ionize the dopants or to break the covalent bonds that give intrinsic carriers. We call this range of temperatures the **freeze-out region**. The carriers freeze out to their dopants and are localized on their specific dopants; there are no free carriers.

As we warm the semiconductor up, we begin breaking the bonds that hold the fifth electron to their donor atoms, and the electron concentration increases. At a high enough temperature, all donors are ionized. This is the **extrinsic region** where the electron concentration equals the doping density. We'll spend most of these lectures in this region.

If we continue to heat up the semiconductor, we can get to the **intrinsic region** where breaking the covalent bonds overwhelms the carriers that were produced by dopants.

Note that semiconductors can be conductive ("metallic") at room temperature and insulating at low temperatures. In fact, the experimental "signature" that the material being measured is a semiconductor is that it becomes an insulator at low temperatures. A plot of hole concentration vs. temperature would look similar to the plot shown above.

6.9 Summary

Doping replaces a few of the atoms in the crystal lattice with atoms from different columns of the periodic table. An ionized donor produces free electrons in the conduction band, and an ionized acceptor produces free holes in the valence band. A shallow dopant is easily ionized at room temperature and is an effective dopant. A deep donor or deep acceptor is harder to ionize at room temperature and is not as effective as a dopant.

The carrier concentration vs. temperature characteristic has a freeze-out region where the carriers freeze out to their individual dopants. It has an extrinsic region where the dopants are fully ionized contributing carriers to the conduction or valence band, and it has an intrinsic region where the breaking of the covalent bonds overwhelms the doping. One of the ways that an experimentalist who is exploring a new material determines whether he or she is dealing with a semiconductor or a metal, is to lower the temperature to near absolute zero, and see the material still continues to conduct electricity. If it does, it's a metal. If it becomes an insulator, it's a semiconductor.

Finally, just a reminder to be careful about units. We will quote dopant and carrier concentrations per cubic centimeter. Proper SI units would be per cubic meter. For example. $N_D = 10^{26}\,\text{m}^{-3} = 10^{20}\,\text{cm}^{-3}$. The safest way to do calculations is to do them in SI units first, and then convert per cubic meter to per cubic centimeter, but we can frequently take shortcuts.

As we wrap up Part 1, it is a good time to review the learning outcomes:

- Explain the terms: "conduction band," "valence band," "forbidden gap" (or band gap), "electron," "hole," and "intrinsic carrier concentration."
- Explain the difference between a metal, an insulator, and a semiconductor in terms of their energy bands.
- Explain the difference between crystalline, polycrystalline, and amorphous semiconductors and be familiar with the crystal structure of Si and GaAs.
- Given the Miller indices, draw the corresponding plane or vector.
- Be able to draw the energy vs. momentum (or energy vs. k) diagram for direct and indirect gap semiconductors.
- Be able to explain how N- and P-type doping is accomplished and the carrier concentration vs. temperature characteristics of semiconductors.

We've covered a lot of ground, but we introduced basic concepts that we will return to, discuss, and extend upon in subsequent lectures.

6.10 Reference

For another treatment of semiconductor doping, see Pierret:

[1] R.F. Pierret, *Advanced Semiconductor Fundamentals,* (2nd Ed.), Pearson Education, Inc., 2003 (see Ch. 4, Sec. 4.3.3).

PART 2
Rudiments of Quantum Mechanics

Lecture 7

The Wave Equation

7.1 Introduction

In Part 1 we discussed electrons and holes in semiconductors and described them as particles that behave according to appropriately defined "Newton's laws." Quantum mechanics tells us that a particle is also a wave, so we will need some basic understanding of quantum mechanics. If you have had a course in quantum mechanics, this lecture will be a refresher. If you have seen little or no quantum mechanics previously, this lecture will introduce you to the basics. By the end of this first lecture in Part 2, you will have some familiarity with the Schrödinger wave equation:

$$-\frac{\hbar}{i}\frac{\partial}{\partial t}\Psi(x,t) = -\frac{\hbar^2}{2m_0}\frac{\partial^2}{\partial x^2}\Psi(x,t) + U(x)\Psi(x,t)$$

Upon completion of Part 2, you will be able to:

- Understand the rudiments of quantum mechanics and how to solve the Schrödinger wave equation.
- Explain key concepts such as phase velocity, group velocity, wavevector, momentum of a plane wave, and the uncertainty relations.
- Explain how quantum confinement affects electrons, what is meant by 1D, 2D, and 3D semiconductors, and what subbands are.
- Understand quantum mechanical tunneling and reflection and how the barrier height and width affect the tunneling probability.
- Explain how $E(k)$ differs for a free electron and for an electron in a periodic crystal potential. Be familiar with $E(k)$ and the constant energy surfaces for common semiconductors. Know what a parabolic energy band is.
- Explain what $D(E)dE$ (where $D(E)$ is the density-of-states) represents and how $D(E)$ varies with energy in 1D, 2D, and 3D.

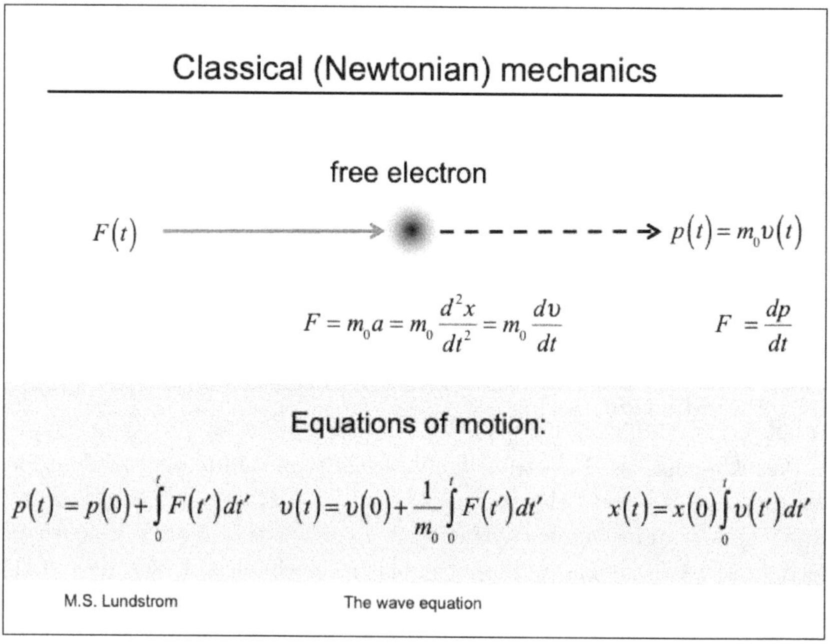

7.2 The need for quantum mechanics

A free electron in vacuum can be thought of as a particle. We apply a force, and the momentum and velocity of the electron change. We compute the velocity and momentum as a function of time from Newton's laws. Force is dp/dt, where p is the momentum. We can compute how the momentum varies as a function of time, how the velocity varies as a function of time, and how position varies as a function of time using equations that we were introduced to in first-year physics. Now we need to go beyond classical physics.

Quantum mechanics was developed in the early part of the 20th century, as scientists worked to explain certain features of nature that were puzzling:

- Black body radiation (Planck, 1901)
- Photoelectric effect (Einstein, 1905)
- Atomic spectra (Bohr, 1913)
- Wave-particle duality (de Broglie, 1924)

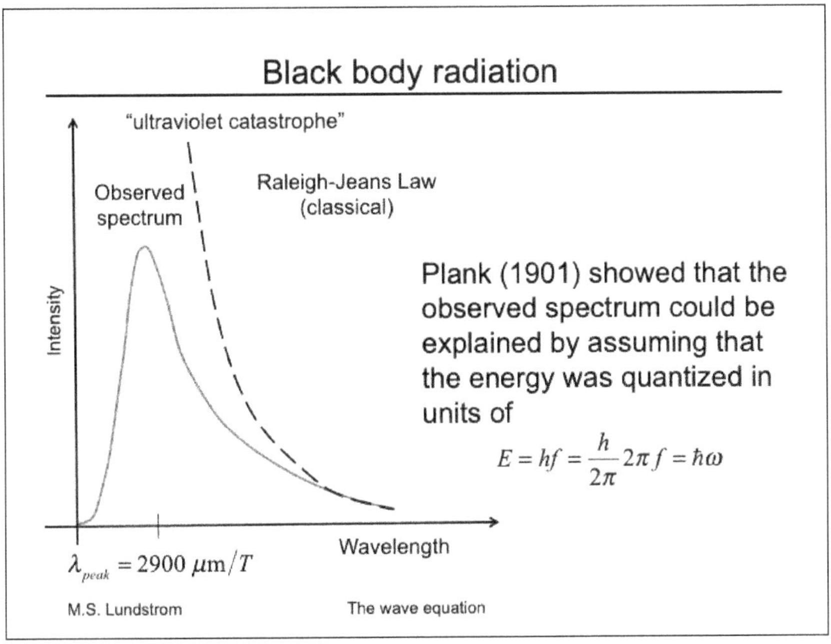

The first puzzle was **black body radiation**. A heated mass emits radiation. A black body emits radiation with an intensity (the number of watts per square centimeter, per unit frequency) vs. wavelength that looks like the solid line shown above. There is a peak intensity at a wavelength of $\approx 2,900\,\mu\text{m}/T$ (where T is the temperature in Kelvin) – no matter what specific material the body is made of. At longer and shorter wavelengths, the intensity decreases towards zero. When physicists tried to compute this spectrum, there was a problem, the so-called "ultraviolet catastrophe" – the intensity diverged at short wavelengths, as shown by the dashed line above. Physicists struggled to explain why the classical calculation failed.

In 1901, Max Planck discovered that he could explain these characteristics if he assumed that oscillating atoms in the black body could only have a series of discrete energies. If an oscillator transitioned from a higher energy to a lower energy, the frequency of the light emitted was given by $\Delta E = nhf = n\hbar\omega$, where n is an integer, h is now called Planck's constant, $\hbar = h/2\pi$, f is the frequency in cycles per second (Hz), and ω is the frequency in radians per second.

It wasn't clear to Planck why the energy should be quantized, but he could explain the measured characteristic if he assumed that it was.

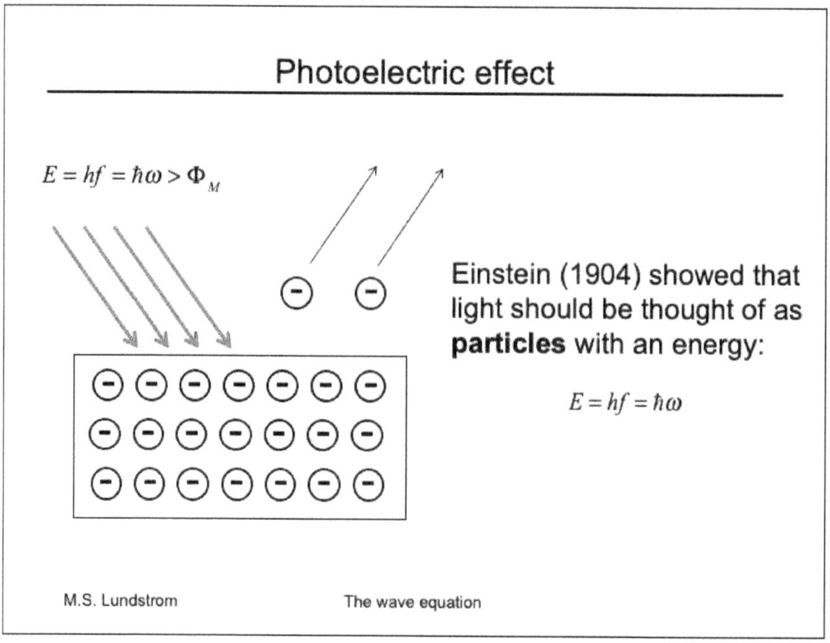

A few years later, Albert Einstein explained the **photoelectric effect**. When we shine light on a metal and then measure the number of electrons that are ejected, it was found that it didn't matter how intense the light was — that is, how many photons per second, per square centimeter were incident on the metal. The light had to be above a certain critical frequency before electrons were ejected from the solid. That critical energy was the **work function**, Φ_M, of the metal. Einstein showed that light could be thought of as particles and that each particle had an energy that was the same energy that Planck had postulated, hf or $\hbar\omega$, and when that energy was higher than the work function of the metal, electrons were ejected.

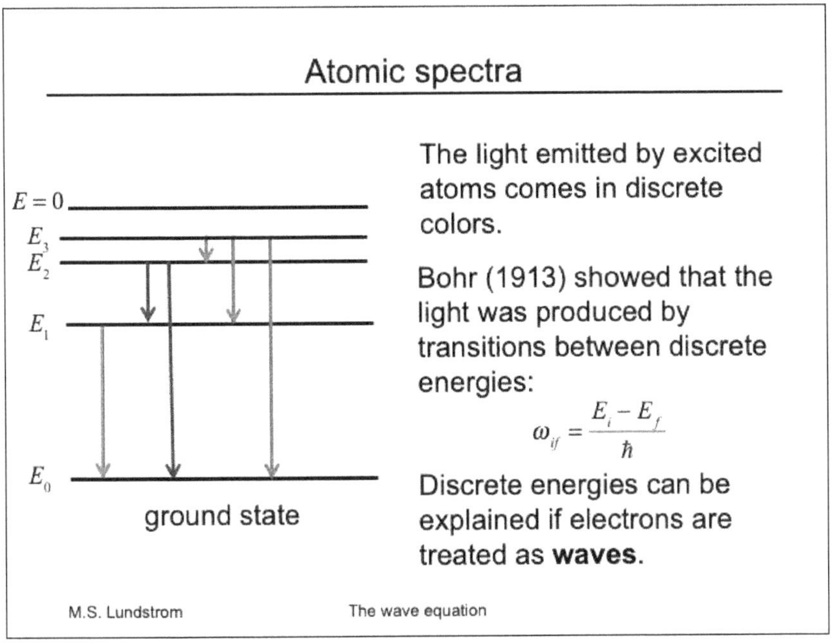

At the same time, people were also measuring atomic spectra, the light given off by molecules or atoms after the atoms had been excited into higher energy states. When allowed to relax back, the excited atoms emitted light, but the light came only in discrete colors, not as a continuous spectrum of colors.

Niels Bohr showed that the discrete energies could be explained if he treated the electrons as waves. The frequencies of the light emitted were given by the difference in the energies of the initial and final electron orbits — the initial energy minus the final energy divided again by Planck's constant. Evidence that particles behave like waves and that waves behave like particles was accumulating.

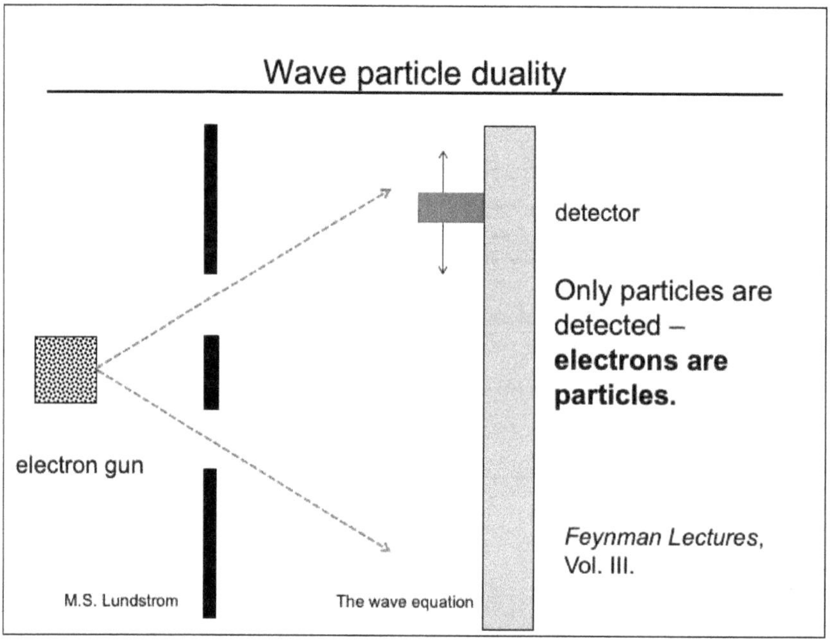

Wave-particle duality is seen more clearly in experiments of the kind shown above in which an electron gun shoots electrons one at a time at two holes in an aperture. Behind the aperture is a screen with a movable detector. When we move the detector up and down, we detect the number of electrons that end up at that position on the screen. What we find is that we always detect either no electron or a single electron. We don't detect parts of electrons, so this tells us that electrons behave as discrete particles.

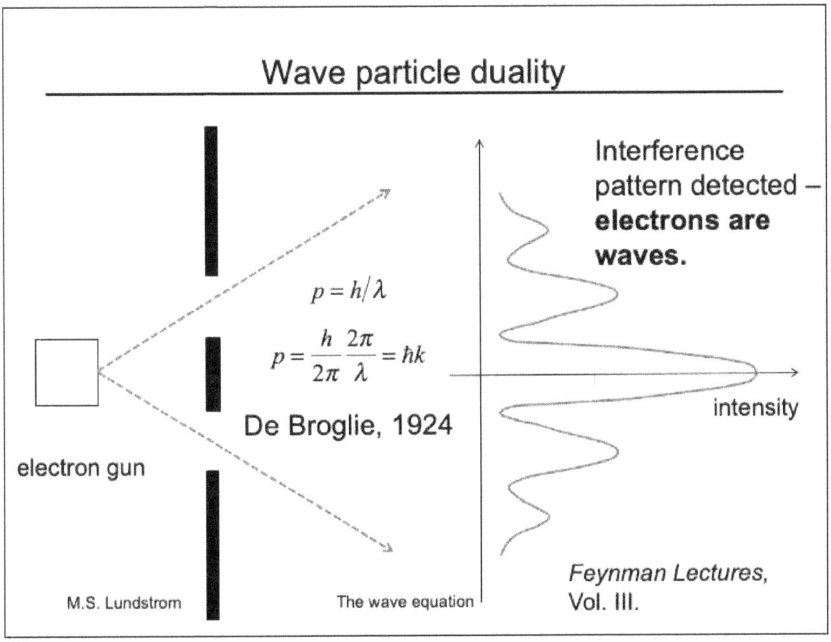

If we do this experiment many times, each time measuring one electron or zero electrons and then add the results, the pattern that we get is shown above at the right. Note that it looks like the kind of interference pattern we see when waves are interfering with each other constructively and destructively. This tells us that electrons behave like waves.

In 1924, Louis De Broglie postulated that any particle with a momentum, p, also behaves like a wave, and the relation between the wave, and the particle is that the particle's momentum is equal to Planck's constant divided by the particle's wavelength, $p = h/\lambda$. The **de Broglie relation** allows us to go back and forth between momentum and wavelength.

Experimental evidence was accumulating that energy is quantized, that waves can behave like particles, and that particles can behave like waves. A new theory to supplant classical physics had to be developed to explain these experimental findings.

It was well understood that waves can show the effects of quantization when boundary conditions are applied. Waves also interfere, and it was also understood that waves can be localized to behave like particles by adding up waves with different wavelengths to produce so-called **wave packets**. This suggested that a wave equation was needed to describe particles.

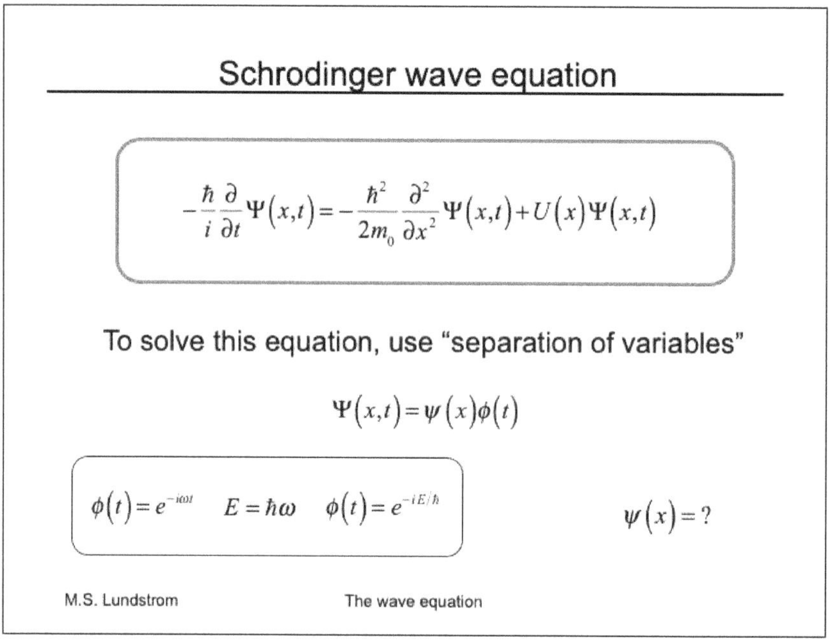

7.3 The wave equation

Shown above is the famous **Schrödinger wave equation**, a function of time and of position. This is not something we can derive, just as Newton didn't derive $F = ma$. The wave equation is simply a description of how nature behaves. It doesn't come from a more fundamental description, at least we haven't yet found a more fundamental description of nature.

One technique to solve a differential equation of this type is to postulate that the time and position dependent parts separate, that the **wave function**, $\Psi(x,t)$, the solution to this equation, is the product of a function of position only, and a function of time only. If you've had a course in differential equations, you may have worked through this technique called **separation of variables**. I won't do this now – I'll simply quote the result. The result is that the time dependent part of this equation goes as

$$\phi(t) = \exp(-i\omega t) = \exp(-iE/\hbar)$$

where the radian frequency ω is related to energy by $E = \hbar\omega$, the relation that we've seen before from Planck and Einstein and others. We know the time dependence of the wave function; how do we find the space dependent part of the wave function?

Time-independent wave equation

$$-\frac{\hbar^2}{2m_0}\frac{d^2\psi(x)}{dx^2}+U(x)\psi(x)=E\psi(x)$$

$$\Psi(x,t)=\psi(x)\phi(t)=\psi(x)e^{-i\omega t} \qquad \omega=E/\hbar$$

The probability of finding an electron between *x* and *x*+*dx*, is:

$$P(x,t)dx=\Psi^*(x,t)\Psi(x,t)dx$$

M.S. Lundstrom The wave equation

To find the space dependent part of the wave function, we solve the equation shown above at the top, which comes from the separation of variables technique. This is the **time-independent Schrödinger wave equation**. The complete space and time dependent wave function is the product of the space-dependent (time independent) part, $\psi(x)$, and the time-dependent (space independent) part, $\phi(t) = \exp(i\omega t)$.

How do we interpret the wave function? Quantum mechanics postulates that the **probability** of finding a particle between x and $x + dx$ is $\psi^*\psi\, dx$. Quantum mechanics is fundamentally about probabilities. It is not a deterministic theory like classical physics.

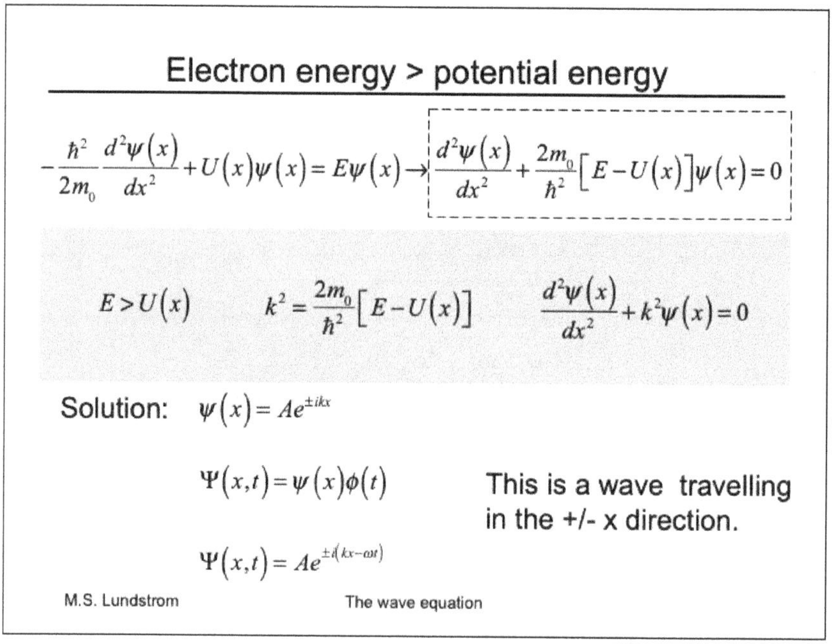

Let's see how this works. Shown on the first line above is the time independent wave equation, the part that describes the spatial dependence of the wave function. If we subtract $E\psi(x)$ from both sides, we can re-write the equation shown on the top line above as shown on the middle line. We will see that the solutions depend on whether the energy of the electron, E, is less than or greater than the potential energy, $U(x)$.

Let's consider first the case where the energy of the electron is greater than the potential energy, $E > U$. In this case, we can lump some things together and call the resulting quantity k^2, where k^2 is a positive quantity. We can then write the wave equation in the simpler form shown above in the right of the gray box.

One way to solve an equation of this form is simply to guess the solution, plug it into the equation, and see if it works. Let's assume that $U(x) = U_0$, a constant. If we guess that the solution is a constant, A, times an exponential, $\exp[\pm ikx]$ and plug it into the equation, we see that it works.

To get the complete solution, we multiply $\psi(x) = \exp(\pm ikx)$ by $\phi(t) = \exp(-i\omega t)$. We'll see next that the solution shown above describes a wave traveling in either the plus or minus x direction.

Waves: phase velocity

$\Psi(x,t) = Ae^{i(kx-\omega t)} = Ae^{i\theta}$ $\quad U(x) = U_0$ $\quad k(x)$ is constant

Follow a point of constant phase:

$$\frac{d\theta}{dt} = 0 \qquad \frac{d(kx-\omega t)}{dt} = 0 = k\frac{dx}{dt} - \omega$$

Phase velocity of the wave: $\upsilon_p = \dfrac{\omega}{k}$

M.S. Lundstrom The wave equation

Let's look at the solution and see what velocity it travels at. Call the argument of the exponential θ and remember that the potential energy, U, is constant, which means that k is constant. Let's follow a point of constant phase in position, and see what velocity that point moves at. The phase $\theta = (kx - \omega t)$ doesn't change with time if we travel along with the wave at the **phase velocity**. Doing the time derivative, we find that the phase velocity is $v_p = \omega/k$.

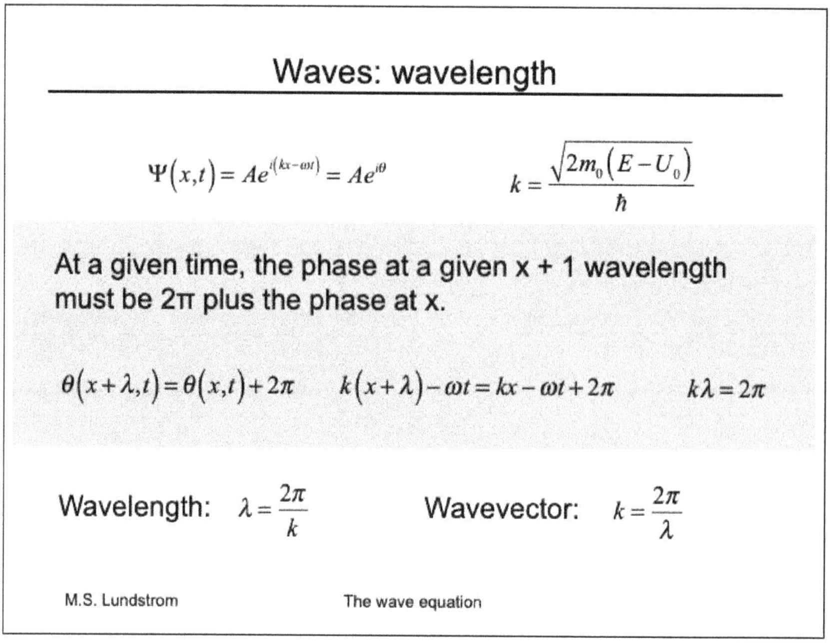

We now have a simple expression for the velocity at which this wave travels. What is its wavelength, λ? If, at a given time, we move one wavelength in position, the wave will accumulate 2π of phase. At position x plus one wavelength, the phase is the phase at position x plus an additional 2π. We find that $k\lambda = 2\pi$ We conclude that the wavelength is simply $\lambda = 2\pi/k$, or we can turn it around and say that the wavevector, k, which we've encountered earlier in Part 1, is simply $k = 2\pi/\lambda$.

So we can understand both the phase velocity and the wavelength of this traveling wave.

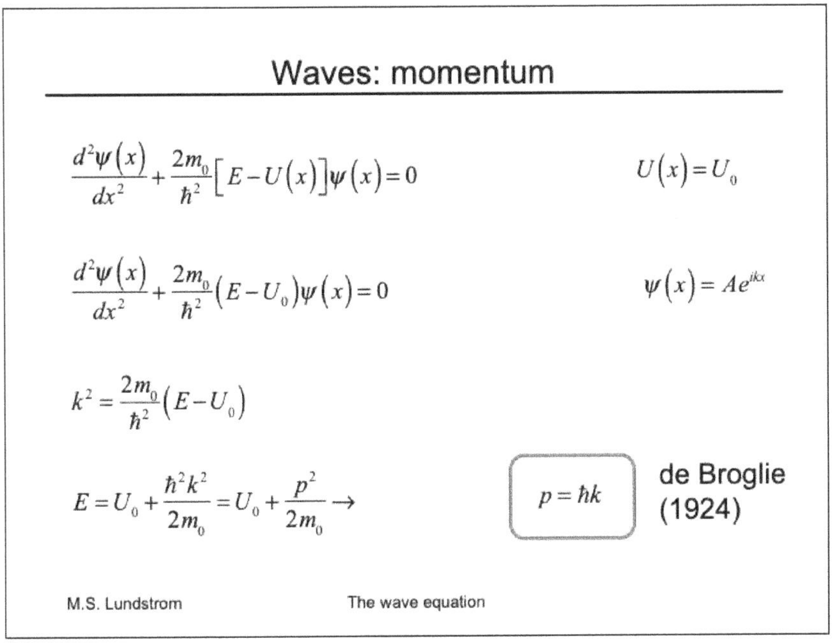

What about the momentum of the wave? The momentum of a particle is mass times velocity, $p = mv$. What is the momentum of a wave?

To answer that question, we start with the wave equation, solve it, and find that the wave function is $\psi(x) = A\exp(ikx)$. We know that k is related to energy, so we can solve for the energy in terms of k. As shown above on the last line at the left, we find that the kinetic energy is $E = \hbar^2k^2/2m_0$, but we also know that the kinetic energy of a particle is $p^2/2m_0$. We conclude that

$$p = \hbar k\,,$$

which is just what Louis de Broglie postulated in 1924. The solution to the wave equation for a free particle gives the same answer.

We also know that the momentum of a particle is mass time velocity, $p = m_0 v$. If we assume that the velocity of the electron is the phase velocity of the electron wave, $v_p = \omega/k$, then we find $p = m_0\omega/k \neq \hbar k$. As we will see, the problem is that the velocity of the electron is not the phase velocity of the wave.

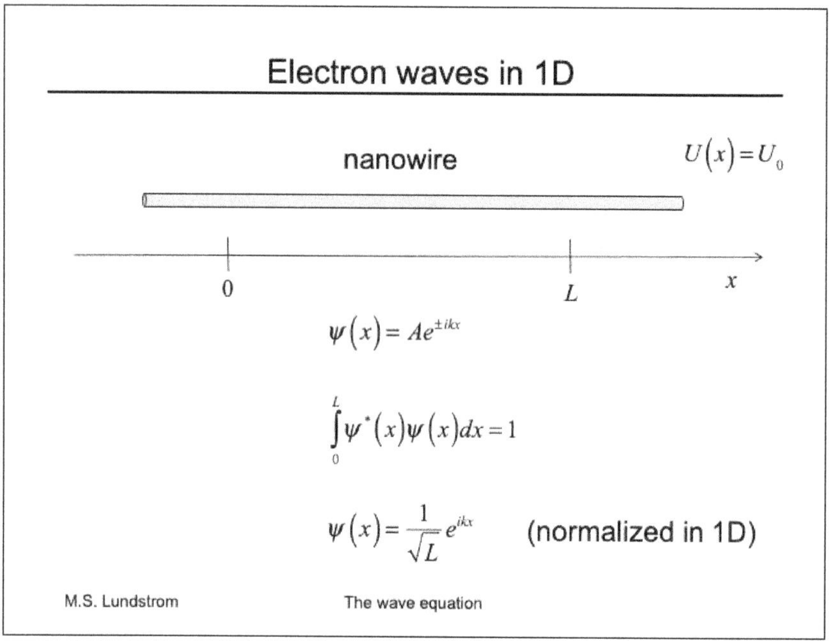

Let's see how things work in 1D. Twenty or thirty years ago, this would have been an academic exercise, but these days, semiconductor **nanowires** in which electrons behave as 1D particles are readily produced.

Assuming a constant potential, U, the solution is just $\psi(x) = A\exp(\pm ikx)$ depending on whether the wave is propagating in the positive or negative direction. Take a section of this nanowire between $x = 0$ and $x = L$ and let's assume that we have one electron in this section. The probability of finding that electron between x and $x + dx$ is $\psi^*\psi\, dx$. Integrating from $x = 0$ to $x = L$, that probability has to be 1 because we are assuming there is 1 electron in the section of length, L. This **wave function normalization** allows us to specify the constant as $A = 1/\sqrt{L}$, where L is an arbitrary **normalization length**. When we're done, we have a wave function that describes the one electron in this region of length, L.

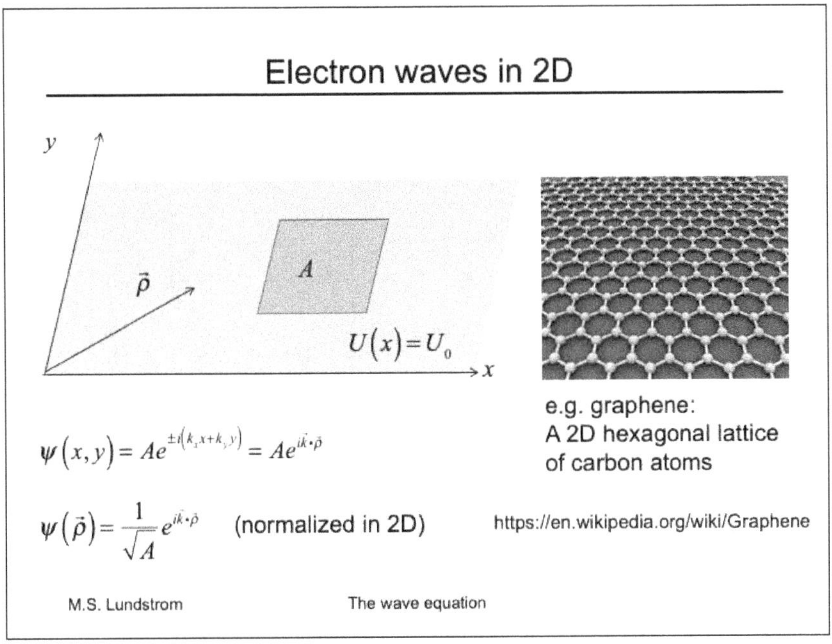

We can also describe electrons in 2D. If we have an infinite 2D sheet in the $x - y$ plane (the lightly shaded region above) and an electron normalization area, A (the darker shaded region above), then we can write the 2D wave function as shown above.

As mentioned earlier, these are not textbook, academic examples anymore, semiconductor technologists work with 1D structures and 2D materials. Shown above on the right is an example, graphene, a 2D material. We won't encounter these structures in the rest of these lectures, but they are the subject of a great deal of current research.

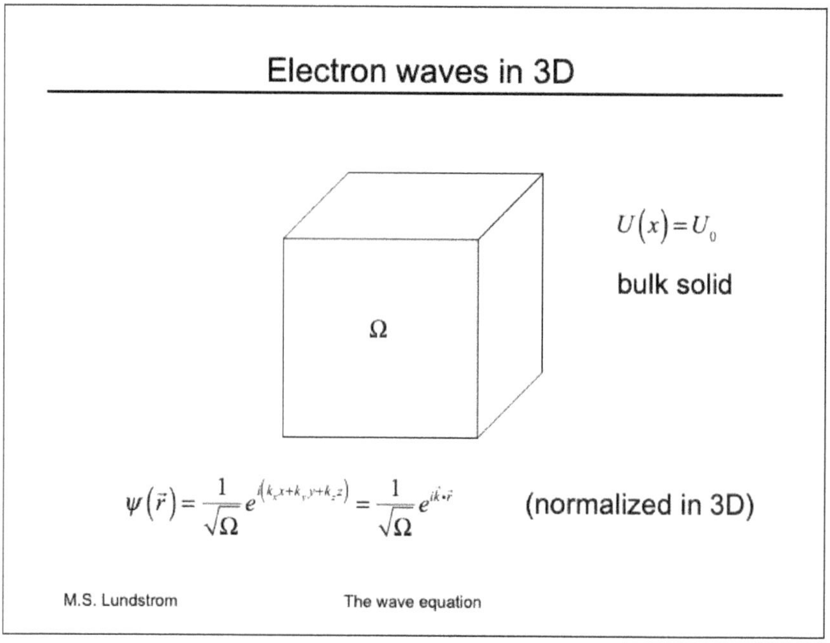

Finally, what about electron waves in three dimensions? For an electron in this 3D volume, Ω, the wave function will be $\psi(\vec{r}) = A \exp[i(k_x x + k_y y + k_z z)]$. Normalizing the wavefunction to describe one electron in this volume, we find that the arbitrary constant $A = 1/\sqrt{\Omega}$

To re-cap, we have spent some time discussing the solutions to the time independent wave equation when the electron's energy is greater than the potential energy, which was assumed to be constant – independent of position. What about the case where the energy is less than the potential energy? What do the wave functions look like then?

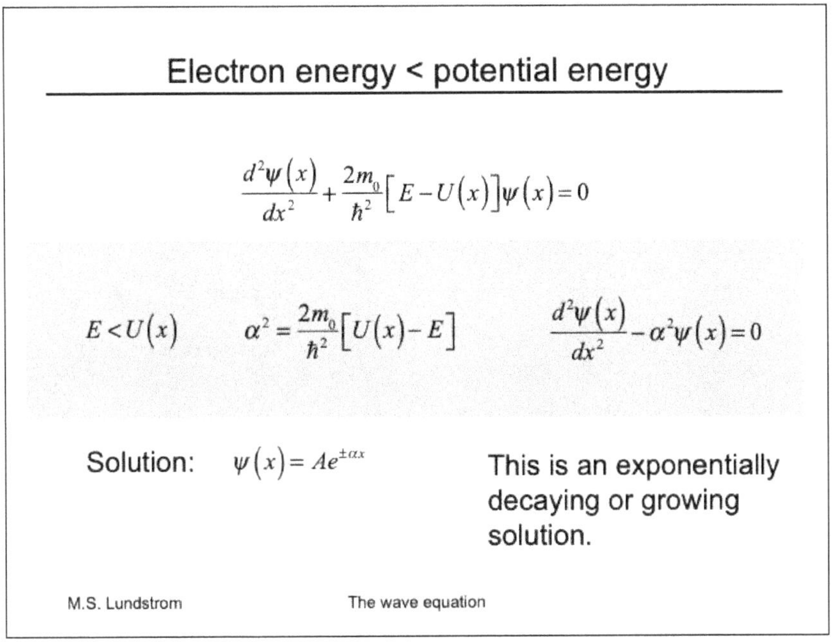

We now understand the solution to the wave equation when $E > U$, what are the solutions when $E < U$? As shown above, we begin again with the time independent wave equation, but in this case, the energy is less than the potential energy, so the term in brackets in the first equation shown above is negative. We define a quantity, $\alpha^2 = U - E$ so that $\alpha^2 > 0$. This allows us to write the wave equation in the form shown at the right in the shaded box. Again, we can guess the form of the solution. The solution is some arbitrary constant A times $\exp(\pm\alpha x)$; either sign satisfies this differential equation. This is a wave that is either exponentially growing with position, or exponentially decaying with position.

We now know the general form of the wave function. When the electron energy is greater than the potential energy, the solution is a traveling wave.

$$\psi(x) = Ae^{\pm ikx}$$

When the energy is less than the potential energy, the solution is a decaying or a growing exponential function.

$$\psi(x) = Ae^{\pm\alpha x}$$

We've been discussing electron waves, but remember, electrons are also particles.

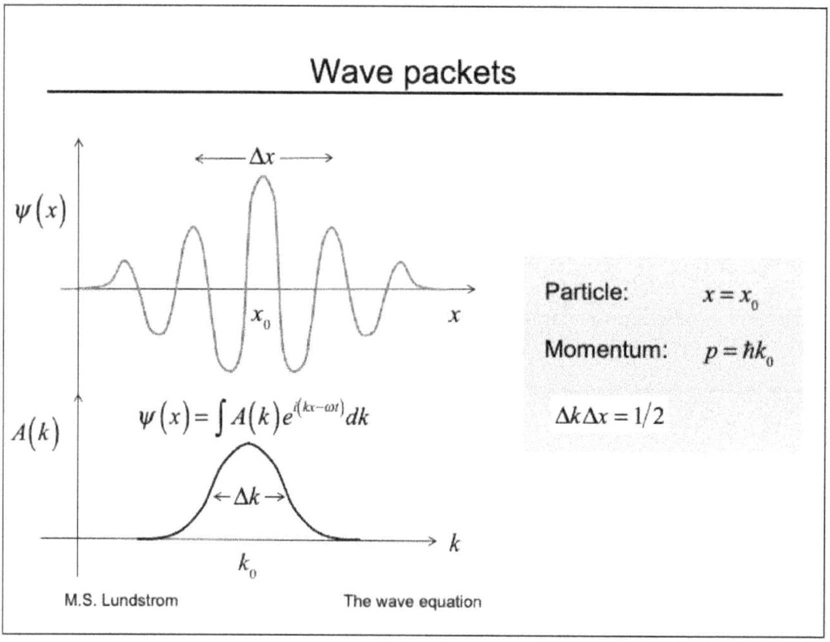

7.4 Particles and wave packets

Consider an electron located somewhere between $x = 0$ and $x = L$ in the nanowire discussed earlier. We describe it by a wave function, $\psi(x) = \exp(ikx)/\sqrt{L}$, but if we take $\psi^*\psi$, we find there's a uniform probability of finding the electron anywhere between 0 and L, $P(x)dx = dx/L$. How do we localize the electron to a specific position?

The slide above shows how we describe an electron that is located near $x = x_0$ – we do this by adding up a group of waves with different wavelengths. If we use one wavelength, the electron is spread out everywhere, but if we have a number of waves with different wave lengths (wavevectors), we can localize the electrons by ensuring that the waves interfere constructively near $x = x_0$, but they interfere destructively when we move away from that particular location. The wider the spread of wave vectors k, that we use, the more tightly we can localize this particle at a particular position. The narrower the spread in wave vectors, the wider it becomes in position.

Forming **wave packet** with different wave vectors is a way to describe a particle that is located near $x = x_0$. If the wave vectors are near $k = k_0$, we are describing a particle with momentum $\hbar k_0$.

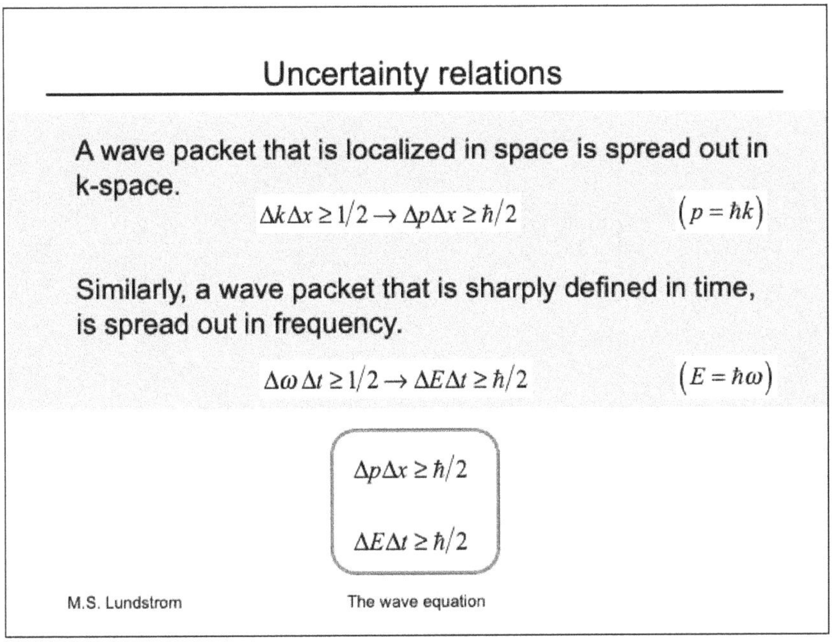

Describing particles with **wave packets** leads to **uncertainty relations**. A wave packet that is very localized in space, is spread out in k-space. A wave vector that is very localized in k-space, is spread out in real space. That leads to uncertainty relations that apply to any wave, phenomenon. If we multiply Δk by $\hbar$, the famous uncertainty relations between momentum and position results. The uncertainty in momentum of an electron and its uncertainty in position has to be greater than or equal to a minimum value, $\hbar/2$.

Similarly, if we have a wave function that's sharply defined in time then it's spread out in frequency. If it's sharply defined in frequency then it's spread out in time. The uncertainty in frequency times the uncertainty in time is also greater than or equal to $\hbar/2$. If we multiply Δt by $\hbar$ the famous uncertainty relations between energy and time result. As wave packets propagate in time, the uncertainties just grow bigger and bigger.

The uncertainty relations shown here describe any wave phenomenon, but they're used in quantum mechanics to describe the quantum mechanical particles that we describe with wave functions. For most of this course, we'll treat electrons as particles, which means that the uncertainty in position, Δx, is assumed to be much smaller than the scale of the device.

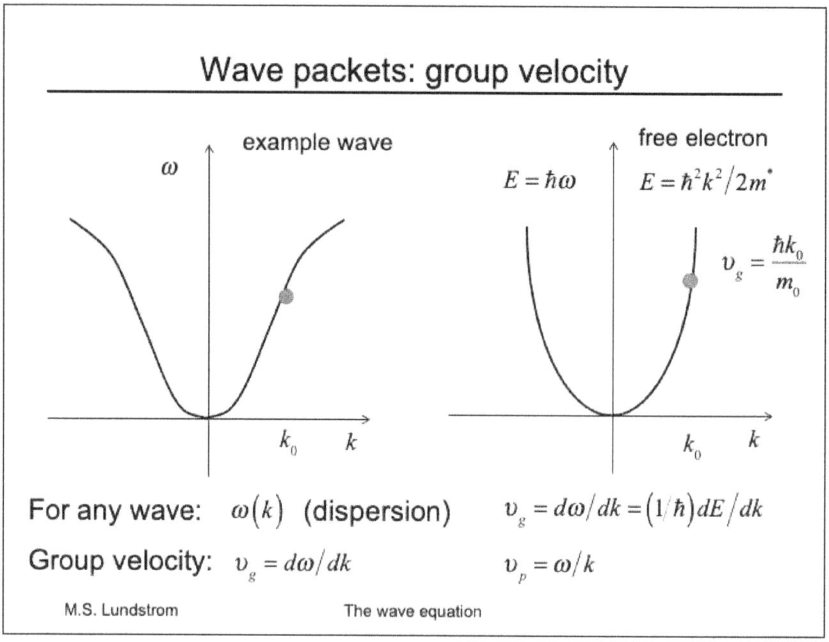

There are two more concepts from quantum mechanics that we need to understand. One is that any wave can be described by its **dispersion**, how its frequency varies with wave vector. (Remember, the wave vector, k is $2\pi/\lambda$, where λ is the wavelength.)

We've seen that the phase of each wave propagates at the phase velocity, $\upsilon_p = \omega/k$, but a wave packet actually propagates at a different velocity. The center of the wave packet propagates at the **group velocity**, $\upsilon_g = d\omega/dk$.

Remember that a free electron is defined by relation $E = \hbar^2 k^2/2m_0$. What is the velocity of the electron wave packet with wave vectors near $k = k_0$? We can simply take the derivative, $\upsilon_g = d\omega/dk = d(E/\hbar)/dk$. Taking the derivative of $E = \hbar^2 k^2/2m_0$, we find that the group velocity of this electron wave is $\upsilon_g = \hbar k_0/m_0$. That makes sense, because $\hbar k_0$ is the crystal momentum associated with the wave packet (i.e. the particle) — the group velocity is momentum divided by mass.

It is important to understand that the phase velocity and the group velocity are different. For this parabolic energy band, the wave packet is propagating at a velocity that is actually twice that of the phase velocity.

Since we are interested in electron particles, it is the group velocity as determined from the dispersion, $E(k)$, that is most relevant.

7.5 Summary

We have covered a lot of ground in this lecture, but we're only going to need a few key concepts from quantum mechanics in this course. Classical mechanics describes the position and momentum of particles with Newton's law, $F = ma$, but nature behaves non-classically as summarized below.

1) $\Psi(x,t) = \psi(x)e^{-i\omega t}$
2) $E = \hbar\omega = hf$
3) $-\dfrac{\hbar^2}{2m^*}\dfrac{d^2\psi(x)}{dx^2} + U(x)\psi(x) = E\psi(x)$
4) $P(x)dx = \psi^*(x)\psi(x)dx$
5) $\Delta p \Delta x \geq \hbar/2 \qquad \Delta E \Delta t \geq \hbar/2$
6) $\upsilon_g = \dfrac{1}{\hbar}\dfrac{dE(k)}{dk}$

(1) We describe quantum mechanical particles with a wave function that has a space dependent part and a time dependent part.
(2) Energy and frequency are related by a fundamental relation, $E = \hbar\omega$ (frequency in radians/s) or by $E = hf$ (frequency in cycles/s).
(3) We find the space dependent part of the wave equation by solving the time-independent Schrödinger equation.
(4) The probability of finding an electron between x and $x + dx$ is proportional to $\psi^*\psi$.
(5) The uncertainty relations arise because particles are described by wave packets formed by a collection of waves with different wavevectors.
(6) Finally, the velocity of the particle is the velocity of the wave packet – the group velocity.

Beginning in the next lecture, we'll explore the implications of quantum mechanics for electrons in semiconductors.

7.6 Reference

For another treatment of the wave equation, see Pierret:

[1] R.F. Pierret, *Advanced Semiconductor Fundamentals*, (2nd Ed.), Pearson Education, Inc., 2003 (see Ch. 2, Secs. 2.1–2.3).

Lecture 8

Quantum Confinement

8.1 Introduction

We won't need much quantum mechanics to understand semiconductors (because others have done the "heavy lifting" for us!), but there are a few cases where quantum mechanics becomes important. One of them is in computing the density of states, which we will discuss in Lecture 11. Another has to do with **quantum confinement**, which is the topic of this lecture. As we saw in Lecture 7, when the electron energy is greater than the potential energy, the time-independent wave equation can be written as

$$\frac{d^2\psi(x)}{dx^2} + \frac{2m_0}{\hbar^2}[E-U]\psi(x) = 0 \rightarrow \frac{d^2\psi(x)}{dx^2} + k^2\psi(x) = 0\,.$$

The solutions were expressed in terms of wave vector, k, as $\psi(x) = A\exp\pm(ikx)$, which is a traveling wave when U is constant. The phase velocity is $\upsilon_p = \omega/k$, and the wavelength, $\lambda = 2\pi/k$. The probability of finding an electron between x and $x+dx$ is

$$P(x)dx = \psi^*(x)\psi(x)dx\,.$$

We multiply $\psi(x)$ by a time-dependent quantity, $\exp(-i\omega t)$, when we need the entire time and space-dependent wave equation for $\Psi(x,t)$.

In this lecture, we will solve the same wave equation but with boundary conditions that confine electrons to a particular region of space. The wave equation is a linear equation with solutions $\exp(ikx)$ and $\exp(-ikx)$, but linear combinations of these solutions are also solutions. In this lecture, it will be convenient to express the solutions as

$$\psi(x) = A\sin(kx) + B\cos(kx)\,.$$

When we impose boundary conditions, we'll see that the solutions are not traveling waves.

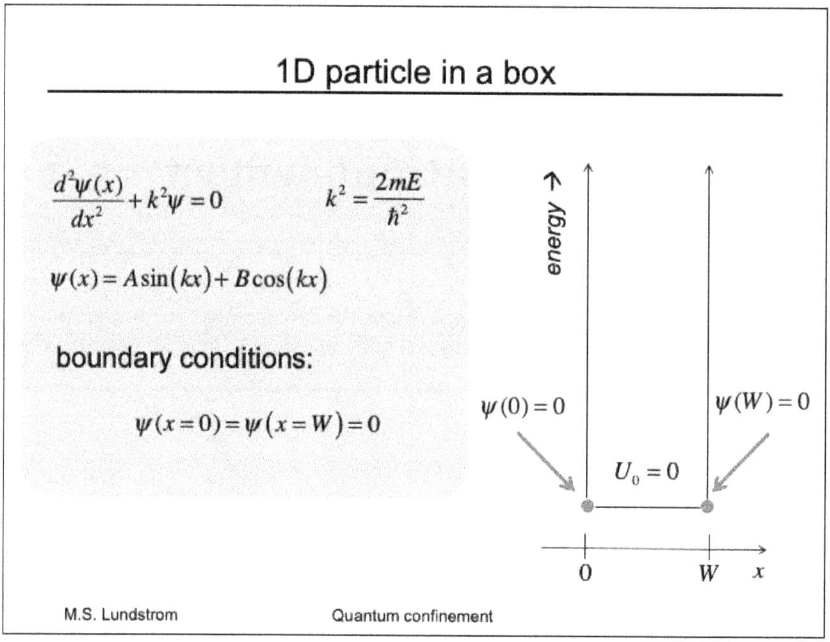

8.2 Quantum mechanical particle in a box

Shown above is a classic quantum mechanics problem, a **particle in a box**, which goes from $x = 0$ to $x = W$. The potential energy is zero inside the box, but it is infinite outside the box. An electron inside the box cannot get out. We'll solve our time-independent wave equation for this problem. Inside the box, $E = \hbar^2 k^2 / 2m$ because $U = 0$, so $k^2 = 2mE/\hbar^2$ inside the box. We need to solve the differential equation shown at the upper left above and then apply the appropriate boundary conditions.

For this problem, it's most convenient to write the solution in terms of sines and cosines. Next, we define the boundary conditions. The electron cannot get out of the box, so there's no probability that it will be found at $x < 0$ or at $x > W$. The appropriate boundary conditions are that the wave function has to be zero at the two ends of the box.

We'll discuss later how to produce quantum wells in semiconductors. When we do that, the mass, m, will be the effective mass, m^*, of the semiconductor.

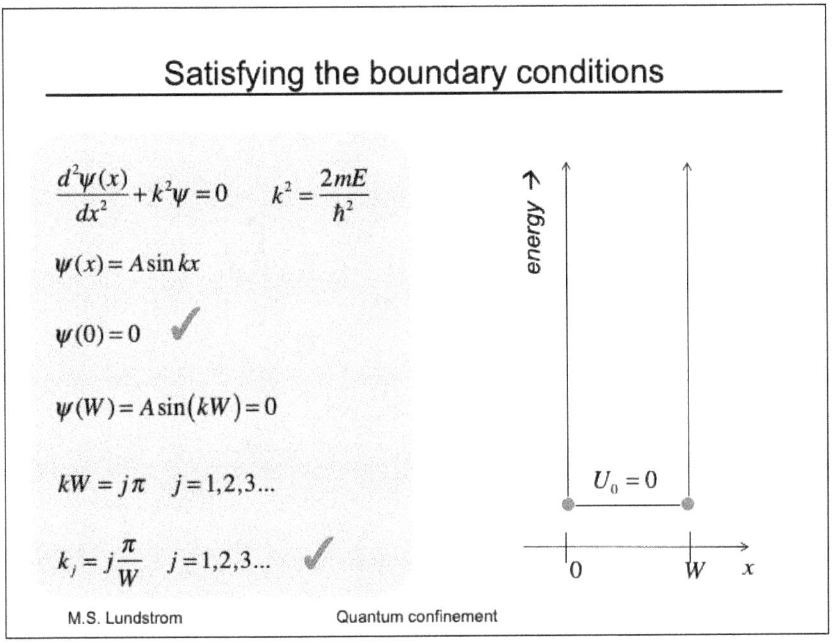

We are ready to solve the particle in a box problem. First, we can throw out the $\cos(kx)$ term because $\cos(0) \neq 0$. The $\sin(kx)$ term is zero when $x = 0$, so it satisfies the first of our boundary conditions, but we have a second boundary condition – we have to ensure that when $x = W$, $\sin(kW) = 0$.

We know that $\sin(\pi) = 0$, $\sin(2\pi) = 0$, the sine of any integer number of pi's is 0, so if kW is some integer number of pi's, then any one of those k's will satisfy the second boundary condition.

We conclude that only a set of discrete wavevectors satisfies the boundary conditions. We have to restrict the wavevector, k, to a set of values given by $k_j = j\pi/W$, where j is an integer. Now we've satisfied both boundary conditions.

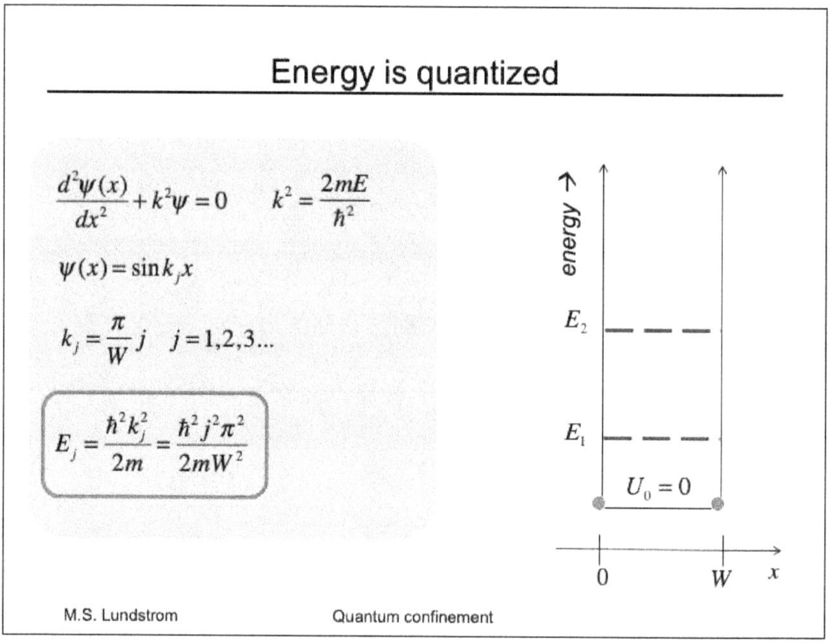

Let's examine the solution. Shown at the top left above is the wave equation, and the relation between k^2 and energy. We've determined that only certain specific values of k satisfy both boundary conditions. This means that there are only specific wave functions and energies are allowed.

When $j = 1$, we get one wave function and one energy. When $j = 2$, we get another wave function and energy, and on and on for $j = 3, 4, 5$ and the other levels. **Confining electrons leads to discrete energy levels.**

This is also what happens in the hydrogen atom. The Coulomb potential confines the electrons to be near the positively charged proton, and that leads to discrete energy levels. The same thing happens for the particle in a box, but the mathematics is a lot simpler.

To summarize, confined electrons have quantized energies. You can see from the relation above that the tighter we confine electrons (the smaller we make W), the higher those energies are. You can also see that if we have a light mass, the energies are larger. The effective mass of electrons in GaAs is much lighter than the effective mass of electrons in Si, so we expect the energy levels to be higher in GaAs quantum wells.

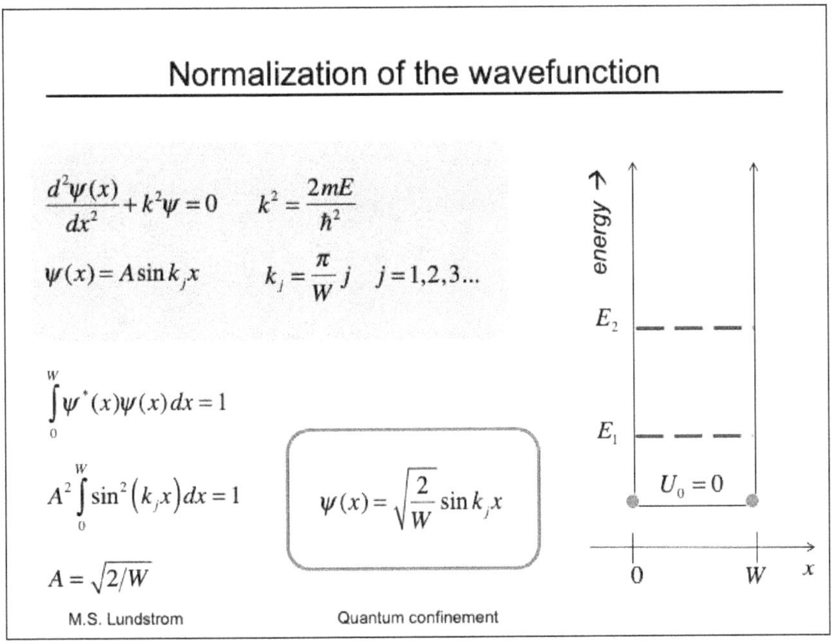

For traveling waves, we talked about normalizing the wave function so that the resulting wave function described finding one electron in a certain region. Let's assume that we have one electron in the quantum well. We need to normalize the wave function properly to describe this electron. When we integrate $\psi^*\psi$ across the width of the well, the integral has to be equal to 1, which allows us to determine the constant A. If you do this calculation, you'll find that $A = \sqrt{2/W}$, so our normalized wave function has the form shown above. The $\sqrt{2/W}$ simply ensures us that we're describing one electron in this quantum well.

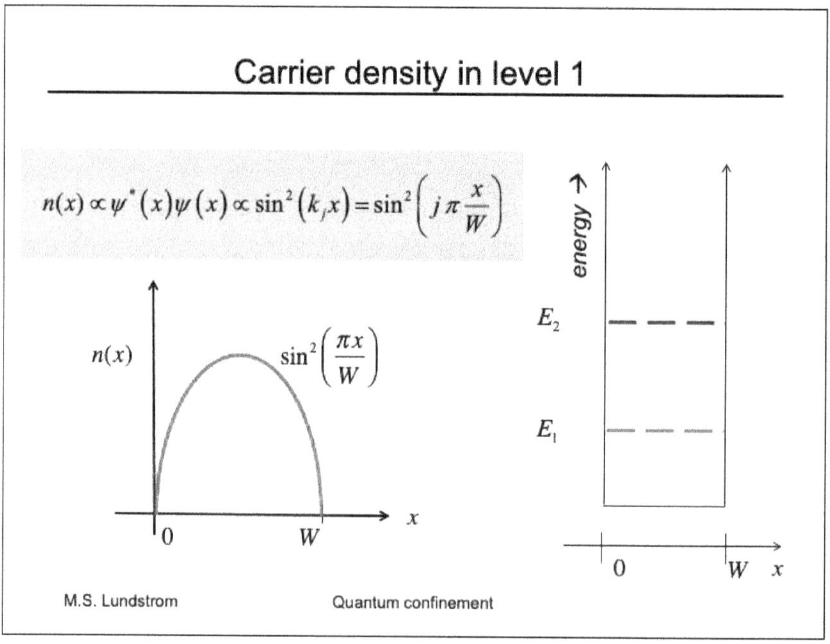

If we have electrons in a quantum well, what is the electron density, $n(x)$? The answer is that it will be proportional to $\psi^*\psi$, which means that means it will be proportional to $\sin^2(k_j x)$. The electron density inside these quantum wells is far from uniform; the electrons will be distributed differently in the first state, the second state, the third state and so on.

Let's assume $j = 1$, which gives the lowest energy level, then $n(x)$ is a sine squared function. As shown above, the electron will have the highest probability of being found right in the middle of the quantum well, and no probability of being at the two edges.

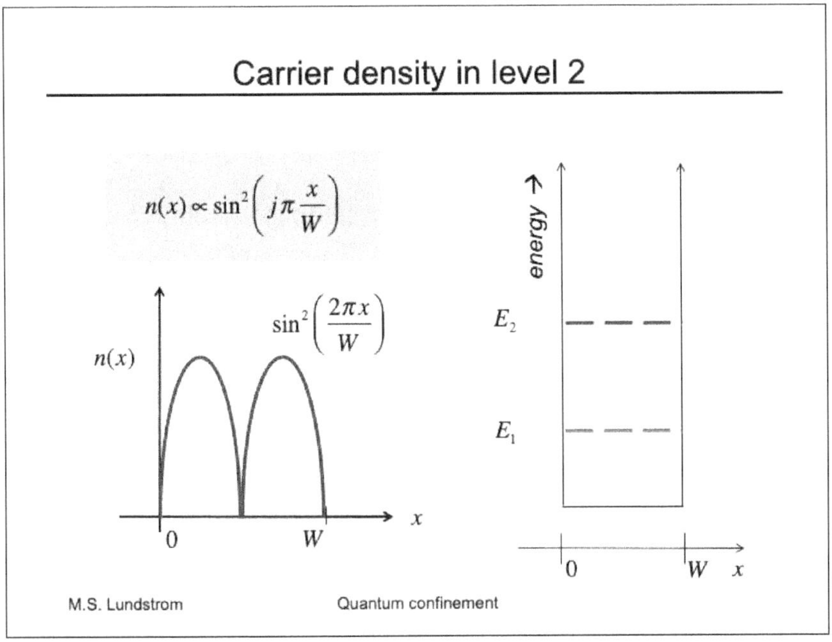

Next, let's look at $j = 2$, the second quantized level. In this case, $n(x) \propto \sin^2(2\pi/W)$. For $j = 2$, the electron will have no probability of being found in the center of the quantum well or at the two ends. We can continue on and on into higher and higher energy levels. To find the total electron concentration in the box, we need to add the contributions from each level.

You might be wondering: "Is this an academic exercise, or are these really effects that we have to worry about?" Or "How narrow do we have to make this well in order to see the effects of quantization?" The answer is that when we confine electrons on the same scale as their De Broglie wavelength, then we should expect these effects to become observable and important. This is what we mean by quantum confinement.

de Broglie wavelength

What is the wavelength of a free (mobile) electron?

$$p = \hbar k = \hbar \frac{2\pi}{\lambda_B} \qquad E = \frac{p^2}{2m} \approx \frac{3}{2} k_B T \qquad \boxed{\lambda_B = \sqrt{\frac{4\pi^2 \hbar^2}{3 m k_B T}}}$$

About 10 nm for electrons in silicon at room temperature

For lighter masses, confinement is felt in wider wells.

M.S. Lundstrom Quantum confinement

What is the wavelength of an electron in the conduction band? We can estimate it if we remember that the relation between momentum and wave number is $p = \hbar k$, where k is 2π over the wavelength. Call this wavelength λ_B, the de Broglie wavelength of the electron. If we can find p for an average electron, then we can estimate the wavelength of an average electron.

To find the average p, we recognize that the average thermal energy of electrons at a temperature, T, is $3/2k_BT$. That kinetic energy is also $p^2/2m$, so if we know the temperature and the mass, we can solve for the thermal average momentum, put the average momentum in the first equation on the left in the slide above, and solve for the average wavelength. The result is an expression for the average wavelength of electrons in terms of the electron mass and the temperature.

Plug numbers in for Si and we'll find that the wavelength of a thermal average electron is roughly 10 nanometers. We would need to make a 10 nanometer wide quantum well in order to see these effects. In fact, it's quite easy to produce 10 nanometer wide quantum wells, so these effects can often be important. For semiconductors with a lighter effective mass (e.g. a III-V semiconductor), the electron's de Broglie wavelength will be even longer, so we'll start to see these effects at even wider quantum wells.

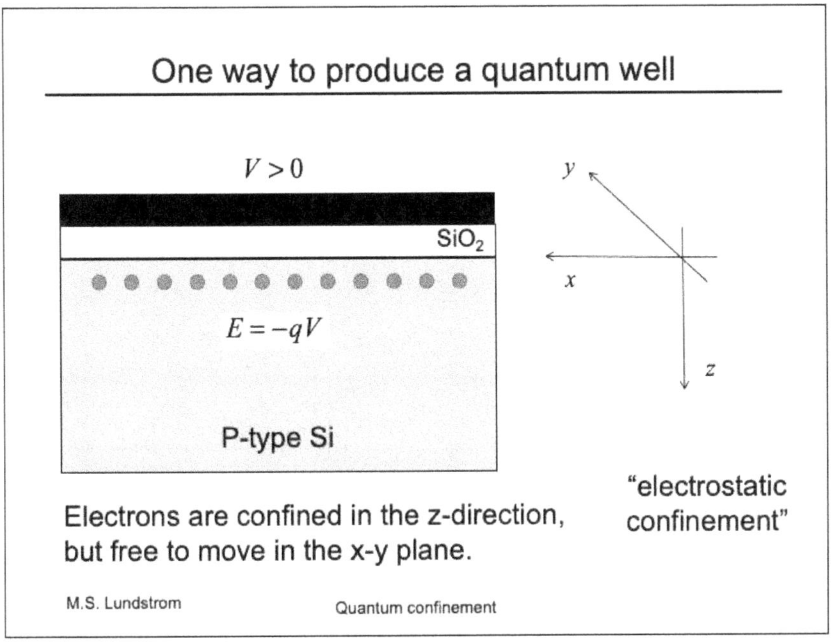

8.3 Quantum confinement in semiconductors

Quantum wells in semiconductors can be produced quite easily. The first way that this was done was by electrostatic confinement. For example, as shown above, assume that we have a silicon wafer with a good insulating layer on top of it and a metal plate on top of that. Apply a positive voltage to the metal plate. You may remember from first-year physics that a positive voltage lowers the potential energy of electrons, so this positive energy on the plate will lower the energy of electrons in the silicon — the positive voltage on the plate attracts negatively charged electrons, and they will pile up in a well, a very thin **quantum well**, near the interface between the Si and the insulating layer. This is, in fact, the first way that scientists observed quantum confinement in semiconductors. It's also important to realize though, that these electrons are still free to move in the x–y plane. They're only confined in the z direction. These are 2D electrons. Actually, they are called **quasi-2D electrons** because they can move a little in the third dimension, z.

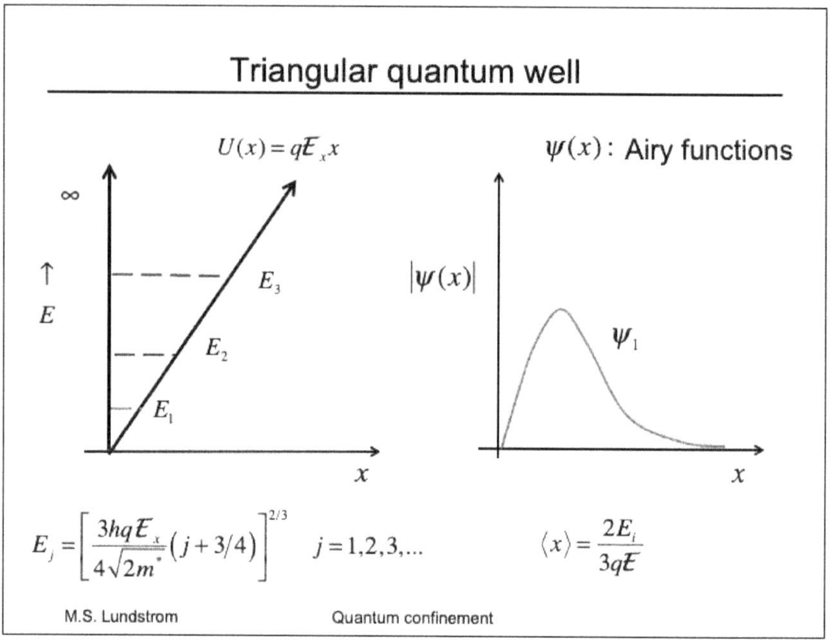

Quantum wells can have different shapes, i.e. different $U(x)$ vs. x profiles. A triangular quantum well is shown above. In this case, the potential is a linear potential as shown above on the left. There will still be a set of quantized levels, but they are not given by the expression for the rectangular quantum well. It is a little more difficult to solve for the energies in this case, but the solutions turn out to be Airy functions. The wave function for the lowest level will look something like that sketched above at the right. The wave function penetrates a little outside the quantum well and peaks inside the quantum well. Things look a little different from the simple quantum box we discussed, but the qualitative features are similar.

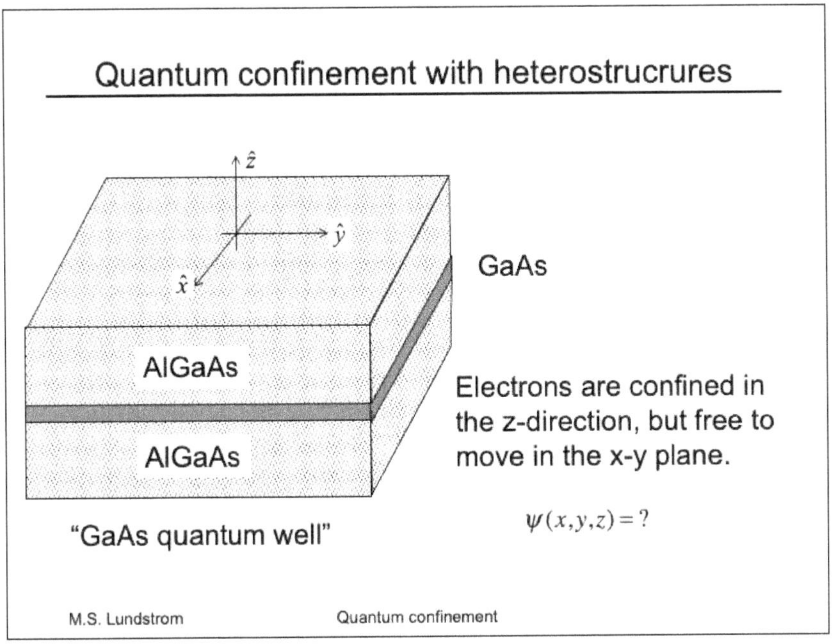

Another way to produce quantum wells is by semiconductor **epitaxy**, which is growing different single crystal layers on top of each other. Aluminum gallium arsenide (AlGaAs) is a semiconducting alloy that has a band gap that is a few tenths of an electron volt larger than a pure gallium arsenide. As shown above, we can grow an aluminum gallium arsenide layer and then a thin gallium arsenide layer and then another aluminum gallium arsenide layer. The two cladding layers have wider band gaps, so they will confine the electrons to be in the GaAs layer. As long as that GaAs layer is thin, 10 nanometers, 20 nanometers, or so, we would produce a quantum well for electrons in the GaAs. Note again that these electrons are confined in the z direction so it's hard for them to move in the z direction, but they are free to move in the x–y plane. These quasi-2D electrons can move only a little in the z direction).

For quantum wells in semiconductors, it can be a little tricky to determine what effective mass to use. For example, if it's a Si quantum well, should we use the transverse or longitudinal effective mass? Reference [2] discusses this issue.

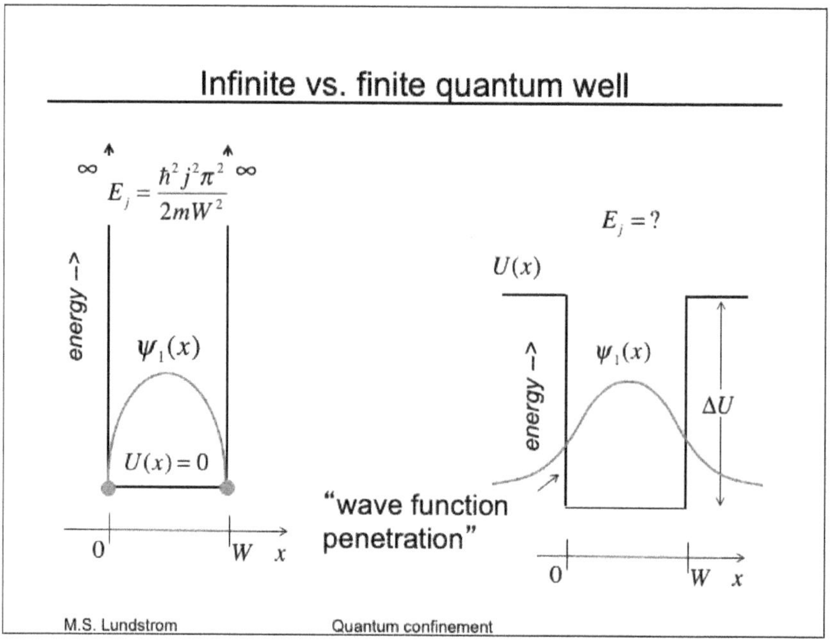

In practice, quantum wells are a little more complicated than our simple examples. For example, it was easy to solve the **particle in a box** problem with infinitely high energy barriers on both sides because that made the boundary condition $\psi = 0$ on both ends of the quantum well. When we actually produce quantum wells, they have a finite barrier, ΔU, as shown in the figure above — not an infinite barrier. Electrons will still have quantized energy levels, but electrons are able to leak out of the well a little bit. The well is effectively a little wider, so we expect the energies to be a little lower.

It takes some mathematics to compute what the energy levels are (see [1]), but it's easy to appreciate qualitatively that they're a little bit less than we would calculate from the formula for an infinite barrier. The spreading of the wave function outside of the well is known as **wave function penetration** into the finite energy barriers.

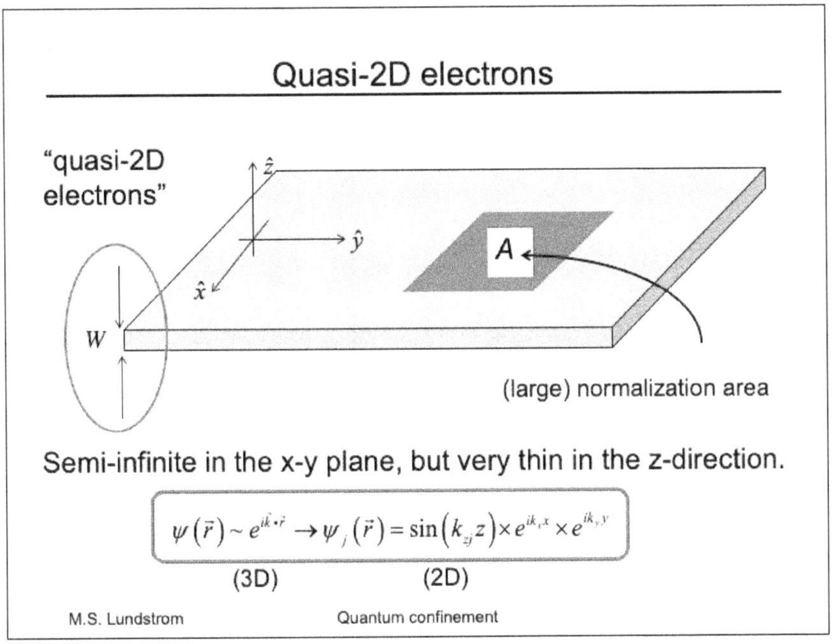

8.4 2D electrons and subbands

Consider the quantum well shown above where electrons are confined to a thin region in the z direction but are free to propagate in the x–y plane. What is the electron wave function in the region with area A? We expect that the solution will look like a particle in a box, solution in the direction of confinement, and a traveling wave in the x–y plane. In the z-direction, the wave function is $\sin(k_z z)$, and in the x–y plane, it is $\exp[(i(k_x x + k_y y)]$. There's no confinement in the x–y plane, so we get our plane waves back.

To actually solve for the wave function, we begin with the 3D Schrödinger equation ($d^2\psi/dx^2$ becomes ∇^2) and solve it using the technique of separation of variables. That is, we postulate that the solution is the product of a function that depends only on x times one that depends only on y, times one that depends only on z. If we plug that assumed solution into our 3D wave equation, we'll see that the equation separates. We solve the separate equations, and find (as shown above) that the wave function is a plane wave propagating in the x direction times a plane wave propagating in the y direction times a different function in the z direction.

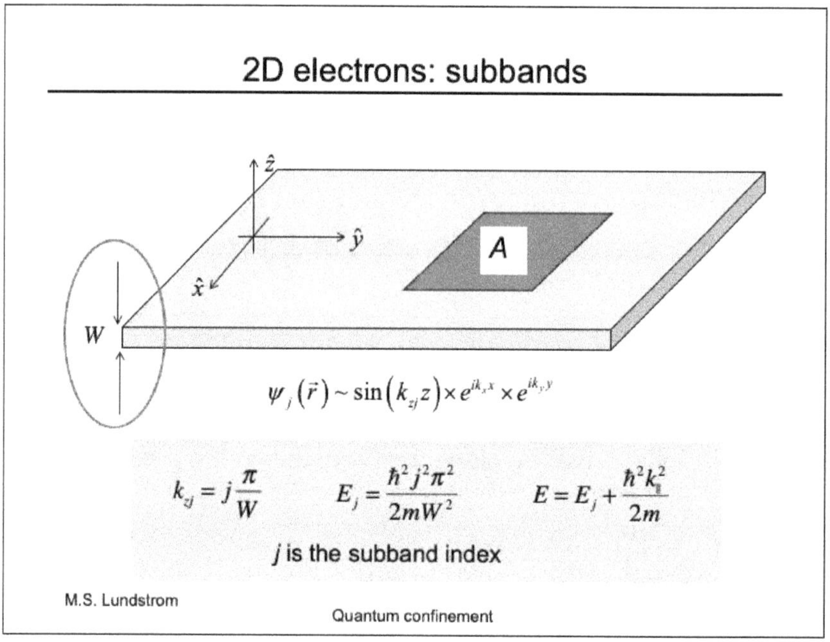

There is an important aspect of quantum wells in semiconductors called **subbands** that we should be clear about. A set of discrete values of k_z satisfy the boundary conditions at the top and the bottom of this quantum well. These discrete $k's$ give rise to a discrete set of energy levels associated with that confinement in the z direction. But remember that electrons are free to move in the x–y plane, so electrons have additional kinetic energy in the x–y plane. For each integer j, the minimum energy, E_j, is the bottom of subband, j, but the total energy consists of that term plus the kinetic energy associated with motion in the x–y plane.

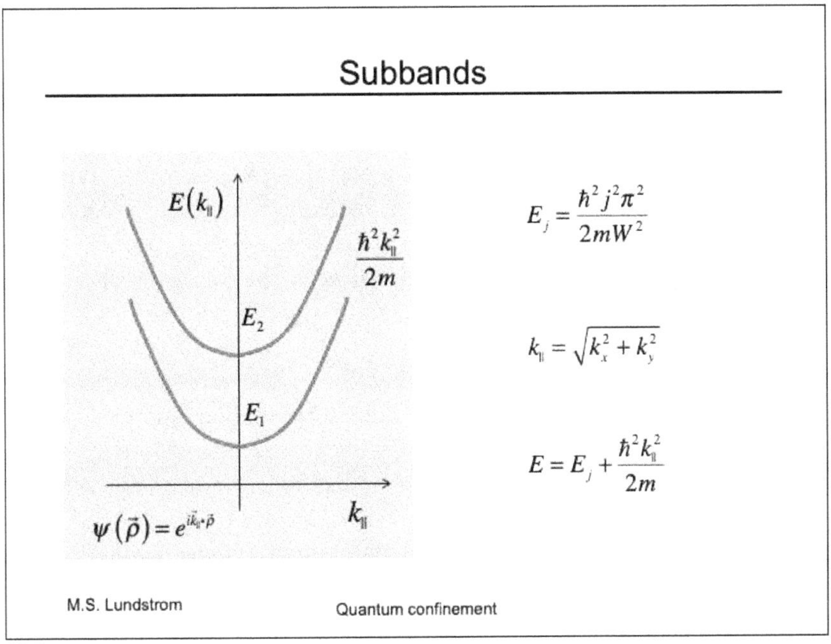

Shown above is a plot of energy vs. the wave vector in the x–y plane. We see that we have discrete levels that come from the confinement in the z direction, but we also have additional energy that comes from the motion in the x–y plane. We call the solutions for each integer j subbands because these are bands that have been induced in the conduction band because of the quantum confinement. The total energy of an electron in a subband is the energy of the bottom of the subband, plus the kinetic energy due to motion in the x–y plane. For each subband, the bottom of the subband is like the bottom of the conduction band for bulk, 3D electrons.

Similar considerations apply for nanowires. In a nanowire, 1D electrons are only free to move in one dimension, and there is a corresponding set of subbands associated with quantum confinement in the other two directions.

8.5 Summary

Quantum confinement, which becomes important from time to time in different types of semiconductor devices, is an example of where quantum mechanics manifests itself in a way that is too important to overlook or to wrap into some effective quantity, like effective masses. The important point to remember is that when electrons are confined, their energy becomes quantized and their wave functions change.

Quantum confinement can be achieved with electric fields or by using epitaxial growth to surround a small band gap semiconductor with larger band gap semiconductors.

Quantum confinement leads to subbands — 2D electrons may be free to move in the x–y plane, but they have different minimum energies depending on which subband they occupy. Similarly, 1D electrons may be free to move in the x-direction, and also occupy different subbands. Subbands arise from the confinement in one direction for 2D electrons and in two directions for 1D electrons. Specifics depend on the shape of the quantum well or quantum wire — the potential vs. position of the confining potential, but the qualitative features can be easily understood in terms of the particle-in-a-box understanding that we've developed here. No matter what the specific shape of the well, narrower wells and lighter effective masses produce higher confined energies.

Quantum confinement is one example of where quantum mechanics becomes important in semiconductors. Another example is quantum mechanical tunneling and reflection, which we'll talk about in the next lecture.

8.6 References

For another treatment of the quantum confinement, including the solution for a quantum will with a finite barrier, see Pierret:

[1] R.F. Pierret, *Advanced Semiconductor Fundamentals*, (2nd Ed.), Pearson Education, Inc., 2003 (see Ch. 2, Secs. 2.3.2 and 2.3.3).

Silicon quantum wells and how the longitudinal and transverse effective masses affect them are discussed the following book.

[2] Mark Lundstrom, *Fundamentals of Nanotransistors*, World Scientific Publishing Co. Singapore, 2018 (see Ch. 9, Sec. 9.3).

Lecture 9

Quantum Tunneling and Reflection

9.1 Introduction

Quantum mechanical confinement is one important way that quantum mechanics manifests itself in semiconductors. Quantum mechanical **tunneling** and quantum mechanical **reflection** are two other important effects. In this lecture, we will be solving the time-independent wave equation,

$$\frac{d^2\psi(x)}{dx^2} + \frac{2m}{\hbar^2}[E - U_0]\psi(x) = 0\,,$$

where U_0 is a constant potential energy. For quantum mechanical reflection, $E > U_0$, and we write the equation as

$$\frac{d^2\psi(x)}{dx^2} + k^2\psi(x) = 0\,,$$

where

$$k^2 = \frac{2m}{\hbar^2}(E - U_0) > 0\,,$$

and for quantum mechanical tunneling, $E < U_0$, and we write the equation as

$$\frac{d^2\psi(x)}{dx^2} - \alpha^2\psi(x) = 0\,,$$

where

$$\alpha^2 = \frac{2m}{\hbar^2}(U_0 - E) > 0\,.$$

In the first case, $E > U_0$, the solutions are

$$\psi(x) = Ae^{+ikx} + Be^{-ikx}\,,$$

and in the second case, $E < U_0$, the solutions are

$$\psi(x) = Ae^{+\alpha x} + Be^{-\alpha x}\,.$$

This is all we need to understand quantum reflection and tunneling.

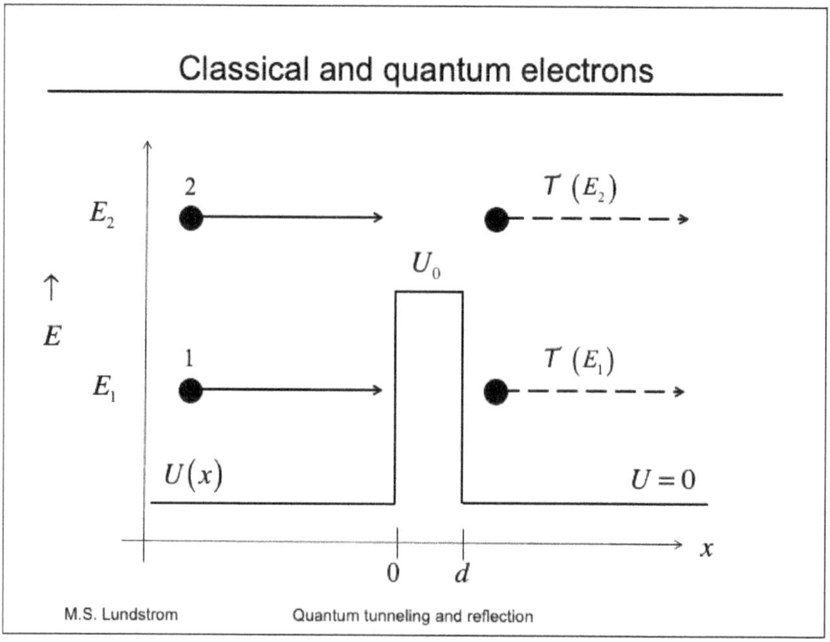

9.2 Quantum tunneling and reflection

Consider two electrons that encounter the energy barrier shown above. Electron 1 has an energy, E_1, and encounters a barrier with a higher energy. If this is a classical particle, there is no chance that electron 1 can get to the other side of the barrier, because it doesn't have enough energy. The probability that it will transmit across is zero, $\mathcal{T}(E_1) = 0$. Electron 2, with a higher energy, E_2, that is above the top of the barrier, can easily cross the barrier, $\mathcal{T}(E_2) = 1$. This is how electrons would behave if they were classical particles.

Electrons are quantum mechanical particles, so they're also waves. If an electron at an energy, E_1, encounters the barrier, there is some probability that the electron will **tunnel** or leak through, $\mathcal{T}(E_1) > 0$. It may be a small probability or might be a significant probability depending on the width of the barrier and the height of the barrier. On the other hand, if the electron has an energy, E_2, that is above the top of the barrier, it might not completely transmit across, $\mathcal{T}(E_2) \leq 1$. Instead of transmitting across, there is some probability that it will reflect and return to the left.

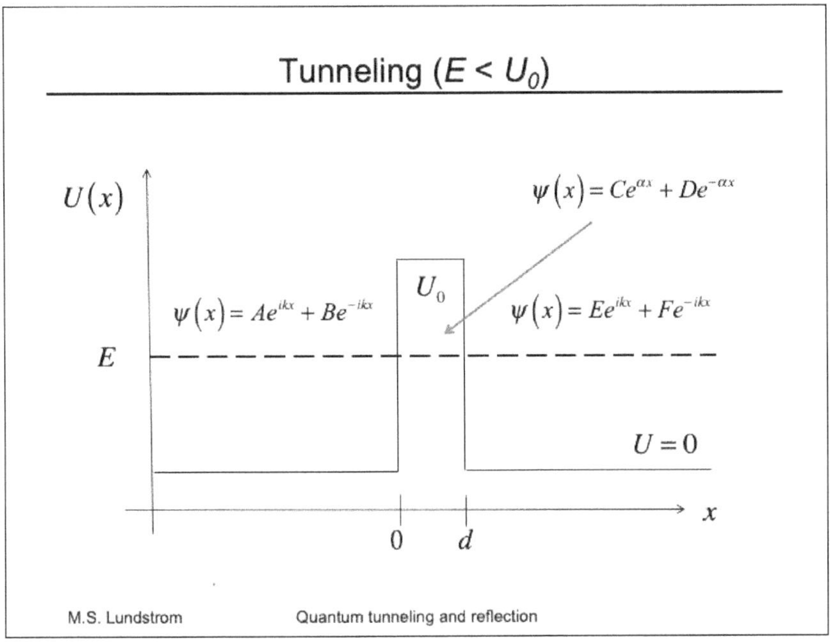

9.3 Quantum mechanical tunneling

Let's first consider tunneling. As shown above, we consider an electron with an energy that is above the potential energy before and after the barrier, but which is below the potential energy inside the barrier. As discussed in Sec. 9.1, we know the form of the solution when $E > U$ and when $E < U$. Before the electron encounters the barrier, $E > U$, and the solution is a wave traveling in the $+x$ direction plus a wave traveling in the $-x$ direction. We also know the form of the wave function after the barrier. It's also a wave traveling in the $+x$ direction and one traveling in the $-x$ direction. Finally, we know the form of the solutions for the electron inside the barrier where $E < U_0$. Inside the barrier, the solutions grow and decay exponentially with position.

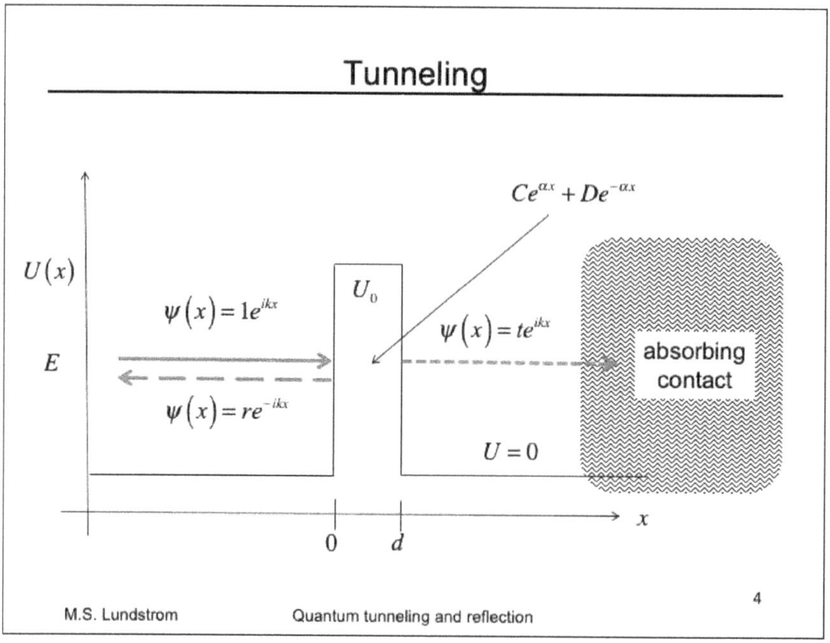

Now let's apply the boundary conditions. As shown above, we've given the incident wave an amplitude of 1 – that is the A in the previous slide. We ask: "What is the probability that it transmits across?" On the right side, we'll assume to have what we call an **absorbing contact**. Any wave that comes across from the left is just absorbed and not reflected. We'll also assume that the absorbing contact does not inject electrons, so there's no wave propagating in the $-x$ direction on the right side of the barrier.

The wave propagating in the $+x$ direction on the right of the barrier is part of the unit amplitude wave injected from the left side of the barrier. Instead of calling the amplitude of the part that transmits across E, as we did on the previous slide, we'll label it with an amplitude, t, that represents the **transmission** amplitude, a complex number.

If the electron that was injected from the left doesn't transmit across, it will be reflected and travel back to the left in the $-x$ direction. Instead of calling its amplitude B, as on the previous side, we'll call it r for **reflection**, also a complex number.

Inside the barrier, we have exponentially growing and decaying solutions with constants C and D that have to be determined.

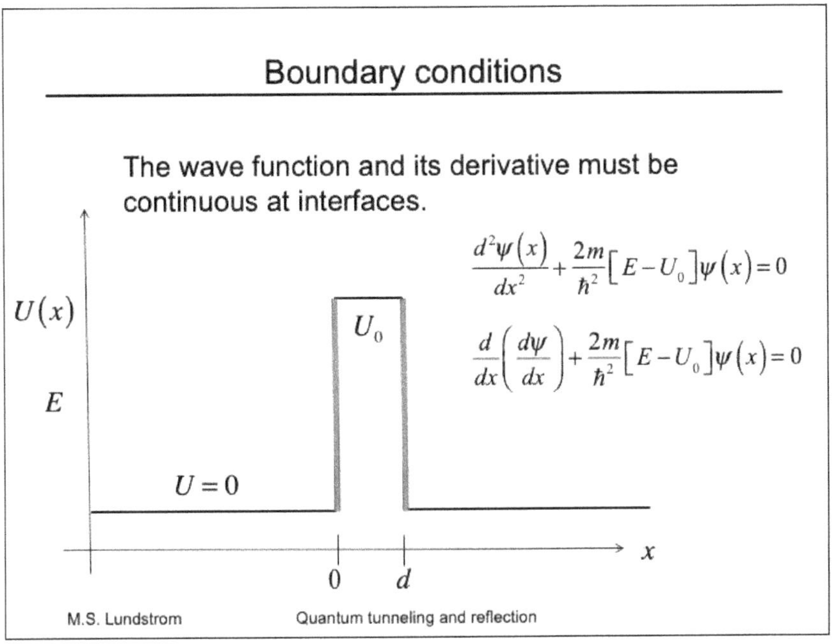

To solve a second order differential equation like the wave equation, we need two boundary conditions. What are those boundary conditions at the two interfaces where the solution changes?

The wave function and its derivative must be continuous at interfaces. You can see why that is necessary by looking at the wave equation and rewriting $d^2\psi/dx^2$ as $d(d\psi/dx)/dx$. This says that the wave function must be continuous, otherwise we'll get an undefined result when we take $(d\psi/dx)$. The slope of the wave function must also be continuous. Otherwise $d(d\psi/dx)/dx$ will be undefined at the interfaces. These are our boundary conditions (or interface conditions in this case) for the wave equation.

Requiring continuity of the wave function and its derivative gives us two boundary conditions at the first interface and two at the second interface.

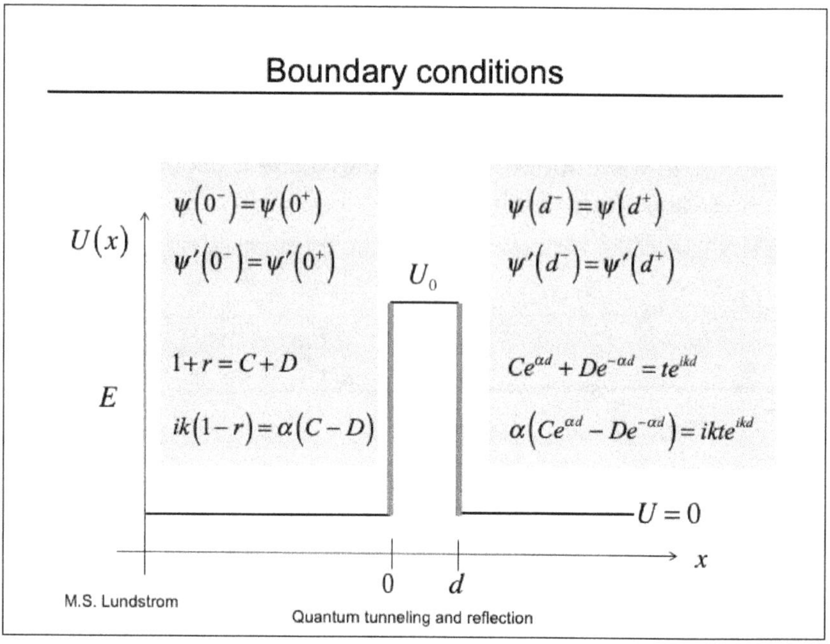

Let's locate $x = 0$ at the first interface. The wave function is continuous there, so $\psi(0^-) = \psi(0^+)$, as shown above. Recall that $\psi(x) = 1e^{ikx} + re^{-ikx}$ on the left and that $\psi(x) = Ce^{\alpha x} + De^{-\alpha x}$ in the barrier. Matching wavefunctions at $x = 0$ gives $1 + r = C + D$.

The derivative of the wavefunction must also be continuous, so $\psi'(0^-) = \psi'(0+)$, and we find $ik\,(1 - r) = \alpha\,(C - D)$. We arrive at the two equations shown above on the left.

We can do the same thing at the second boundary. The result is the two equations on the right.

Now that we have four equations in four unknowns, it's simply a matter of doing some algebra and solving these equations for the four unknowns, t, r, C, and D.

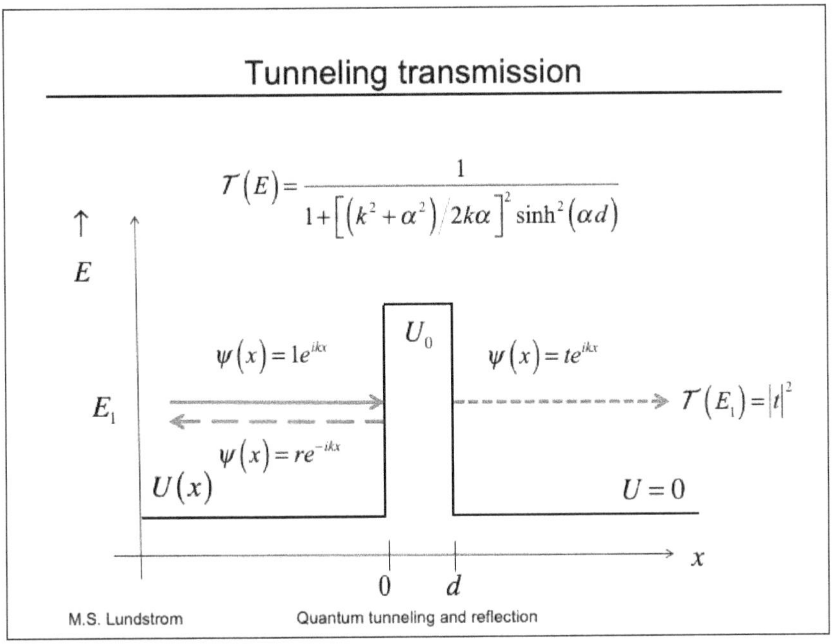

We won't go through that algebra here, but it is a good exercise. We'll just look at some of the features of the solution. Be sure you understand the process we went through to get the equation shown above on the top.

The incident wave has an amplitude of 1. The reflected wave has an amplitude of r, and the transmitted wave an amplitude, t. The quantities t and r are complex numbers. We're looking for the probability that the flux of electrons is transmitted, which is a real number between 0 and 1. So we're interested in the transmission, $\mathcal{T}$, which is $|t|^2$. The resulting equation for the transmission is the rather complicated looking expression on the top. Let's take a closer look at it.

Examine solution

$$\mathcal{T}(E)=\frac{1}{1+\left[\left(k^2+\alpha^2\right)/2k\alpha\right]^2\sinh^2(\alpha d)}$$

$$\sinh(\alpha d)=\frac{e^{\alpha d}-e^{-\alpha d}}{2}\approx\frac{e^{\alpha d}}{2} \qquad \alpha^2=\frac{2m}{\hbar^2}\left[U_0-E\right]$$

$$\sinh^2(\alpha d)\approx\frac{e^{2\alpha d}}{4}>>1 \qquad k^2=\frac{2m}{\hbar^2}\left[E\right]$$

$$\mathcal{T}(E)\approx\left(\frac{16k\alpha}{k^2+\alpha^2}\right)^2 e^{-2\alpha d} \qquad \boxed{\mathcal{T}(E)\propto\exp\left[-2d\sqrt{2m^*\left(U_0-E\right)/\hbar^2}\right]}$$

M.S. Lundstrom Quantum tunneling and reflection

Shown above is the expression that we arrived at for the probability that an electron flux with an energy less than the height of the barrier will transmit through the barrier and come out the other side. The algebra is a little complicated, but it is straightforward. Be sure that you understand how we set up the problem and imposed the boundary conditions.

If the barrier is just a little bit thick (it doesn't take very much thickness), then we get a simplified expression for the transmission shown above on the bottom left.

Remember that we know α and k. The quantity, α, is related to the electron energy, E, and the potential inside the barrier, U_0. The potential outside the barrier is 0, and k is the wave vector of the traveling waves outside the barrier. We know k because we know the energy of the incident electron.

The formula for the transmission through a barrier shown above on the lower right provides some insights into quantum mechanical tunneling.

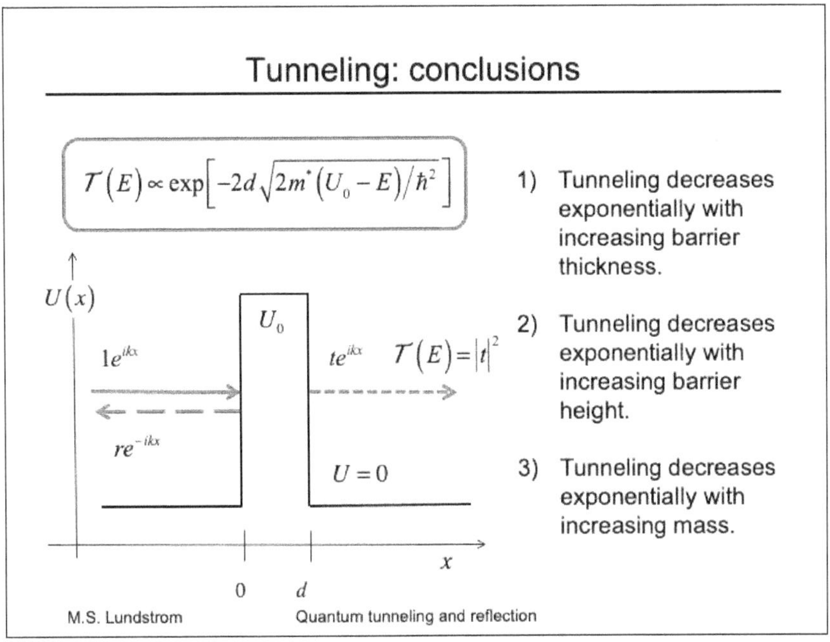

Let's examine the result and identify a few points to remember.

The first thing the expression above tells us is that the transmission probability decreases exponentially with the thickness of the barrier. Stated differently, the transmission increases exponentially as the barrier gets thinner. That's something very important to remember about quantum mechanical tunneling.

The tunneling expression shown above also says that the transmission probability decreases exponentially with the height of the barrier, U_0. That's also very important to remember.

It's also worth remembering that the transmission probability decreases exponentially as the mass of the particle increases. That's important to remember because if we have a semiconductor with a light effective mass, we will expect more tunneling than for semiconductors with a heavy effective mass.

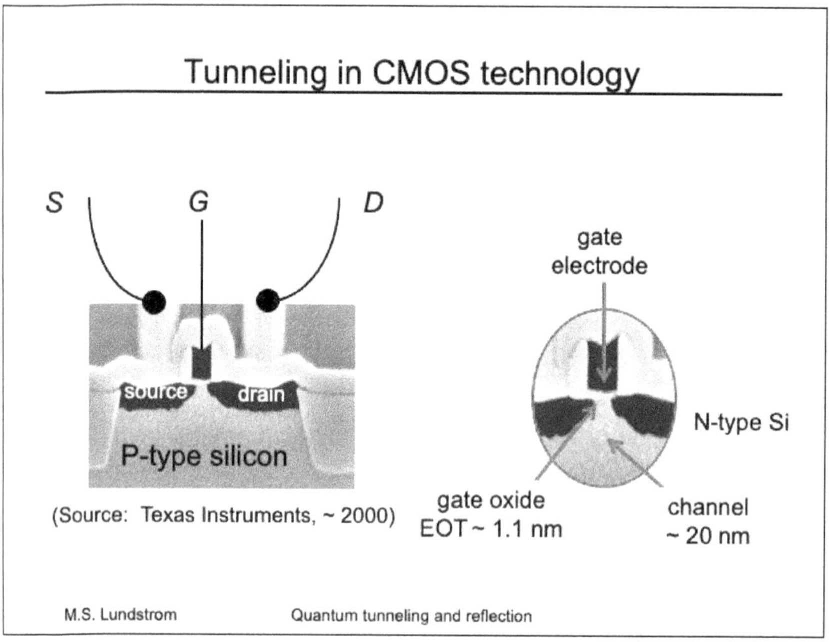

Tunneling can play a role in semiconductor devices. Shown above is a cross section of a silicon MOSFET. Let's look at the heart of the MOSFET and point out where quantum mechanical tunneling can become a problem.

There's a very thin insulating layer (the gate oxide) that separates the metallic gate electrode (labeled G above) from the underlying silicon. We apply voltages to the gate and influence the potential in the silicon, but we don't want current to flow from the gate into the silicon. The function of the insulating layer is to block current. For many years, the gate oxide was scaled down in thickness each technology generation until it eventually reached a thickness of about 1.1 nm — only four atoms or so thick. As we now know, the transmission, and therefore the tunneling current, depends exponentially on the thickness of the insulator. The silicon dioxide thickness could be thinned to just a little more than one nanometer thick, but not thinner because the exponentially increasing gate current led to unacceptable performance of nanoscale MOSFETs. This necessitated a change to thicker insulators with a higher dielectric constant.

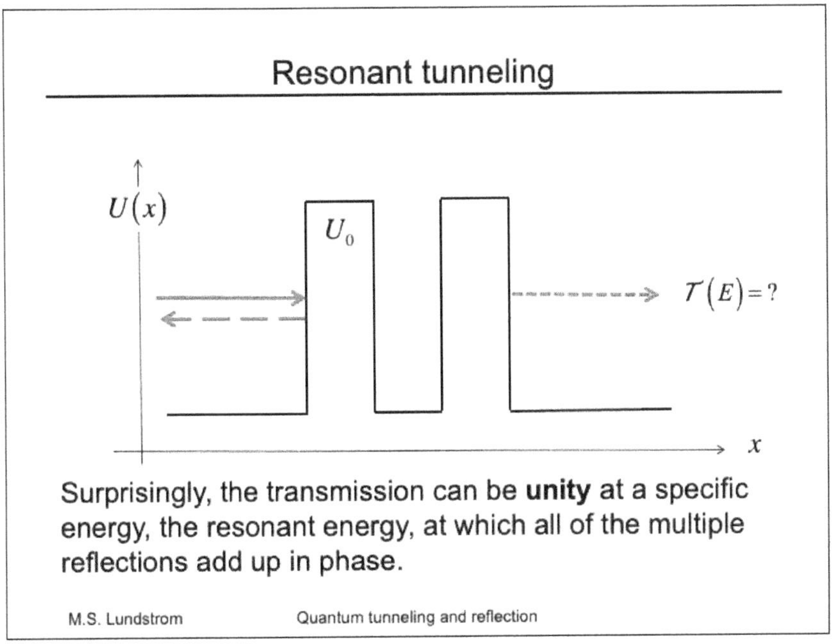

Some curious things can happen when you think about electrons as waves. For example, shown above is our barrier with a modest thickness. There is some probability that electrons can tunnel across, but that probability is generally much less than 1.

What happens if we put two barriers together? You might expect that the probability of transmitting across two barriers in series would be even smaller than the probability of transmitting across one. But there's a phenomenon known as **resonant tunneling**. The reflection and transmission of electron waves back and forth in the cavity between the two barriers can add up in a way that at certain particular energies (certain particular wavelengths) constructive interference occurs. At some specific energies, the transmission can actually be one although the transmission across any one individually is very small. Semiconductor researchers actually found applications for semiconductor devices based on this kind of phenomena. Google "resonant tunneling diode" to find out more.

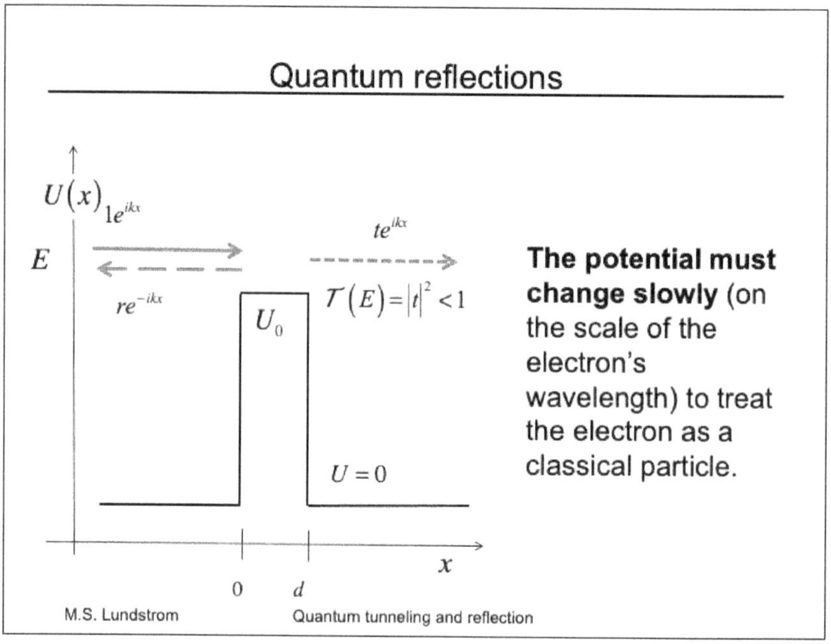

9.4 Quantum mechanical reflection

Quantum tunneling can be very important, usually in a detrimental way. Quantum mechanical reflections can also occur, and it's worth knowing about them too.

If a classical particle has an energy high enough to get over the barrier, it just goes over, but for a quantum mechanical particle, there is a probability that the electron might reflect off the barrier. To compute the reflection coefficient, $R = |r|^2$, we would match the wave function and its derivative at the two interfaces, just as we did for the tunneling problem. The only difference is the energy, $E > U_0$, so we get traveling wave solutions everywhere.

To avoid reflections, we should vary the potential slowly on the scale of the electron's de Broglie wavelength. We saw that that was roughly 10 nanometers or so in silicon. So we'd have to vary our potential slowly on the scale of 10 nanometers if we don't want quantum mechanical reflections to occur.

9.5 Summary

Quantum mechanical tunneling and reflection both occur in semiconductor devices. Tunneling typically degrades the performance of a device, but it can occasionally be exploited (e.g. in resonant tunneling diodes).

It's important to remember three things about tunneling.

(1) Tunneling probabilities decrease exponentially with increasing thickness of a barrier.
(2) Tunneling probabilities decrease exponentially with increasing height of the barrier.
(3) Tunneling probabilities decreases exponentially with the increasing mass of the particle doing the tunneling.

Remembering these three features of tunneling will help us understand the effects that occur in modern semiconductor devices.

Quantum mechanical reflections usually degrade device performance. To avoid reflections, we engineer the structure so that the potential changes slowly on the scale of the electron's wavelength.

9.6 Reference

Pierret briefly discusses quantum mechanical tunneling and reflection.

[1] R.F. Pierret, *Advanced Semiconductor Fundamentals*, (2nd Ed.), Pearson Education, Inc., 2003 (see Ch. 2, p. 48).

Lecture 10

Electron Waves in Crystals

10.1 Introduction

The topic for this lecture is waves in crystals, specifically semiconducting crystals. Our goal is to develop some comfort with a few key concepts that will help us understand free carriers in semiconductors. This lecture can also serve as a starting point for those who wish to dive deeper into the quantum theory of semiconductors.

We begin with the time-independent wave equation for a constant potential energy, U_0,

$$\frac{d^2\psi(x)}{dx^2} + \frac{2m_0}{\hbar^2}\left[E - U_0\right]\psi(x) = 0\,.$$

When $E > U_0$, ,we define a positive quantity, k,

$$k = \frac{\sqrt{2m_0\left[E - U_0\right]}}{\hbar}\,,$$

and write the wave equation as

$$\frac{d^2\psi(x)}{dx^2} + k^2\psi(x) = 0\,,$$

which has solutions

$$\psi(x) = Ae^{\pm ikx}\,.$$

The full time-dependent wave function is a traveling wave,

$$\Psi(x,t) = \psi(x)\phi(t) = Ae^{\pm(ikx-\omega t)}\,.$$

We saw that the quantity, k, is related to energy, so we can write the energy in terms of k as

$$E(k) = U_0 + \frac{\hbar^2k^2}{2m_0}\,.$$

The momentum of the electron is

$$p = \hbar k \,.$$

We describe electrons as particles by forming wave packets with waves having different $k's$. The velocity of the wavepacket is the group velocity, which is the velocity of the electron particle.

$$\upsilon_g = \frac{d\omega}{dk} = \frac{1}{\hbar}\frac{dE}{dk} = \frac{\hbar k}{m_0} = \frac{p}{m_0}\,,$$

so the velocity of the electron is its crystal momentum divided by its mass.

The equations above describe a free electron with mass, m_0, in the presence of a constant potential energy, U_0. In this lecture, we'll be discussing a much more complicated problem, an electron in the presence of a spatially varying, periodic potential, $U(x) = U_c(x)$, where U_c, the **crystal potential**, has the periodicity of the lattice. We'll find, however, that in the end, the key results can be expressed with equations very similar to those summarized above.

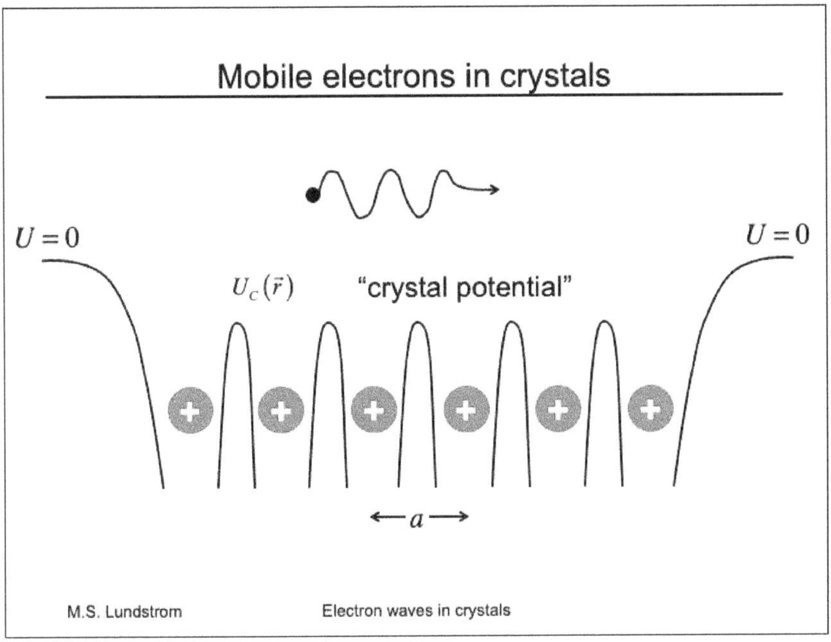

10.2 Free electrons in the crystal potential

When we treat atoms, we must use the correct potential energy, U. For the hydrogen atom, the potential energy is the Coulomb potential,

$$U(\vec{r}) = -\frac{q^2}{4\pi\epsilon_0 r}.$$

The solutions to the wave equation with this potential give us the s, p, d orbitals and the quantized energy levels that we saw earlier for the hydrogen atom.

For a crystal, the potential is even more complicated. At the two ends of the crystal shown above, we have the vacuum where the energy is arbitrarily set to zero. This is our reference energy. Inside the crystal, the energy is negative; the lower energy inside confines the electrons to stay inside the crystal.

Associated with each atom, there's a potential due to all of the positive charges on the protons and due to the negative charges of the core electrons and the valence electrons too. That can be a very complicated potential, but it does have one simplifying feature – this crystal potential has the periodicity of the lattice.

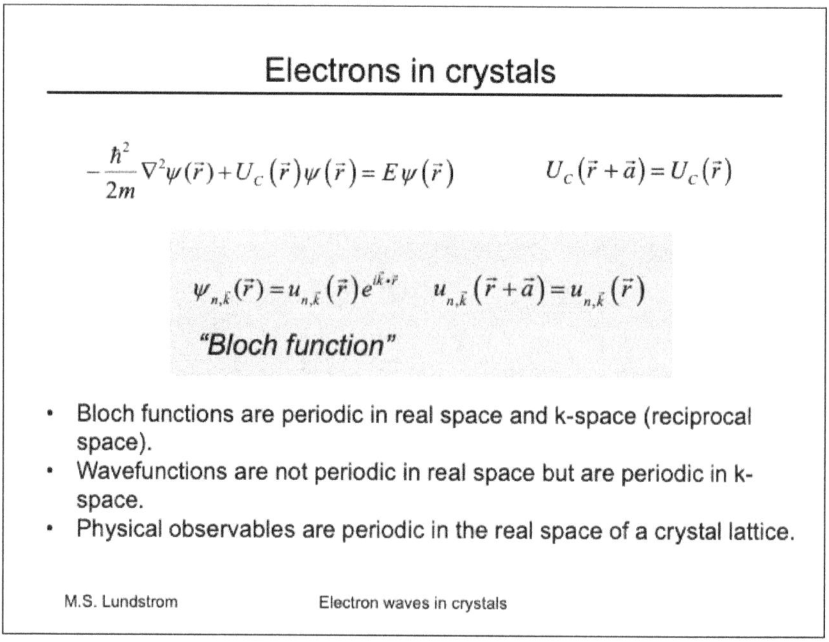

To understand electrons in semiconductors, we need to solve a wave equation that includes the periodic crystal potential. The solutions to equations like this have been studied in detail by physicists and mathematicians. The solution is shown in the shaded box above. It consists of a component with the periodicity of the lattice times the traveling wave that we've seen earlier. The subscript, n, is a quantum number that labels the various "bands" (e.g. conduction band, valence band), and k is the wavevector we have seen before. The periodic component, $u_{n,k}(r)$, has the periodicity of the lattice; it is also periodic in k-space, which is called **reciprocal space**. The overall wave function is not periodic in space, but it is periodic in k-space. The quantity, $\psi^*(x)\psi(x)$ is periodic in space as are all physical observables.

The Fourier transform of a function of time, describes the same function in frequency space; frequency has units of 1/time. Similarly, the spatial Fourier transform of a function of position, or real space (e.g. a crystal lattice) can be described in k-space (reciprocal space), where the wave vector k has units of 1/distance.

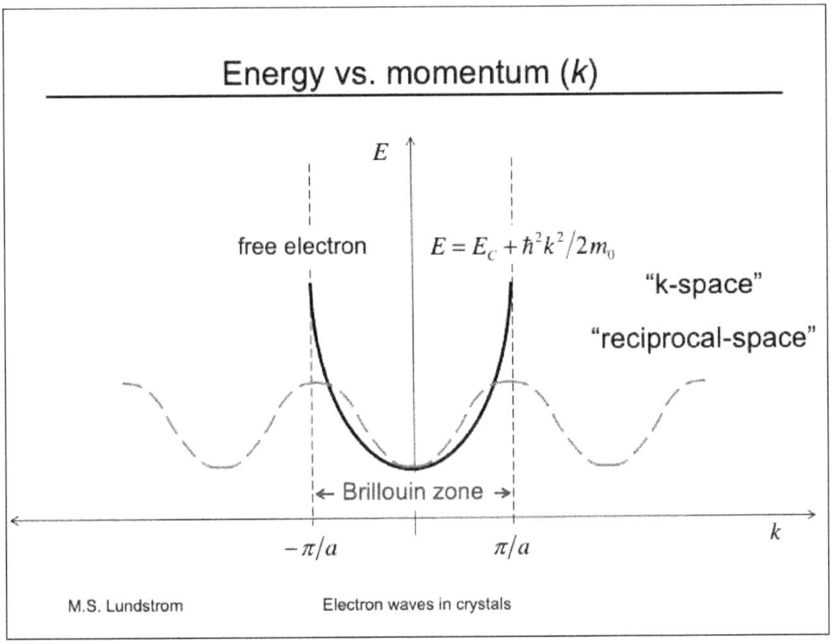

10.3 Energy vs. momentum

The energy vs. momentum (or E vs. k) relation is an important relation for electrons in semiconductors. For free electrons $E = p^2/2m_0$. The momentum is $\hbar k$, so $E = \hbar^2k^2/2m_0$. This energy vs. momentum relation is shown above as the solid line.

In the presence of a periodic crystal potential, $E(k)$ becomes periodic in k, as shown in the dashed line above. At the energy minima, the curvature might be a different from the curvature for a free electron in vacuum. The difference in curvature means that there's a difference in the mass of the particle in these two cases. This is where the **effective mass** comes from.

As seen in the figure above, $E(k)$ is periodic in k. The distinct values occur over a limited region of k-space. In this simple 1D example that region goes from $k = -\pi/a$ to $k = +\pi/a$, which is called the **Brillouin zone**. Since the solution is periodic, we only need to consider the solution within one Brillouin zone. In 3D crystals, the Brillouin zone can assume a complicated shape, but the basic idea is that the solutions repeat themselves in reciprocal space (i.e. k-space).

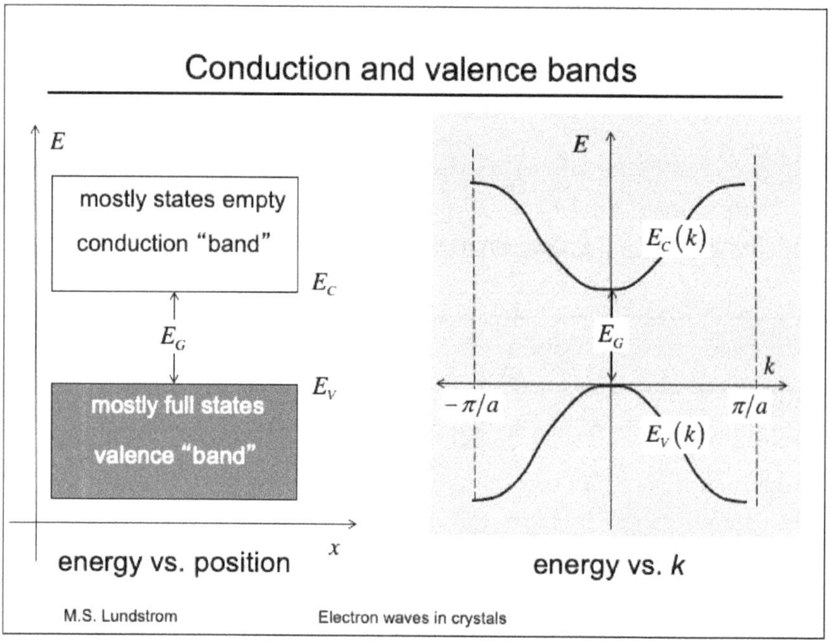

As shown above on the left, the valence band is mostly filled, and the conduction band is mostly empty. We have previously talked about the $E(k)$ relation. A typical conduction band is sketched above on the right. There's a minimum (at $k = 0$ in this case), and $E_c(k)$ looks like a parabola near the minima. The valence band has a maximum (also at $k = 0$ in this case), and near the maximum, $E_v(k)$ looks like a parabola. These solutions come from solving the wave equation with a crystal potential.

The **crystal momentum** is an important quantity. The momentum of an electron in a crystal is not $p = \hbar k$ because the electron is continually being acted on by the electrostatic forces of the atoms in the crystal lattice. The crystal momentum does, however, behave much like the momentum of a free particle. For example, when electrons collide in semiconductors, crystal momentum is conserved just as when free particles collide, momentum is conserved. When we talk about "momentum" for electrons in semiconductors, we will always be talking about the crystal momentum.

The key quantity that we get by solving the wave equation is $E(k)$, which we refer to as the **band structure** of the semiconductor or sometimes as the **dispersion**. Every wave has a dispersion, $\omega(k)$ — a relation between frequency (or energy because $E = \hbar\omega$) and wave vector, k.

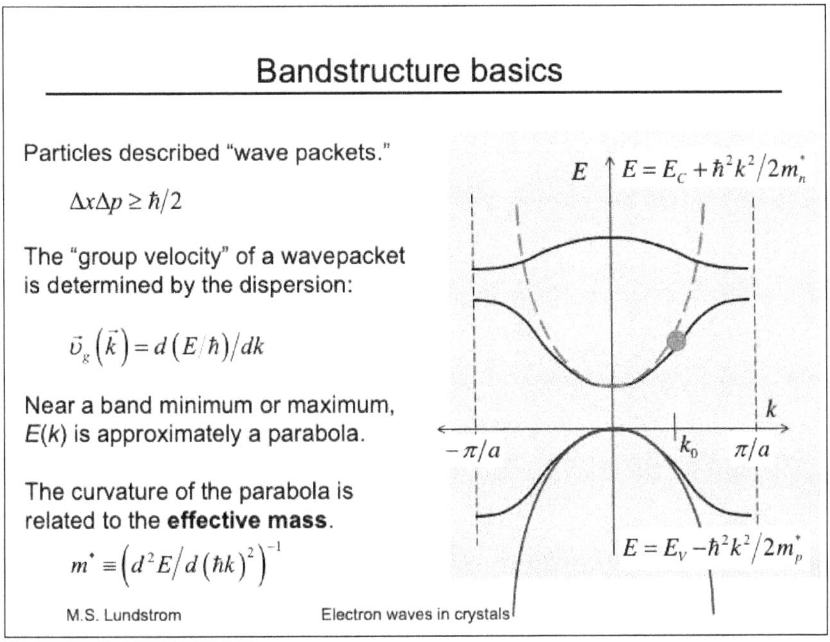

10.4 Bandstructure

Shown above are valence and conduction bands, $E_v(k)$, and $E_c(k)$. There are other bands, but we don't usually need to concern ourselves with them, but as discussed in Sec. 18.5, under some conditions, they can become important.

We describe electrons with wave packets. The uncertainty in the location and momentum of the particle is given by $\Delta x \Delta p \geq \hbar/2$. The electron wave packet moves at the group velocity, $\upsilon_g = (1/\hbar)(dE/dk)$ at $k = k_0$, where k_0 is the center wave vector of the wave packet.

Electrons in the conduction band are near the bottom of $E_c(k)$, and most of the holes in the valence band are near the top of $E_v(K)$. Near the minimum, $E_c(k)$ looks like a parabola, so we can approximate $E_c(k)$ by a parabola — as long as we use the correct curvature (i.e. the correct effective mass). From the slope of $E_c(k)$ we find that for a parabolic band, the group velocity is crystal momentum divided by effective mass, i.e. $\upsilon_g = (1/\hbar)(dE/dk)|_{k=k_0} = \hbar k_0/m^*$.

Similarly, near the top of the valence band, we can approximate $E_v(k)$ with a parabola with the appropriate curvature (i.e. hole effective mass). This **parabolic band assumption** generally (but not always) works well for semiconductors and greatly simplifies calculations.

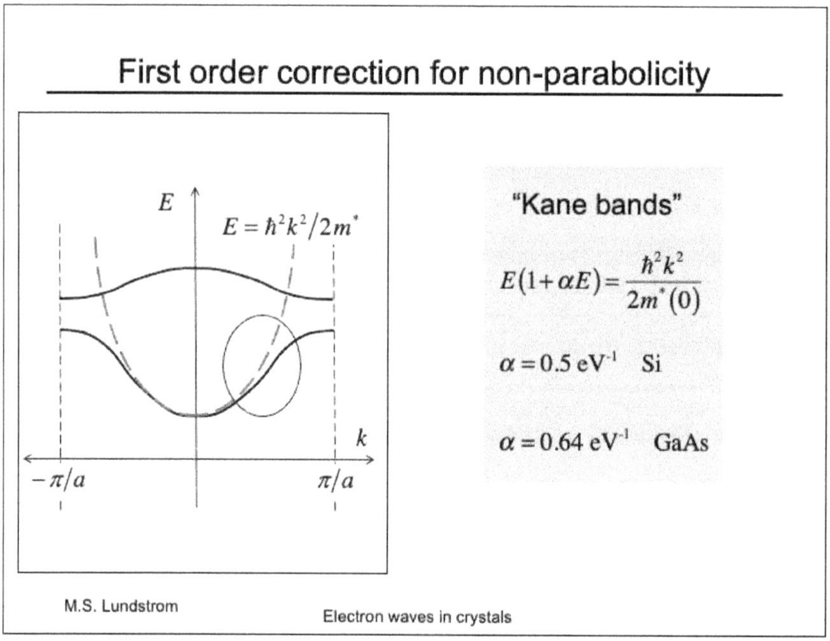

Because electrons are near the bottom of the conduction band and holes are near the top of the valence band, the parabolic band assumption often works well, but there are times when we need to do a little better. Sometimes a first order correction for nonparabolicity is needed. For example, you can see from the figure above, for energies a little above the minimum, there is some difference between the parabolic approximation (dashed line) and the actual band structure (solid line). Shown at the right is a first order correction known as **Kane bands**. The quantity $m^*(0)$ is determined from the curvature of $E(k)$ at $k = 0$. You can see that for a given energy, k for the nonparabolic band is larger than for the parabolic band, so this equation describes a "flattening" of $E(k)$. The **nonparabolicity parameter**, α, is a material dependent parameter that we can obtain from numerical solutions to the wave equation.

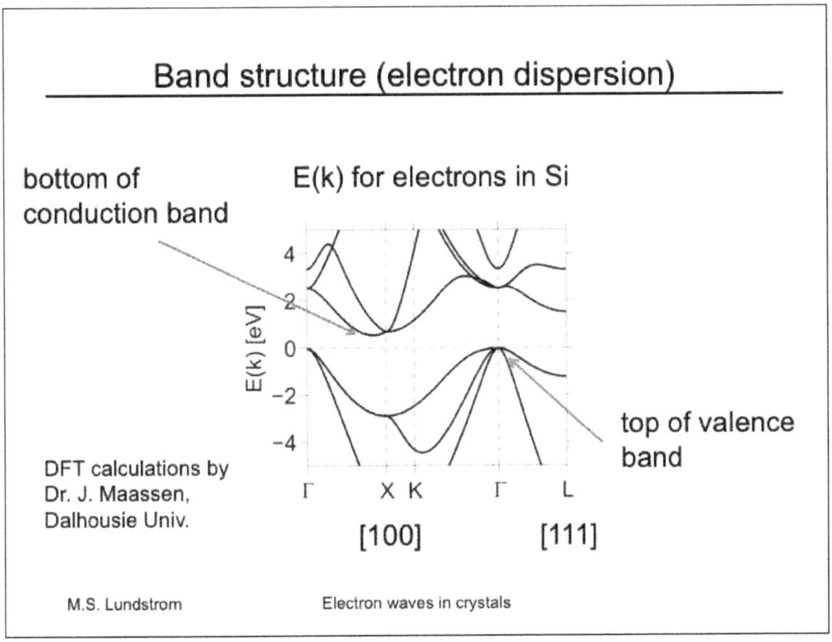

We have talked about band structure and reciprocal space (k-space) using simple, 1D examples, but real crystals are 3D, which means that we need to plot $E(k)$ in 3D. The Brillouin zone is a 3D volume in k-space.

Shown above is a numerical solution to the wave equation in three dimensions along various high symmetry directions for a Si crystal. The Γ point is the origin, in k-space. From Γ to X is a [1 0 0] direction. From Γ to L is a [1 1 1] direction. You can see that the bands have a complicated shape, but if you look carefully, you can see that the lowest conduction band minima is along a [1 0 0] direction. The minimum does not occur at $k = 0$, but is near the edge of the Brillouin zone along a [1 0 0] direction.

If we look at the valence band we can see that the maximum is at $k = 0$, the Γ point, so Si is an indirect band semiconductor. There are actually three valence bands and they all have a maxima at the Γ point, $k = 0$. (Note that when spin-orbit coupling is included in the calculations, the third valence band is actually split-off from the other two.)

This complicated band structure can be calculated numerically, but we are primarily interested in the region of lowest energy in the conduction band and highest energy in the valence band.

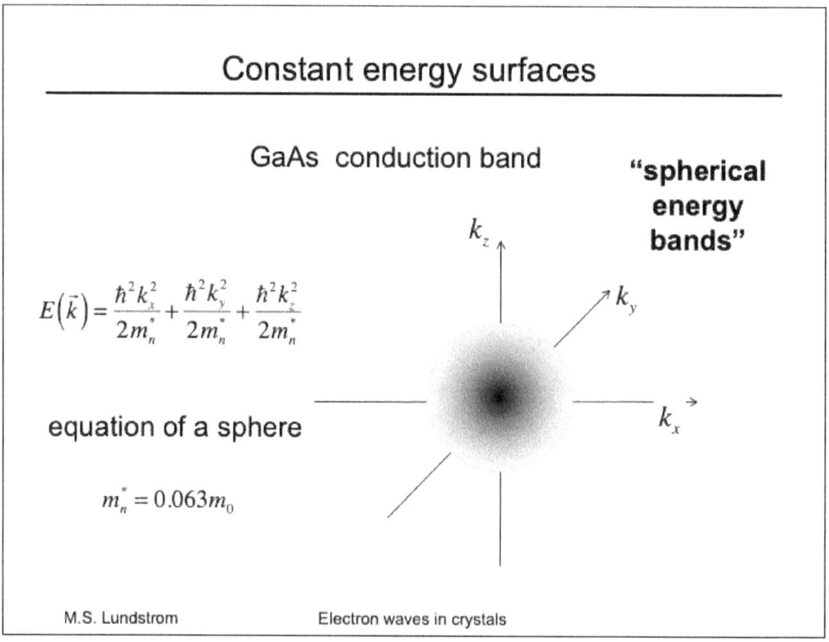

10.5 Constant energy surfaces

We will also talk from time to time about **constant energy surfaces**. For a parabolic band, the energy is $E(k) = \hbar^2 k^2/2m^*$, or in 3D, $E(k) = \hbar^2(k_x^2 + k_y^2 + k_z^2)/2m^*$ where I have assumed that the effective mass (band curvature) is the same in all directions. For a constant energy, this is the equation of a sphere in k-space. We will call energy bands like this **spherical energy bands**. The conduction band of GaAs is well approximated as a spherical band for energies close to the minimum.

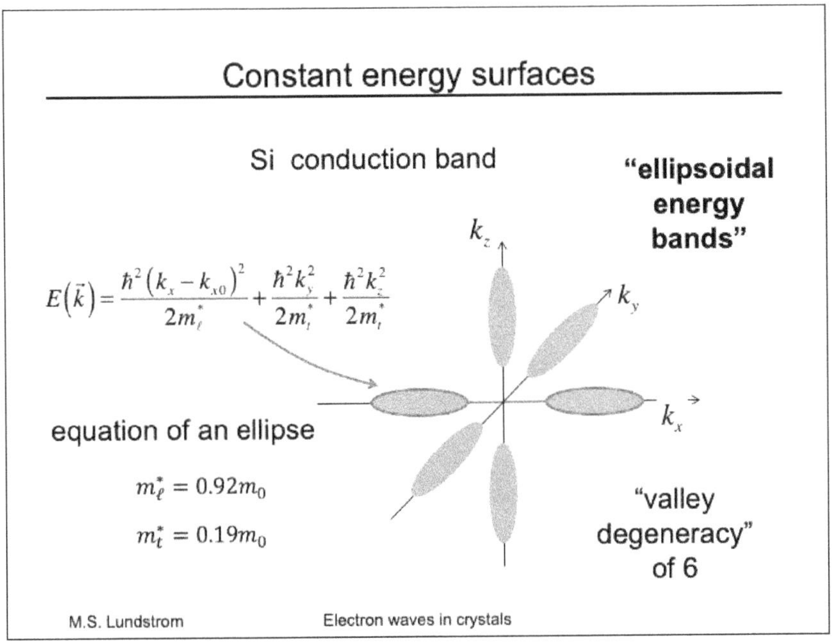

In some materials, silicon is an example, things get a little more complicated. The curvatures in the x direction, the y direction, and the z direction are different. If we plot the constant energy surfaces for electrons in silicon, there will be six of them along the six [100] directions. The constant energy surfaces are six ellipsoids. The curvature in the long direction is related to the **longitudinal effective mass**. The curvature in the transverse direction is related to the **transverse effective mass**, and they can be quite different. For example, in Si, the longitudinal effective mass is $m_\ell^* = 0.92m_0$, and the transverse effective mass is only $m_t^* = 0.19m_0$. We will call energy bands like this **ellipsoidal energy bands**. There are 6 different minima (or **valleys**) that occur along three coordinate axes. We say that there is a **valley degeneracy** of 6.

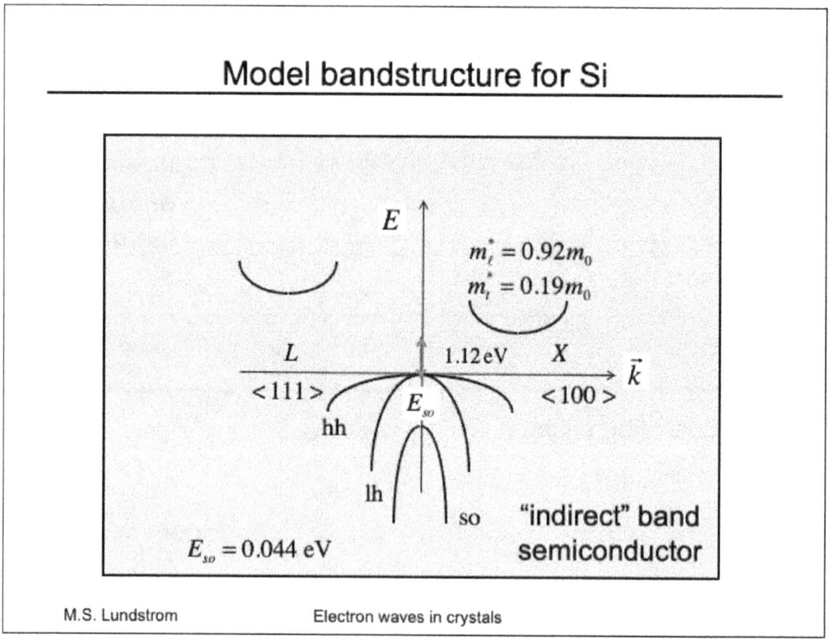

10.6 Model bandstructures

Shown above is a simple, "model band structure" for Si that is informed by measurements and detailed calculations.

In Si, the lowest conduction band minima are along [1 0 0] directions with two different effective masses that describe the curvature in different directions. (There's a set of higher valleys along the direction L too, but they don't typically come into play.) The conduction band is parabolic and can be described by $E_c(k) = (\hbar^2/2)\left[(k_x - k_{x0})^2/m_{xx}^* + (k_y - k_{y0})^2/m_{yy}^* + (k_z - k_{z0})^2/m_{zz}^*\right]$, where m_{xx}^*, m_{yy}^*, and m_{zz}^* are either the longitudinal or transverse effective mass depending on which of the 6 ellipsoids we are considering.

The three valence bands have their maxima at $k = 0$. The **heavy hole band** has very little curvature. There is also a **light hole band** with much larger curvature and a **split-off band** that sometimes comes into play. To first order, we can describe the valence bands as spherical, but there are different effective masses for the three different bands.

Silicon is an indirect semiconductor; the minima of $E_c(k)$ and the maxima of $E_v(k)$ occur at different locations in k-space.

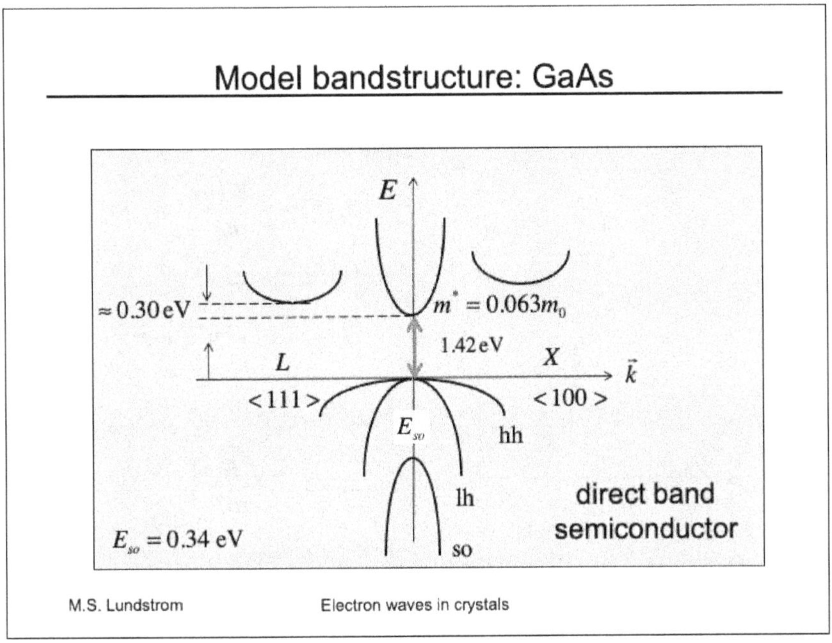

In gallium arsenide (GaAs), a III-V semiconductor, the minimum of the conduction band occurs at $k = 0$. It has a very high curvature, which means a light effective mass. If we look out along the [1 1 1] directions, we see that there are conduction band minima that are a little higher in energy. They are close enough in energy to the minima of the first band so that they sometimes become important and can affect charge carrier transport (see Sec. 18.5). There's yet another set of conduction band minima if we look along the [1 0 0] directions, and these "upper valleys" can also come into play under high electric fields where the electrons gain kinetic energy. The valence bands for GaAs are similar to the valence bands in silicon.

The simplest description of gallium arsenide is the model band structure shown above. There is a spherical conduction band with its minimum at $k = 0$. Three spherical valence bands have their maxima at $k = 0$. There is also a set of ellipsoidal upper conduction band valleys along L and another set along X.

Finally, note that gallium arsenide is a direct gap semiconductor because the minimum of the conduction band and the maximum of the valence band both occur at $k = 0$.

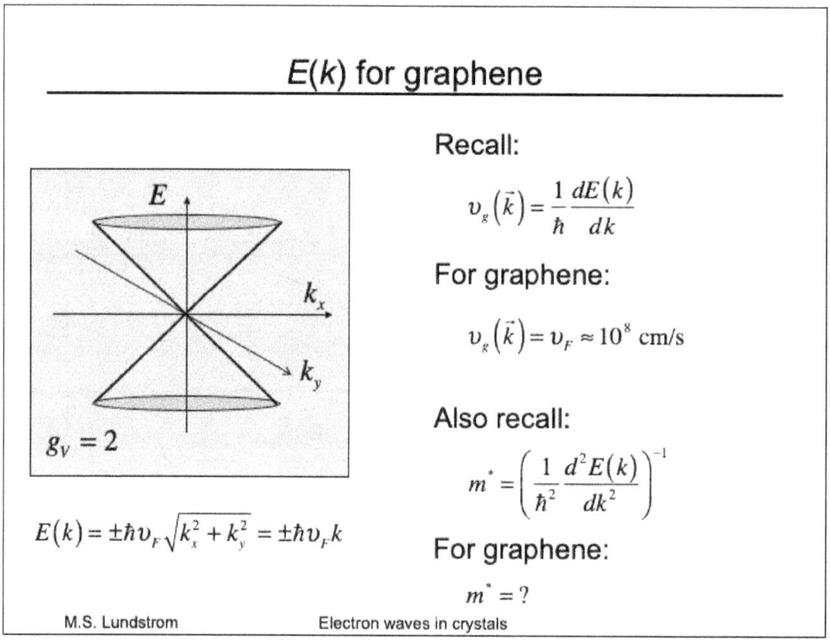

Graphene is an interesting 2D material with conical constant energy surfaces as shown above. There is no band gap; graphene is a **semi-metal**. Its $E(k)$ is not complicated, but it's much different from the parabolic relation for common semiconductors. By taking the slope of $E(k)$, we find that the group velocity is independent of momentum. It is about $\upsilon_F = 10^8$ cm/s, which is a very high velocity for electrons.

Things get a little confusing if we ask about the effective mass. The definition of the effective mass is that it's proportional to one over the curvature of the band. There is no curvature of these bands, so you might say that the hole effective mass is infinite. On the other hand, this dispersion is very much like that of photons. Instead of $\omega = ck$, where c is the speed of light, we have $E/\hbar = \omega = \upsilon_F k$. Photons are massless, so maybe the effective mass of electrons in graphene should be zero.

While it is convenient to introduce an effective mass when dealing with parabolic band semiconductors, there is no need to do so. If we know $E(k)$, we can analyze materials by dealing directly with the $E(k)$ relation. In these lectures, we'll stick to parabolic band semiconductors, but you may encounter materials where the bands have a much different shape, and then you will have to deal directly with the material's $E(k)$.

10.7 Summary

As discussed in this lecture, we can often treat electrons and holes as semi-classical particles with the quantum mechanics being embedded in the effective mass. The key points can be summarized as follows.

(1) Electrons in a solid behave as both particles and as waves.
(2) Electron waves are described by a band structure, $E(k)$.
(3) Because the crystal is periodic, the dispersion, $E(k)$, is periodic in k (Brillouin zone).
(4) Electrons and holes in semiconductors have a crystal momentum that we will treat like the actual momentum of a particle.
(5) Since electrons and holes are typically near the band minima and maxima, we can usually treat the bands as parabolic.
(6) Electrons and holes in a parabolic band semiconductor can be treated as semi-classical particles with an effective mass and crystal momentum.

There are times where we do have to include quantum mechanics explicitly. Examples are quantum confinement and quantum mechanical tunneling. In those cases, we must solve the wave equation summarized in Sec. 10.1, but the mass will be the effective mass of electrons or holes.

Finally, this lecture, even more than others, may leave readers with questions. Exactly how is the bandstructure computed, why can the crystal momentum often be treated as the real momentum, etc. Questions like these require a deeper understanding of solid state theory, but if we accept for the time being the key points summarized above, we can learn much about the behavior of free carriers in semiconductors.

10.8 References

For those who seek a deeper understanding of energy bands, Pierret presents a simple introduction and Tiwari a more advanced treatment.

[1] R.F. Pierret, *Advanced Semiconductor Fundamentals*, (2nd Ed.), Pearson Education, Inc., 2003 (see Ch. 3).

[2] Sandip Tiwari, *Semiconductor Physics: Principles, Theory, and Nanoscale*, Oxford Electroscience Series, Vol. 3, 2020 (see Chapters 3 and 4).

Lecture 11

Density of States

11.1 Introduction

We are now ready to discuss an important topic – the **density of states**. As you know, the valence band is mostly filled with electrons. There's a top of the valence band, E_v, and a bottom of the valence band that is so low in energy that we generally don't need to worry about it. We also have a conduction band that consists of states that are mostly empty. There's a bottom of the conduction band, E_c, and there's a top of the conduction band that we generally don't need to worry about because it's so high in energy.

Within the valence and conduction bands, the states are not uniformly distributed in energy. It will be important to know how the states are distributed within the conduction and valence bands. Most of the empty states in the valence band (the holes) are near the top of the valence band, and most of the filled states in the conduction band (the electrons) are near the bottom of the conduction band, so we only need to know how the states are distributed in energy near the edges of the bands. That's our focus in this lecture.

The quantity that we will discuss is the **density of states** (DOS), $D(E)$. The DOS is simply a way to count the number of states between a certain energy, E and $E + dE$. The number of states in that range is $D(E)dE$.

We will make extensive use of the DOS in the next unit. In this lecture, we want to understand the DOS and how to calculate it for a given semiconductor. We'll make use of a simple, parabolic $E(k)$ which generally works well near the top of the valence band and near the bottom of the conduction band.

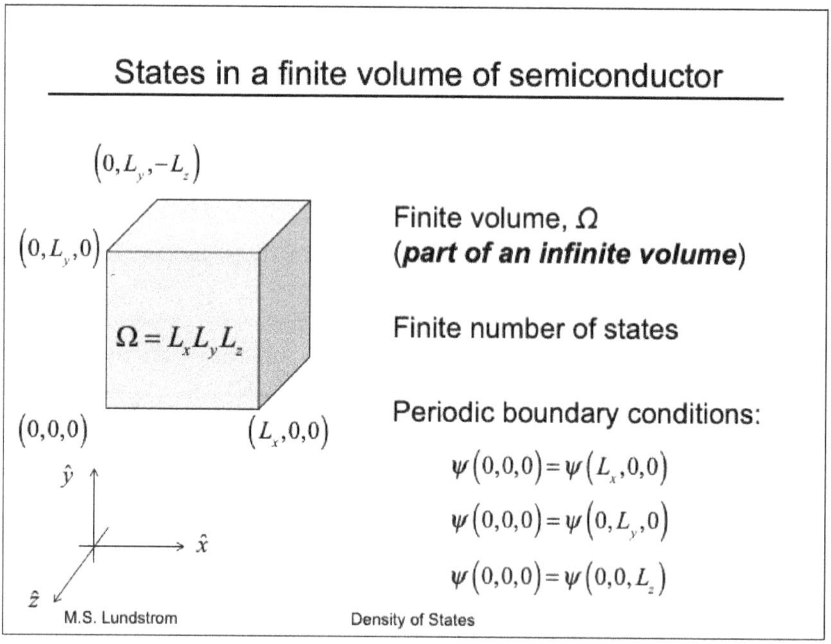

11.2 Density of states in k-space

Consider the large volume, Ω, of a semiconductor crystal as shown above. The length is L_x in the x direction, L_y in the y direction, and L_z in the z direction. Consider this to be a large but finite volume carved out of an infinite, bulk semiconductor. Because the volume is finite, there are a finite number of states, which come from the finite number of atoms in the volume. We need to count these states.

Since this is a piece of an infinite solid, we will assume that at $x = 0$ and at $x = L_x$ the wave function is the same. (The same goes for y and z directions.) That is, we assume **periodic boundary conditions**.

Of course, a real semiconductor will have a finite volume, and the crystal lattice will terminate at the boundaries. There will be **dangling bonds** because atoms on the surface will be missing atoms to bond with. The dangling bonds produce **surface states**, which generally lie within the forbidden gap. Surfaces states can have important consequences for devices, but our focus here is on the density of states inside the semiconductor, far from the surfaces.

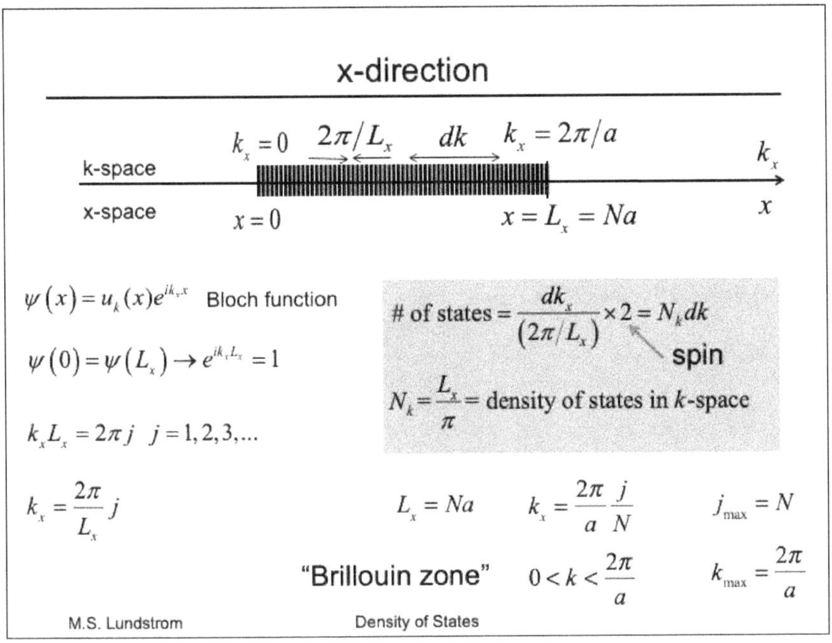

Let's look in the x direction and measure things in units of the lattice constant; N is the number of atoms in the length, L_x, and a is the lattice constant. We know the solutions to the wave equation in this region. They're Bloch functions as shown above at the left in the first equation — a periodic part times a wave solution.

At $x = 0$ the wave function is the same as at $x = L_x$. The periodic part will be the same because it repeats itself at every lattice constant. To make $\psi(x) = \psi(x + L_x)$, we require that $\exp[ik_x L_x] = 1$, so $k_x L_x = j(2\pi)$, where j is an integer. Only a discrete set of k_x's is allowed, $k_x = j(2\pi/L_x)$. These are the discrete k_x's shown in the figure at the top. When we apply boundary conditions to a wave function, we get a set of discrete states, but this is a particle in a very large box, so the allowed k_x's are very finely spaced.

Next, we ask: "In some range, dk, how many states are there?" Well, we know that the states are spaced by $2\pi/L_x$, so the number of states is the width of the region, dk, divided by the spacing between states, then we multiply by 2 because each state can hold an electron with spin up or with spin down. This quantity, N_k in the shaded box above is the density of states in k-space. The number of states in dk is $N_k dk$.

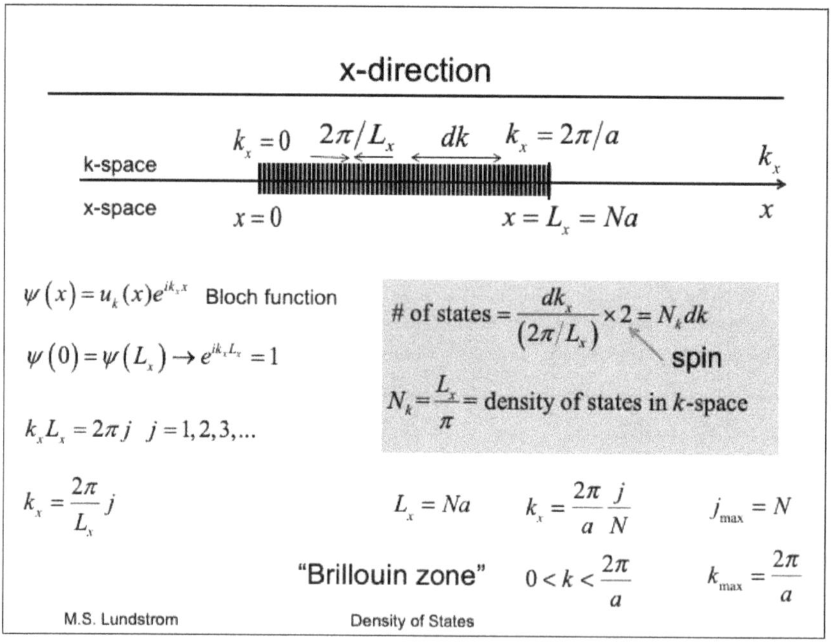

To continue our discussion, let's look at the result more closely.

As shown above below the shaded box, the length, L_x, is some integer number of lattice constants, which means $L_x = Na$, and the allowed values of k_x are $k_x = (2\pi/a)(j/N)$. As j increases, it eventually reaches N. When that happens, the phase of the exponential has gone from zero to 2π. If we continue to let j increase, we're just going around the circle again, and we get another set of results that are completely equivalent. So, the unique solutions are between $k_x = 0$ and $k_x = 2\pi/a$ which is just the **Brillouin zone** that we talked about in the previous lecture. The Brillouin zone is a region in k-space in which the unique solutions are found. We can shift this region in k-space, and it's usually convenient to shift it so that the unique solutions are between $-(\pi/a) < k_x < +(\pi/a)$, The width of the region over which we have a unique set of solutions is $2\pi/a$, the Brillouin zone.

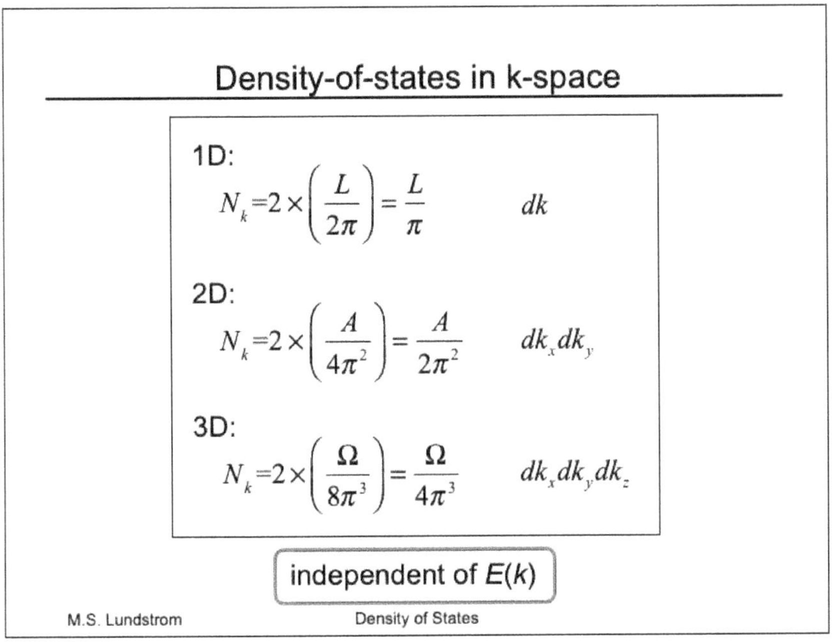

We've gone through the counting in 1D, and we've seen that the density of states in k-space is 2 for spin times $L/(2\pi)$, so $N_k = L/\pi$ in 1D.

If we have a 2D material, the density of states would be 2 for spin but now they're spaced $L_x/(2\pi)$ in the x direction and $L_y/(2\pi)$ in the y direction, so we get $L_x L_y$, which is the area, A, times 2 over $(2\pi)^2$, so $N_k = A/(2\pi^2)$ in 2D.

In 3D, we have the factor of 2 for spin, but we now have states spaced in x, y, and z, so the factor (2π) in 1D, $(2\pi)^2$ in 2D, becomes $(2\pi)^3$ in 3D. and $N_k = \Omega/(4\pi^3)$ as shown above. The important thing to point out is that the 1D, 2D, and 3D DOS's in k-space are independent of $E(k)$. The density of states in k-space is the same — no matter what the material is.

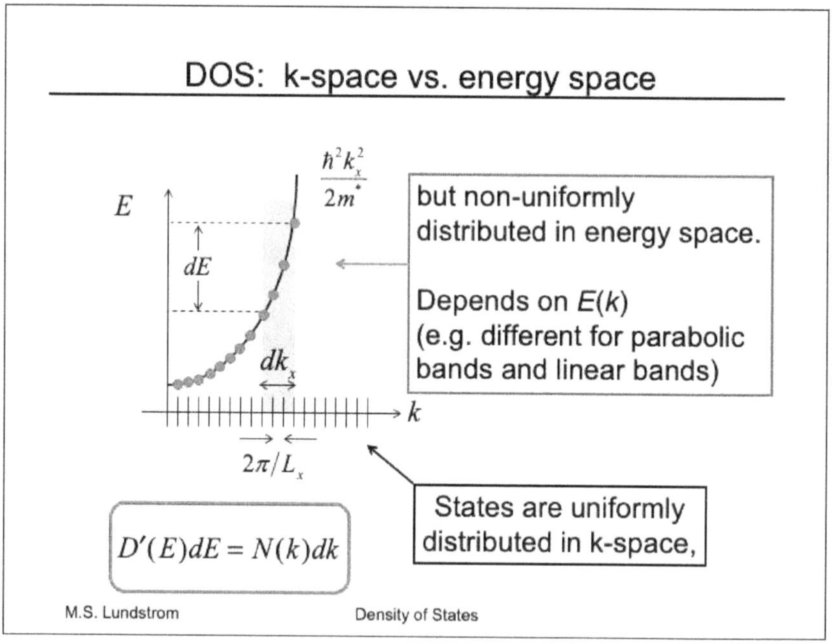

11.3 Density of states vs. energy

We can also ask: "How are the states distributed in energy?" As in k-space, the DOS in energy space depends on the dimension being considered, but in contrast to the DOS in k-space, $D(E)$ also depends on the $E(k)$.

The figure above shows the bottom of a parabolic conduction band. We've seen that the states are spaced uniformly in k. At any particular k, there is a corresponding energy given by $E(k)$. If we map the states that are uniformly distributed in k-space to their corresponding energies, you can see from the figure above, that they are not distributed uniformly in energy. Exactly how they are distributed in energy will depend on $E(k)$.

To derive $D(E)$ in 1D, we'll count the states in dk, and then map those states into a corresponding region in energy, dE. As shown by the equation at the lower left, the number of states in the energy range, dE, must be the same as the corresponding number of states in dk. This will allow us to define a density of states in energy, which we will write as $D'(E)dE$. By using the relation on the lower left, we will be able to deduce from the known DOS in k-space, the corresponding DOS in energy space, if we know $E(k)$. Let's see how this works in 1D.

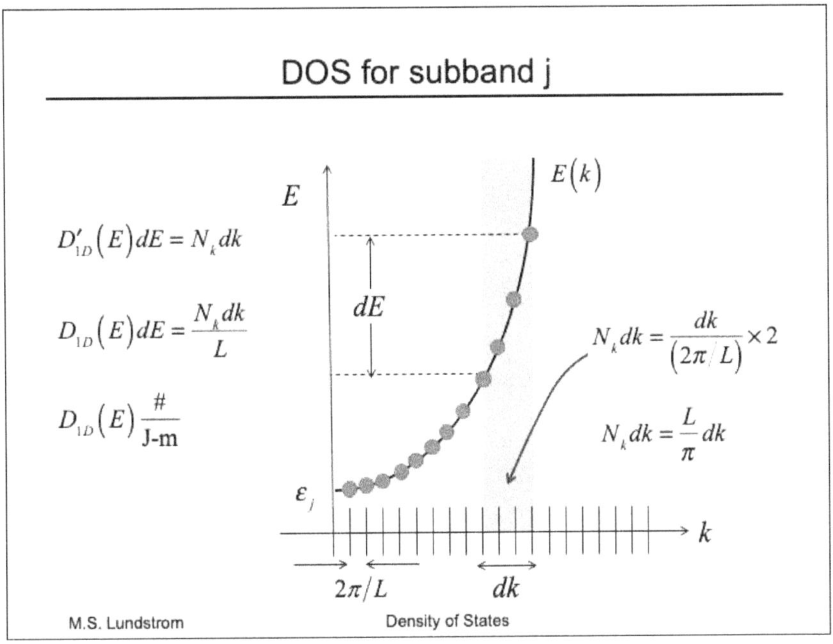

Consider a 1D nanowire, once academic exercise, but today we can readily produce semiconducting nanowires. We'll assume a single subband (recall that there are multiple subbands due to quantum confinement). If we assume a single subband with an energy ϵ_j, then we need to add the kinetic energy associated with motion in the x direction, $\hbar^2 k_x^2/2m^*$, so $E = \epsilon_j + \hbar^2 k_x^2/2m^*$.

Now let's compute the DOS in 1D, $D_{1D}(E)$. We'll map the k-states in the region, dk, to the corresponding region in energy space, dE. The number of states in the region, dk, is $N_k dk = (L/\pi)dk$, which has to be the same total number of states in the corresponding dE, $D'_{1D}(E)dE$, because these are the same states.

We'll drop the prime on $D'(E)$ now. Typically when we talk about density of states we are interested in the density of states per unit length. We're not interested in the actual number of states because it depends on the arbitrary length, L, that we chose. If we double the length we'll get twice the number of states. So, in 1D, we are really interested in the density of states per unit length. That's why we divide by L; $D'_{1D}(E)$ is the density of states, and $D_{1D}(E)$ is the density of states per unit length, and $D_{1D}(E)\,dE$ is the number of states in the energy range, dE, per unit length.

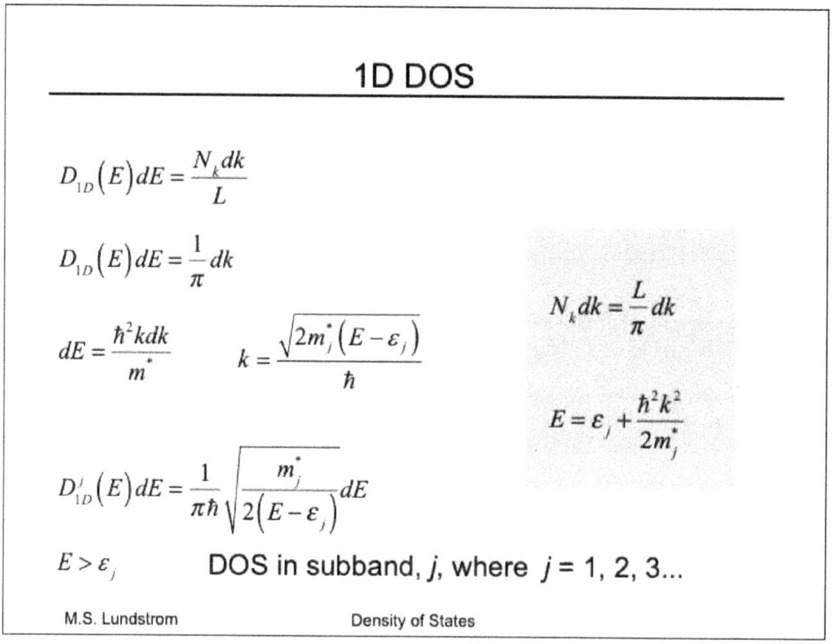

We are now prepared to calculate the 1D density of states, $D_{1D}(E)$. As shown on the first line above, we have a relation between the number of states per unit length in k-space and the corresponding number in energy space. We are also assuming a parabolic band for $E(k)$. The calculation is shown on the left. Note that we compute a DOS for each subband separately, and that the subbands are labeled by the integer, j.

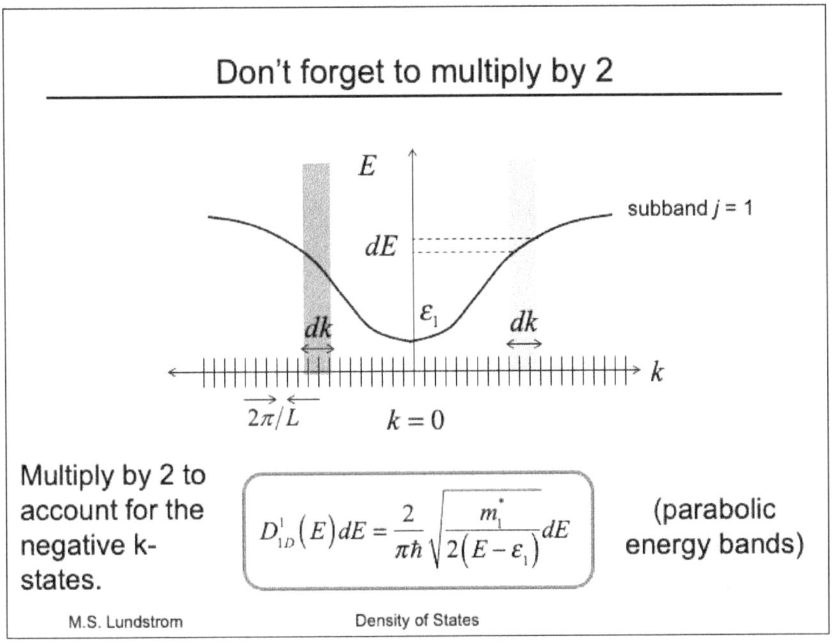

We've done the calculation for $+k$, and we found the number of states that are in that corresponding energy range. The band structure is symmetric, so there's an equal number of states along $-k$ that map into the same energy range. We need to multiply by 2, so the final result is as shown above at the bottom of the slide above. Remember that we've assumed parabolic bands; the answer would be different if we had a different band structure.

In general, there are multiple subbands. For the jth subband.

$$E = \epsilon_j + \frac{\hbar^2 k^2}{2m_j^*},$$

and we'll have a DOS for each one of them,

$$D_{1D}^{j} = \frac{1}{\pi\hbar}\sqrt{\frac{2m_j^*}{(E-\epsilon_j)}},$$

Note that the density of states goes to infinity at the bottom of the subband when $E = \epsilon_j$. There is a density of states associated with each subband, and the total density-of-states is the sum of the contributions from each subband.

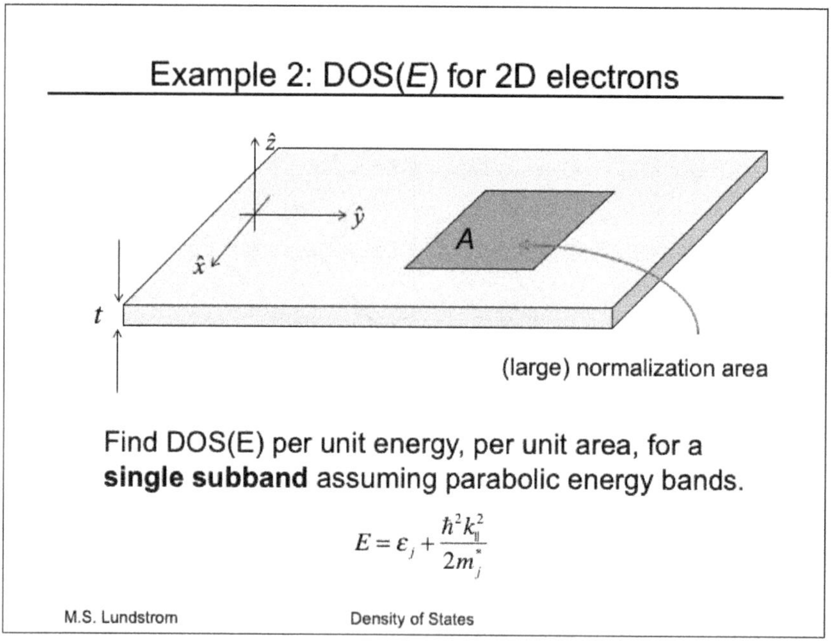

Next, let's do the calculation for 2D electrons. We can produce 2D electrons with semiconductor heterostructures (i.e. wide band gap/small band gap/wide band gap layers), and there are also 2D materials. One example is the semiconductor, MoS_2, which we can approximate with a parabolic energy band. Another is the 2D semimetal, graphene, which has a non-parabolic (conical) energy band.

Consider a thin semiconducting slab of thickness t as shown above. The electrons are confined in the z direction, so there will be particle-in-a-box subbands associated with this confinement. The electrons are free to move in the x–y plane, and we would like to count the states within the area, A, shown above. We'll assume again that the bands are parabolic so the electron energy is $E = \epsilon_j + \hbar^2 k_{||}^2/2m^*$, where $k_{||}$ lies in the x–y plane.

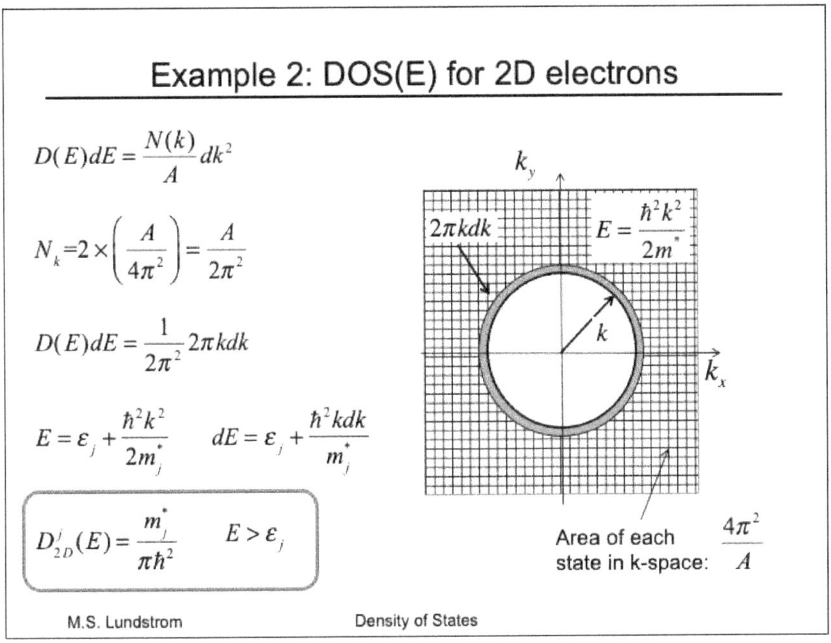

We'll approach this problem in the same way we did for the 1D problem by beginning with $D(E)dE = N_k dk/A$. The red ring shown above has a width dk. The area of the region that we are counting the states in is the circumference of the circle, $2\pi\hbar$, times dk.

We're asking how many states are there in this ring. Well, each state occupies an area of $(2\pi)^2/A$, as we've seen when we applied periodic boundary conditions. The calculation is shown in the slide above on the left, and the result for the 2D DOS is shown on the bottom line at the left. Note that the 2D density of states is effective mass divided by $\pi\hbar^2$. Of course these states only exist for energies above the bottom of the subband, ϵ_j.

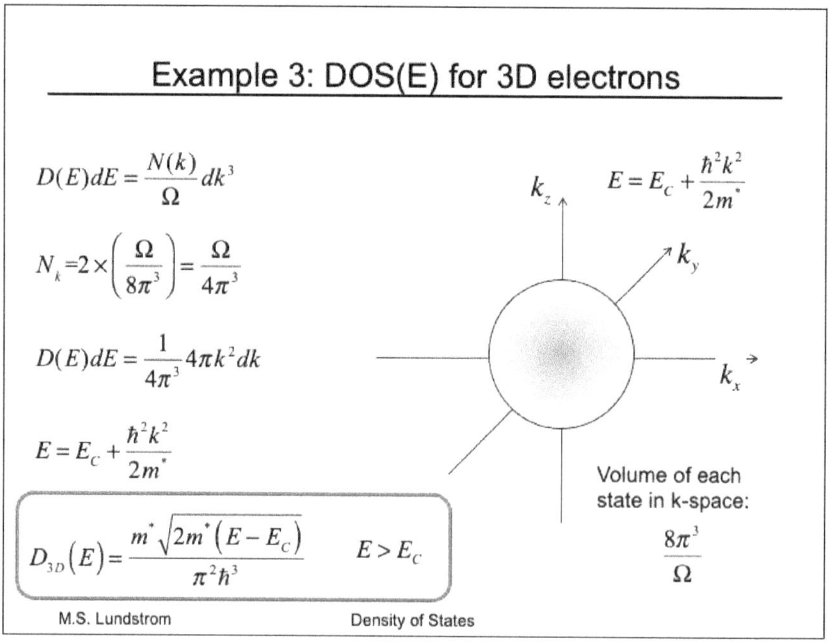

As we've seen, the density of states in 2D is quite different from the density of states in 1D. The 3D density of states is different yet. Let's do the 3D calculation because most of the course will be concerned with 3D semiconductors. We assume parabolic bands again, so the constant energy surfaces are spheres. We equate the number of states in the energy range, dE to the number of states in the corresponding range dk. We count k-states in a thin shell of volume, $4\pi k^2 dk$. Next, we make use of the dispersion to express $k^2 dk$ in terms of energy.

When we do the algebra, we end up with the result on the lower left in the slide above. We find that the density of states is proportional to $(m^*)^{3/2}$, and that it varies as the square root of energy. This is worth remembering.

We've now calculated the density of states for 1D, 2D, and 3D electrons all under the assumption that the bands are parabolic.

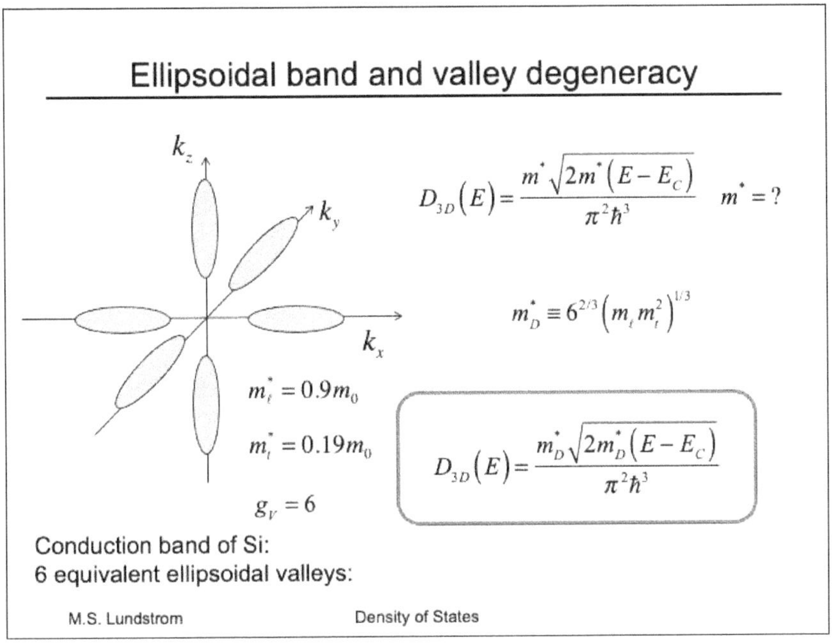

A spherical energy band makes the calculations easy, but things can get more complicated. For example, you will recall that the conduction band of silicon is parabolic, but ellipsoidal because there are different curvatures in different directions, and there are six minima at the same energy. The constant energy surfaces for the conduction band of Si are shown above.

As shown in the slide above, the minima of the conduction bands are located along, $\langle 100 \rangle$, $\langle 010 \rangle$ and $\langle 001 \rangle$ directions. We have computed $D(E)$ assuming a spherical band centered at $k = 0$. We could compute the density of states for Si with its ellipsoidal conduction band, but the calculation is a little more complicated (see [1]).

After doing the calculation, it turns out that we can wrap everything together and express $D(E)$ for the Si conduction band in a way that looks like our simple result for a spherical band with one effective mass [1]. To do this, we define a quantity, m_D^*, called the **density of states effective mass**. This is not really an effective mass — real effective masses are related to the curvature of $E(k)$; it is just a definition such that if we use it in our simple spherical band result, we'll get the correct DOS for Si.

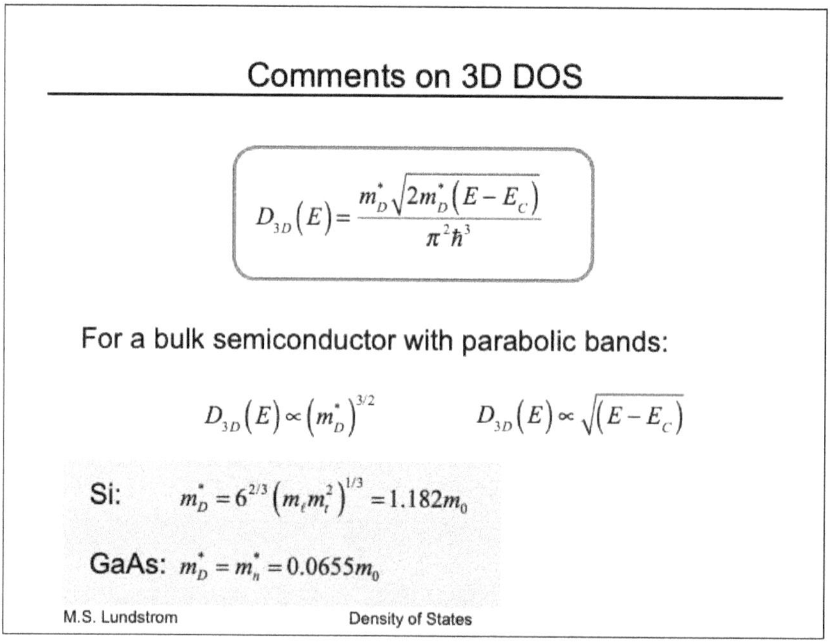

11.4 Density of states discussion

The expression for the density of states for a 3D semiconductor with parabolic energy bands is shown above. It's worth remembering that the 3D density of states varies as effective mass to the 3/2 power, so if I'm dealing with a semiconductor that has a light effective mass, I know it's going to have a small density of states. For example, at a given energy, Si has a much higher density of states effective mass than GaAs. It's also worth remembering that the 3D DOS varies as the square root of energy. As energy increases, the number of states increases.

When computing the density of states for a semiconductor, we have to be careful to use the appropriate density of states effective mass. If it's a simple spherical band semiconductor, the DOS effective mass is just the effective mass associated with the curvature of the band, but if the band is more complicated, like the conduction band of Si, it is a more complicated quantity. As shown above, the DOS effective mass for Si depends on longitudinal and transverse band curvature effective masses and on the number of valleys.

Remember to use the DOS effective mass when computing the DOS.

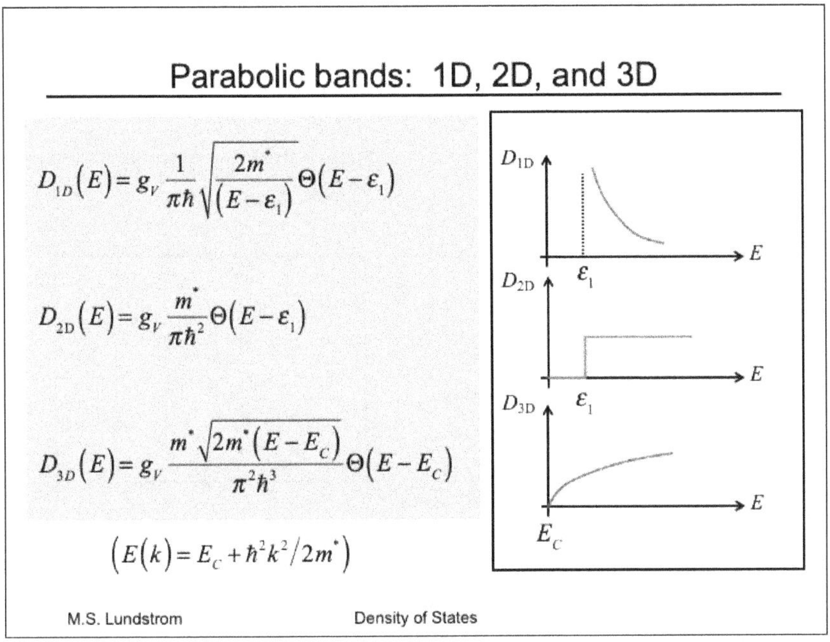

We have derived the density of states for 1D, 2D, and 3D semiconductors assuming parabolic bands. From the plot at the top in the slide above, we see that the 1D density of states goes as $1/\sqrt{E-\epsilon_1}$; it approaches infinity at the bottom of the subband. In 2D, the density of states is independent of energy above the bottom of the subband. In 3D, the density of states is zero at the bottom of the conduction band and increases as the square root of energy above the bottom of the conduction band. We will make significant use of the density of states in Part 3.

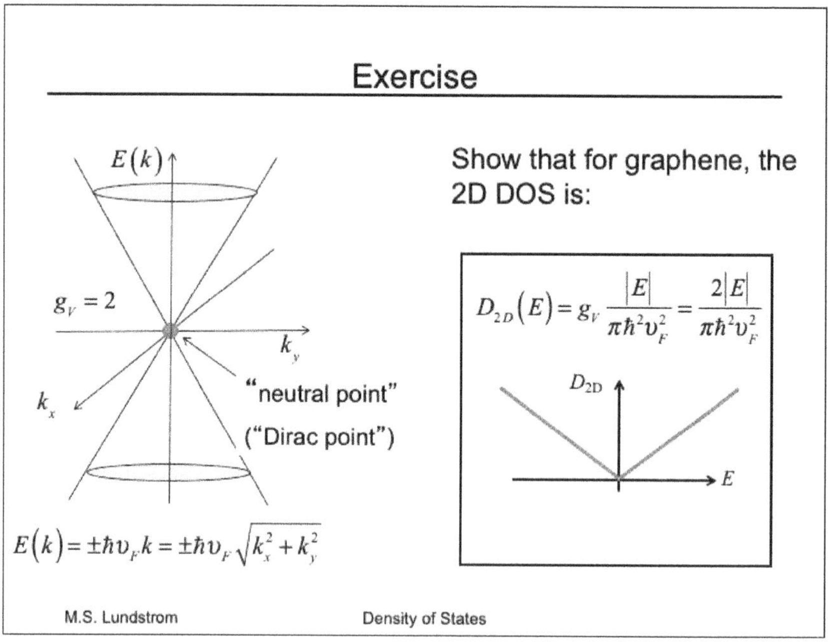

To test yourself and see if you understand the basic concepts, compute the density of states for graphene. Graphene does not have a parabolic band structure. It is a 2D material, but it has a density of states that is significantly different from the 2D density of states for parabolic bands that we have seen. The result is shown above. The density of states increases linearly with energy as we go above and below $E = 0$. See if you can derive the result shown above. If you can do that, then you understand the process for deriving a density of states in any dimension and for any band structure.

11.5 Summary

In this lecture, we've introduced an important quantity – the density of states. The important points to remember are:

(1) That the density of states depends on the dimension of the semiconductor. We have different density of states in 1D, 2D, and 3D.
(2) The density of states in k-space is constant and does not depend on band structure but the density of states in energy space does depend on band structure and can be quite different for different band structures.
(3) For 3D bulk semiconductors with parabolic bands, which will be our focus for the rest of this course, the density of states is proportional to $(m_D^*)^{3/2}$ and to the $\sqrt{(E-E_c)}$ above the bottom of the conduction band.

Quantum mechanics helps us understand the density of states and how to compute it. We will make significant use of $DOS(E)$, especially the 3D $DOS(E)$, in the next unit.

Finally, as we wrap up Part 2, let's review the learning outcomes:

- Understand the rudiments of quantum mechanics and how to solve the Schrödinger wave equation.
- Explain key concepts such as phase velocity, group velocity, wavevector, momentum of a plane wave, and the uncertainty relations.
- Explain how quantum confinement affects electrons, what is meant by 1D, 2D, and 3D semiconductors, and what subbands are.
- Understand quantum mechanical tunneling and reflection and how the barrier height and width affect the tunneling probability.
- Explain how $E(k)$ differs for a free electron and for an electron in a periodic crystal potential and be familiar with $E(k)$ and the constant energy surfaces for common semiconductors. Know what a parabolic energy band is.
- Explain what $D(E)dE$ (where $D(E)$ is the density-of-states) represents and how $D(E)$ varies with energy for parabolic energy bands in 1D, 2D, and 3D.

With this brief introduction to quantum mechanics, and especially with the discussion about the density of states, we are ready to proceed.

11.6 References

Those who seek a deeper understanding of energy bands in semiconductors should consult the follow texts. Pierret presents a simple introduction and Tiwari a more advanced treatment. Pierret also has a good discusison of the density of states and the density of states effective mass,

[1] R.F. Pierret, *Advanced Semiconductor Fundamentals*, (2nd Ed.), Pearson Education, Inc., 2003 (see Ch. 3 and Ch. 4, Sec. 4.1).

[2] Sandip Tiwari, *Semiconductor Physics: Principles, Theory, and Nanoscale*, Oxford Electroscience Series, Vol. 3, 2020 (see Chapters 3 and 4).

PART 3

Equilibrium Carrier Concentrations

Lecture 12

The Fermi Function

12.1 Introduction

The first two units of these lectures provided the background needed to dive into real semiconductor problems. The topic of Part 3 is understanding how to compute the concentrations of electrons and holes under equilibrium conditions. Before we can do this, we need to understand the Fermi function, which is the subject of this lecture.

Upon completion of Part 3, you will be able to:

- Explain what the Fermi function is, how it is used, and the effect of the location of the Fermi level and temperature on the Fermi function.
- Understand the relation between the Fermi level, E_F, and the equilibrium carrier concentrations for degenerate and nondegenerate semiconductors.
- Relate the intrinsic carrier concentration to the effective densities of states and to the band gap and understand what the intrinsic Fermi level, E_i, is.
- Relate the equilibrium carrier concentrations to the location of the Fermi level relative to the conduction and valence band edges and to the intrinsic level.
- Compute the equilibrium carrier concentrations given the donor and acceptor concentrations and the temperature.

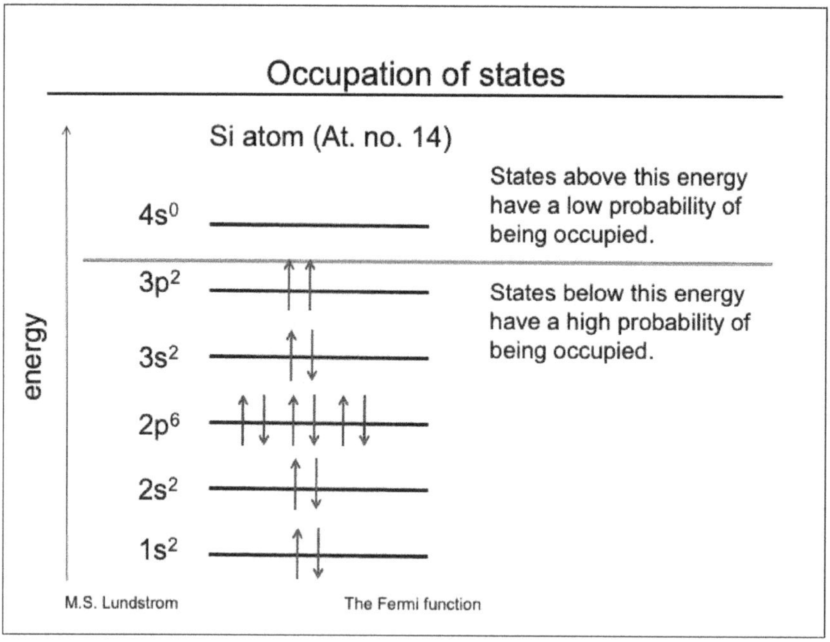

12.2 The Fermi function

To begin, let's go back to an isolated silicon atom with its discrete energy levels. We have 14 electrons in the various s, p, d, and f orbitals associated with the principal quantum numbers $n = 1, 2$, and 3. As shown above, we can draw a line, an energy, such that below the line the states are filled and above the line, the states are empty. Of course, there is no need to draw this line for the Si atom, because we can simply put the 14 electrons in, and we know exactly which energy levels are filled, which are partially filled, and which ones are empty, but it will turn out that this line, known as the **Fermi energy**, is very useful when the states are distributed in energy, as in a semiconductor.

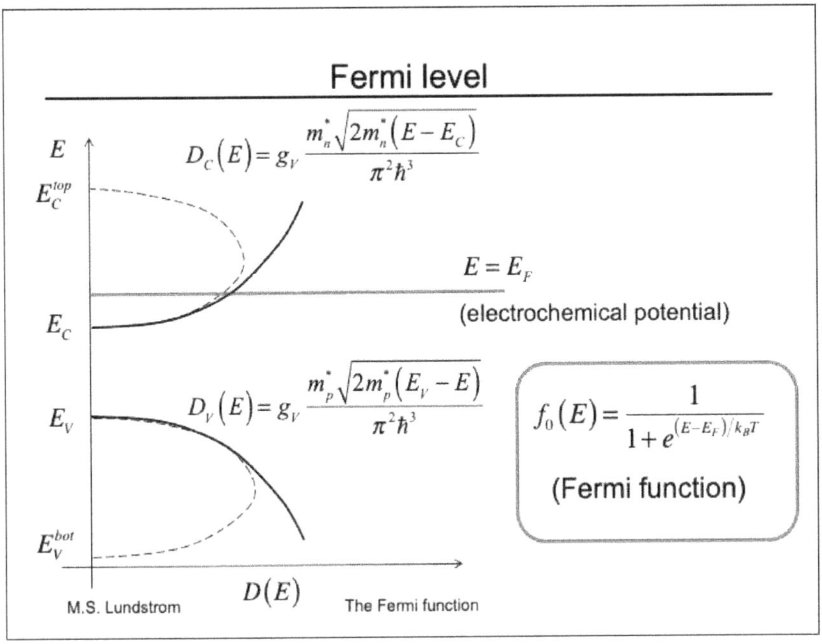

Shown above is the density of states, $D(E)$, vs. energy near the top of the valence band and the bottom of the conduction band. As discussed in the previous lecture, near these band edges, $D(E)$ goes as the square root of energy. The conduction and valence band densities of states actually have more complicated shapes and go to zero at the top of the conduction band and at the bottom of the valence band, but that doesn't need to concern us here because in equilibrium and near equilibrium, the carriers are near the band edges.

The horizontal line shown above is called the **Fermi level** (or Fermi energy). It's sometimes also called the **electrochemical potential**, or sometimes just the **chemical potential**. Below the line, the states are mostly filled, and above the line, the states are mostly empty. It turns out that there is a very simple function that describes the probability, a number between 0 and 1, that a state at an energy, E, is occupied by an electron. That function is the **Fermi function**, f_0 shown above. The subscript zero is to remind us that this function applies to equilibrium conditions. As we will discuss later in these lectures, we can also compute this probability out of equilibrium, but in Part 3, we assume that the semiconductor is in equilibrium, and discuss only the equilibrium Fermi function, f_0.

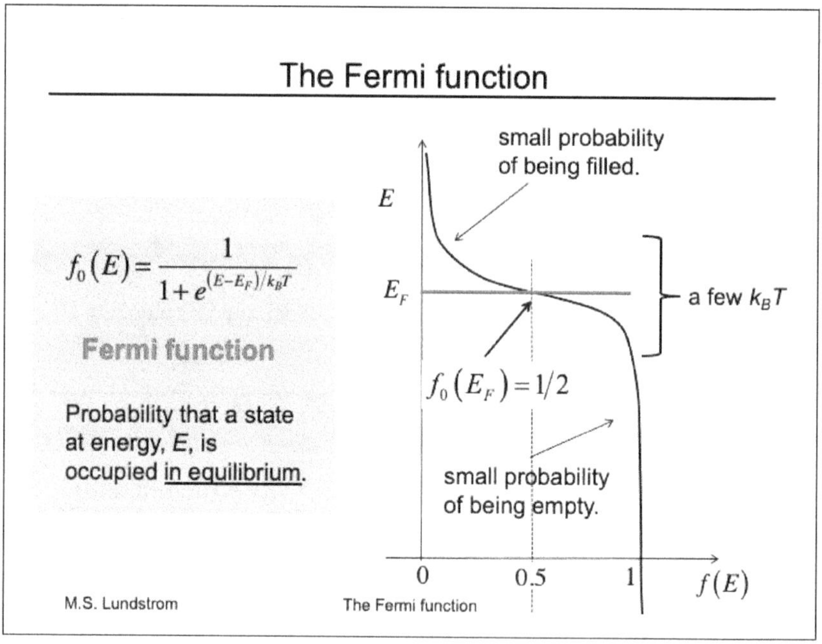

We won't derive the Fermi function, but you can see that it makes sense. Electrons like to occupy the lowest energy states, so it makes sense that the states with low energy should have a high probability of being occupied. States with high energies are less likely to be occupied.

To get comfortable with the Fermi function, we can plot it as shown above on the right. The Fermi function goes between 0 and 1, as a probability should. At energies much lower than E_F, we have the exponential of a large negative number in the denominator, which is close to 0, so f_0 approaches 1. At energies much higher than the Fermi energy, we have the exponential of a large number, so f_0 approaches 0. For a state located at the Fermi energy, the probability that the state is occupied is 1/2.

To summarize, states below the Fermi energy have a small probability of being empty, and states above the Fermi energy have small probability of being filled. At the Fermi energy, the probability that a state (if a state exists) is occupied is one-half. As shown above, the Fermi function transitions between 1 and 0 in a range of energies a few k_BT wide, which is centered at E_F.

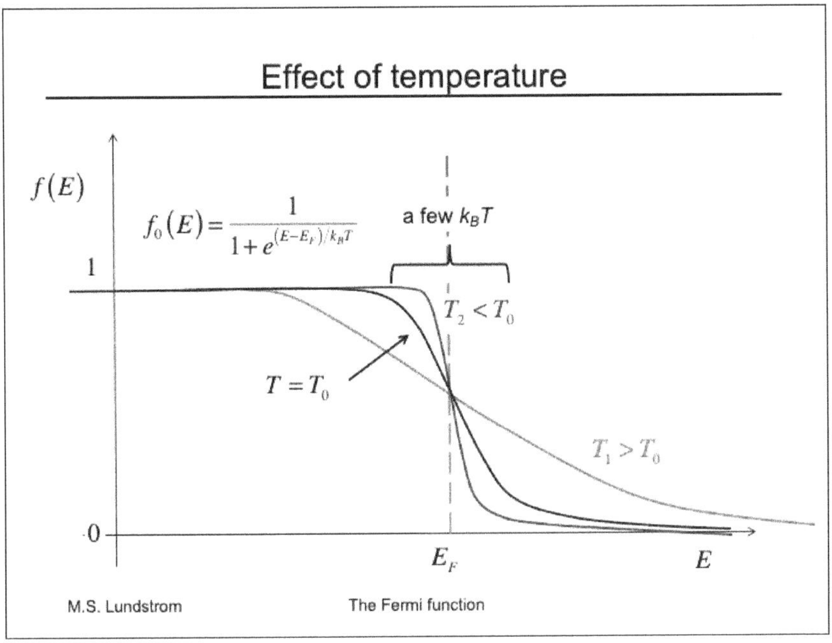

12.3 The effect of temperature

Let's flip the Fermi function on its side and plot $f_0(E)$ vs. E as shown above. At the Fermi energy, the probability that a state (if a state exists) is occupied is one-half. The Fermi function transitions between 1 for $E \ll E_F$ and 0 for $E \gg E_F$ in a range of energies that is a few k_BT wide. What happens if we increase the temperature? If we decrease the temperature?

Shown above are sketches of the Fermi function vs. energy at three different temperatures. If we increase the temperature, the width in energy of the transition from 1 to 0 gets larger. As we decrease the temperature, the width of the transition region gets narrower. If we were to go all the way to $T = 0$ K, the Fermi function would become a step function. All the states below the Fermi level are filled at $T = 0$ K, and all the states above the Fermi level are empty at $T = 0$ K.

Temperature has important effects on the Fermi function.

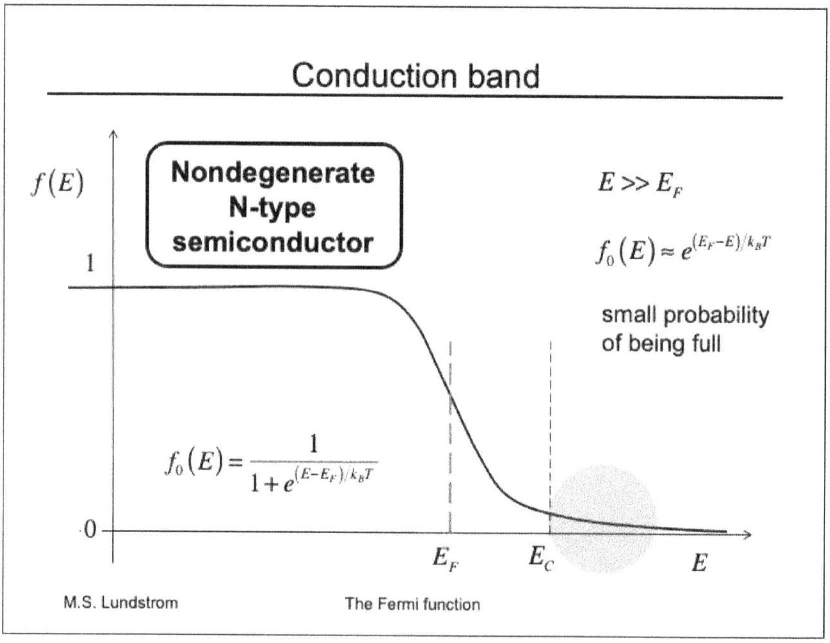

12.4 Nondegenerate semiconductors

Let's apply the Fermi function to the conduction band, and assume that the Fermi energy is located in the forbidden gap, as shown above. The probability that a state at E_F is occupied is one-half, but there are no states in the forbidden gap so there are no electrons there. You can see that the probability that a state in the conduction band is occupied is small because all of the states in the conduction band are well above E_F.

There is an important approximation to the Fermi function when the probability that states are occupied is small (i.e. when the energies of the states are well above E_F). A **nondegenerate** N-type semiconductor is one in which the conduction band states have a small probability of being filled. As shown above, for a nondegenerate semiconductor, $f_0(E)$ simplifies to an exponential, $f_0(E) \approx \exp\left[E_F - E)/k_B T\right]$. The order of the arguments is easy to remember. The higher the Fermi energy moves, the higher the probability that states in the conduction band are occupied, so it's $(E_F - E)$ in the argument of the exponential. This is an approximation that we're going to make use of frequently as we work out semiconducting problems.

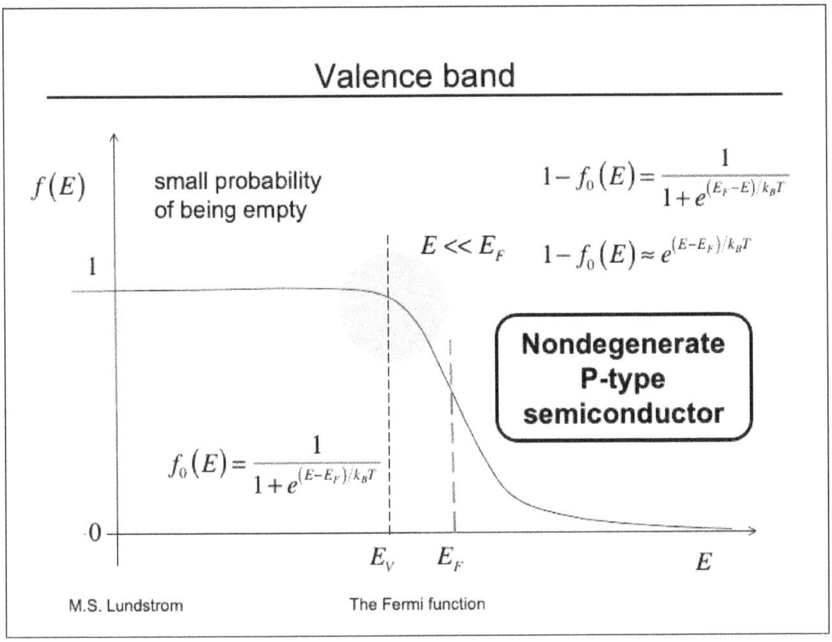

There's a similar approximation that we can make for nondegenerate P-type semiconductors. For a nondegenerate P-type semiconductor, there is a small probability that the states in the valence band are empty.

What we're interested in now is not the probability that the state is filled, but, rather, the probability that the state is empty. As shown above, as long as the Fermi energy is well above the top of the valence band, then $(1 - f_0) \approx \exp[(E{-}E_F)/k_BT]$. The order of the arguments is easy to remember. The lower the Fermi energy moves, the higher the probability that states in the valence band are empty, so it's $(E{-}E_F)$ in the argument of the exponential.

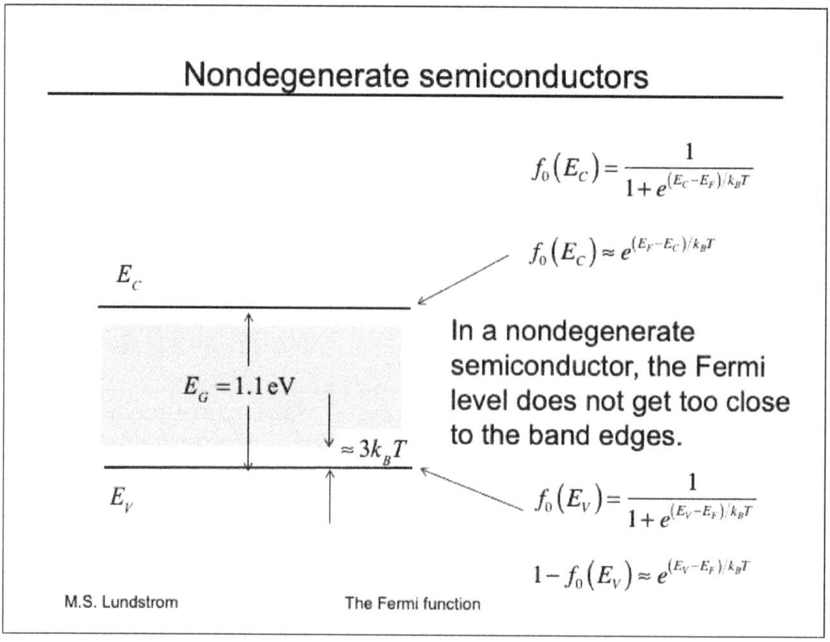

In calculations, we will frequently assume nondegenerate semiconductors. A nondegenerate semiconductor is one in which the states in the conduction band are always well above the Fermi level, and the states in the valence band are well below the Fermi level. So as long as the Fermi level lies in the forbidden gap and doesn't get too close to the conduction band or too close to the valence band, this nondegenerate assumption will be a good approximation. When we make the nondegenerate assumption, then the probability that a state at E_c is occupied is just $\exp\left[(E_F - E_c)/k_BT\right]$. The probability that a state at the top of the valence band is empty is just $\exp\left[(E_v - E_F)/k_BT\right]$.

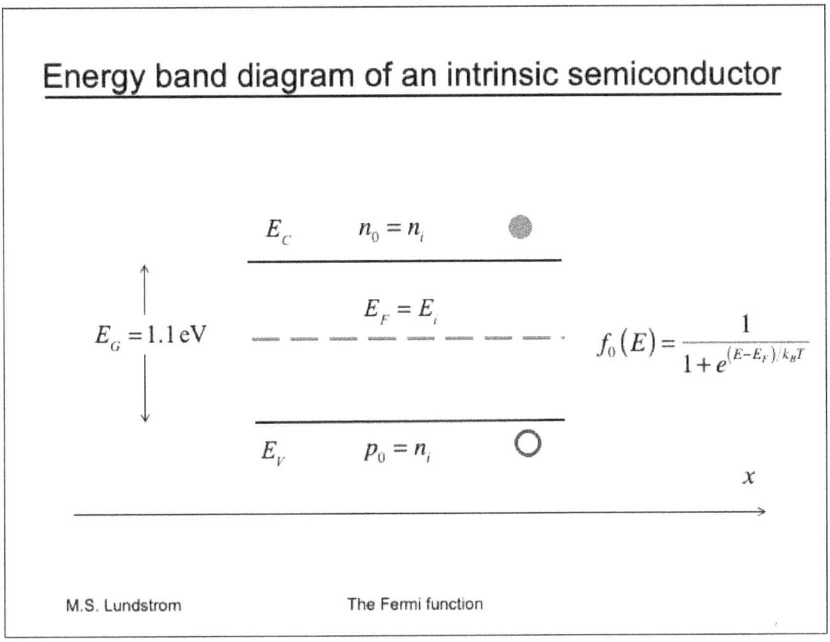

12.5 Fermi levels and energy band diagrams

A question that we will often ask is: "Where is the Fermi level?"

Shown above is an energy band diagram for an intrinsic semiconductor. In this case, there are equal numbers of electrons and holes, so we expect the Fermi level to be near the middle of the gap so that there's an equal probability of the states in the conduction band being filled and the states in the valence band being empty. It turns out, however, that E_F is not exactly in the middle because the density of states in the valence band is slightly different from the density of states in the conduction band.

Generally, the Fermi level will lie inside the bandgap or slightly above E_c or slightly below E_v. If it lies inside the bandgap but not too close to E_c or E_v, then we are dealing with a nondegenerate semiconductor. If it lies above E_c, then we are dealing with a degenerate N-type semiconductor, and if is lies below E_v, then we are dealing with a degenerate P-type semiconductor.

States high in the conduction band, way above E_F will be empty, and the states deep in the valence band, way below E_F, will be filled. States near E_c have a small probability of being filled, and states near E_v have a small probability of being empty.

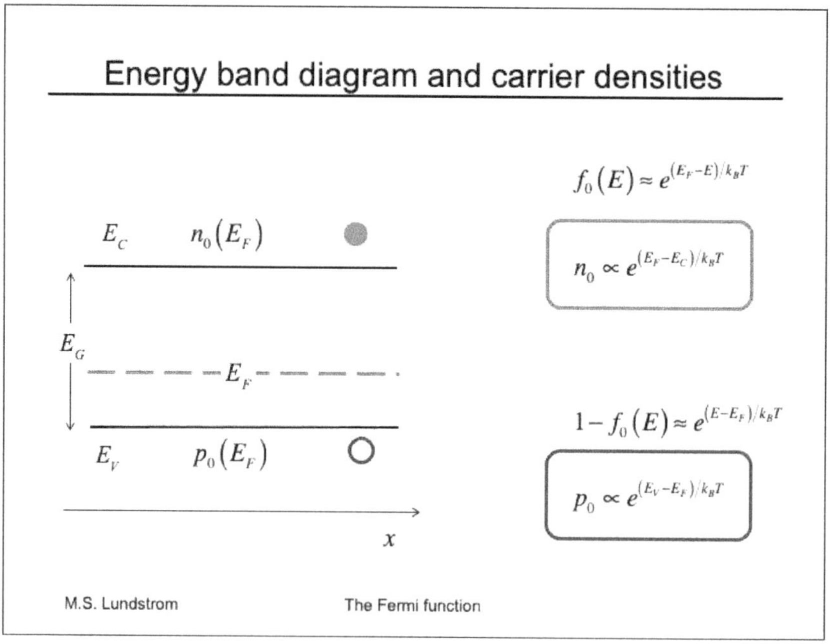

If we see an energy band diagram with a Fermi level, our question will be: "What are the equilibrium electron and hole densities, n_0 and p_0?" The nondegenerate approximation says that the probability of states in the conduction band are filled is given by an exponential, as shown above. Most of the electrons in the conduction band are near the bottom of the conduction band, near $E \approx E_c$, so we expect the electron concentration to be proportional to $\exp\left[(E_F - E_c)/k_BT\right]$.

In the valence band, we're interested in the probability that the state is empty. The nondegenerate approximation says that the probability that states in the valence band are empty is given by the exponential shown above. Most of the holes in the valence band are near the top of the valence band at $E \approx E_v$, so we expect the hole concentration to be proportional to $\exp\left[(E_v - E_F)/k_BT\right]$.

When we do the math in the next lecture, we'll see that both the electron and hole concentrations are related exponentially to the location of the Fermi level relative to the band edges. When we see the Fermi level above the middle of the bandgap on an energy band diagram, we are dealing with an N-type semiconductor $n_0 > p_0$, and when we see the Fermi level below middle of the bandgap, we are dealing with a P-type semiconductor, $p_0 > n_0$.

12.6 Summary

This lecture was an introduction to the Fermi function,

$$f_0(E) = \frac{1}{1 + e^{(E-E_F)/k_bT}},$$

which gives the probability that a state at a certain energy, E, is occupied in equilibrium (if a state exists at that energy). There are two parameters in the Fermi function. One is the Fermi level, E_F, and the second is the temperature, T.

When we compute carrier concentrations in the next lecture, we'll see how to make use of the Fermi function.

12.7 References

Both Pierret and Tiwari derive the Fermi function.

[1] R.F. Pierret, *Advanced Semiconductor Fundamentals,* (2nd Ed.), Pearson Education, Inc., 2003 (see Ch. 4, Sec. 4.2).

[2] Sandip Tiwari, *Semiconductor Physics: Principles, Theory, and Nanoscale,* Oxford Electroscience Series, Vol. 3, 2020 (see Appendix E).

Lecture 13

Fermi-Dirac Integrals

13.1 Introduction

In the previous lecture, we discussed the Fermi function, which is used to compute carrier concentrations in semiconductors. When we do the calculation, we'll encounter an integral known as the Fermi-Dirac integral. That integral is the subject of this lecture. We'll also discuss how to simplify the Fermi-Dirac integral for nondegenerate semiconductors to obtain simple expressions that relate the electron and hole concentrations to the location of the Fermi level. Before we begin, remember that the Fermi function gives us the probability that a state at energy, E, is occupied in equilibrium, if that state exists. The Fermi function is

$$f_0(E) = \frac{1}{1 + e^{(E-E_F)/k_BT}} ,$$

where E_F is the Fermi level and T is the temperature.

The Fermi function, which we saw in the last lecture, and the density-of-states, which we discussed in Lecture 11, will allow us to develop expressions for the equilibrium electron concentration, n_0, and the equilibrium hole concentration, p_0, in terms of the Fermi level and temperature.

There are two ways to do the calculation; we can do the calculation in k-space or we can do it in energy space. We'll first discuss how we do it in k-space, but we won't actually do that calculation. Instead, we'll discuss how to do the calculation in energy space using the density of states, $D(E)$.

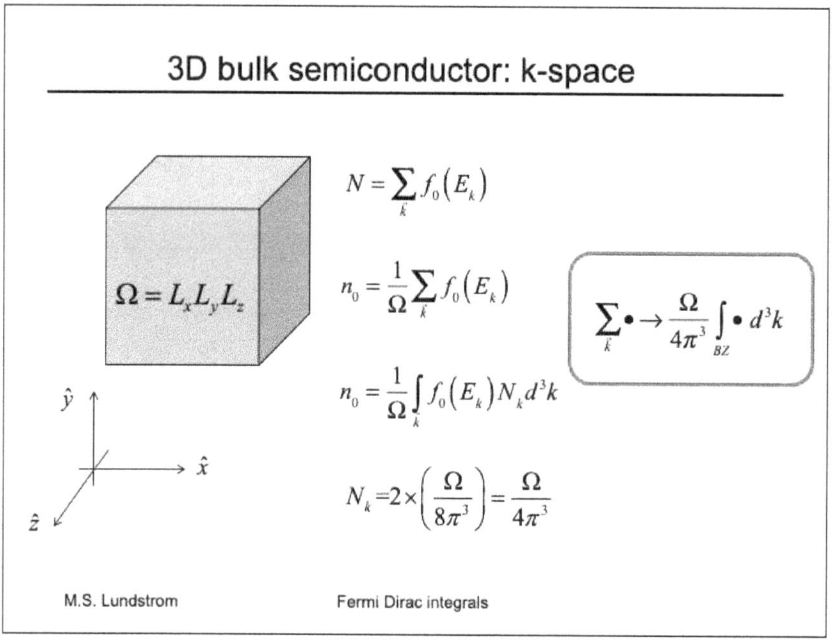

13.2 Carrier concentration calculation in k-space

Consider again a volume, Ω, of semiconductor, a cube shown above, which is carved out of an infinite semiconductor. We want to calculate the number of electrons in this volume. We saw earlier that when we apply periodic boundary conditions, we get a finite number of closely spaced k-states. To find the total number of electrons in this volume, we just add up the probability that each of those states is occupied in equilibrium. This is the first sum shown in the slide above. The volume itself is arbitrary, so we're really more interested in the carrier concentration — the number of carriers per unit volume, so we would divide the result by volume to find $n_0 = N/\Omega$. That is the second equation shown in the slide above.

Since we are dealing with a significant size volume, the k-states are very finely spaced, so we can convert the sum to an integral as long as we have the correct number of states in each element of k-space, dk. The correct number will be given by $N_k dk$. where N_k is the density of states in k-space, which is 2 for spin times $\Omega/(2\pi)^3$. If we put this all together and perform the integral shown above, we will get the answer that we're after.

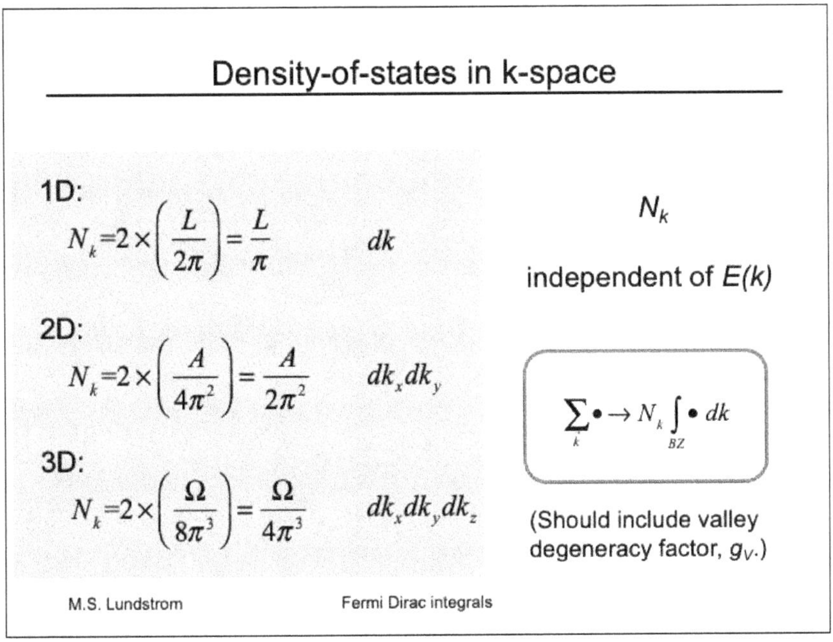

The k-space calculation proceeds straight forwardly using the density of states in k-space as shown above and then converting the sum over the finely spaced states in the volume, Ω, into an integral. The DOS in k-space is the same for every material, but when doing the calculation, we will need the $E(k)$ for the specific material because the Fermi function depends on energy. Be sure to include the valley degeneracy — e.g. remember that Si has six equivalent valleys in the conduction band along the six directions, $+k_x, -k_x, +k_y, -k_y, +k_z$, and $-k_z$. We won't do the k-space calculation here, but you might give it a try and see if you get the same answers we're going to get by doing the calculation in energy space.

When we do the calculation in energy space, we will need the density of states in energy. As discussed in Lecture 11, we know the DOS for parabolic energy bands 1D, in 2D, and in 3D. We'll do the calculation in 3D using a DOS that goes as $D_{3D}(E) \propto (m_D^*)^{3/2}\sqrt{E - E_c}$. To test your understanding, you can work out the corresponding results in 1D and 2D.

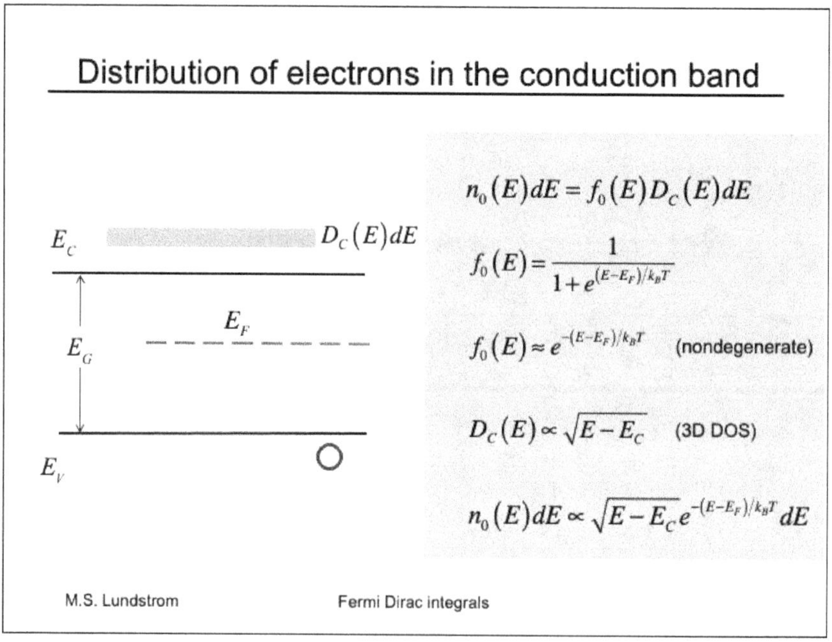

13.3 Carrier concentration calculation in energy space

The calculation can also be done in energy space. To see how, let's first take a look in the conduction band and ask: "How many electrons are there in the shaded energy range shown above?" The number of states in that region, is $D_c(E)dE$, where $D_c(E)$ is the 3D conduction band density of states. To find the number of electrons in this portion of energy space, we take the number of states, $D_c(E)dE$, and multiply by the probability that those states are occupied, $f_0(E)$.

For now, let's make the nondegenerate approximation to the Fermi function to get the expression shown above at the bottom of the slide. We conclude that the number of electrons in dE, $n_0(E)dE$, is proportional to the square root of energy times an exponential factor.

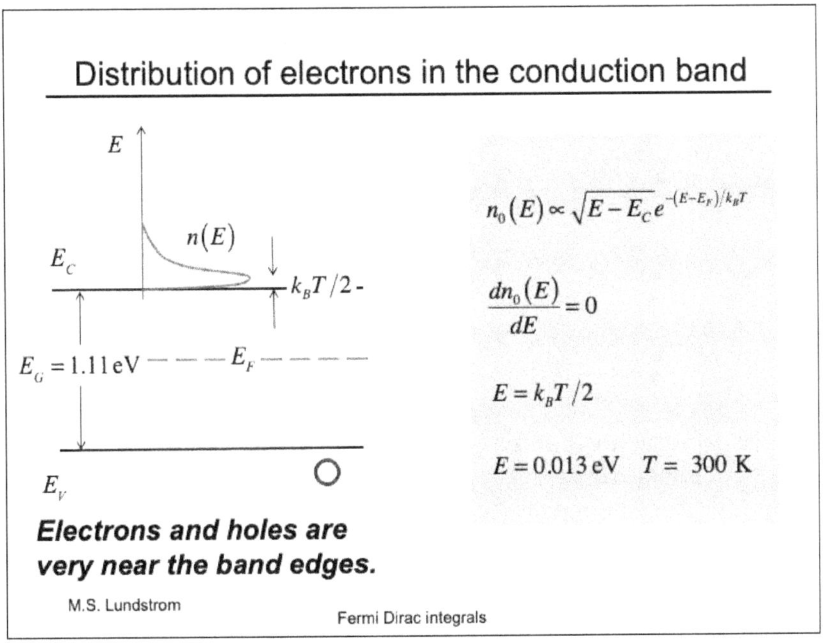

If we plot $n_0(E)dE$, we get the result shown above, which tells us where the electrons are located in the conduction band. As the energy increases, the square root starts to increase, so $n_0(E)dE$ increases, but as we continue to increase the energy, the exponential factor dominates and $n_0(E)dE$ begins to decrease. There is a peak in $n_0(E)dE$. To find the energy of maximum $n_0(E)$, we differentiate it with respect to energy and set it equal to zero. If you do that calculation, you'll find that the peak occurs at an energy that is $k_BT/2$ above the bottom of the conduction band. (Remember that we have assumed a nondegenerate semiconductor.)

At room temperature, k_BT is about 0.026 eV, so the peak $n_0(E)$ occurs at 0.013 eV above the bottom of the conduction band. The band gap of silicon is 1.11 eV, so this verifies our assumption that most of the electrons in the conduction band are near the bottom of the conduction band. Similarly the holes are located within about $k_BT/2$ from the top of the valence band. We conclude that under equilibrium conditions for a nondegenerate semiconductor, electrons and holes are located very near the band edges.

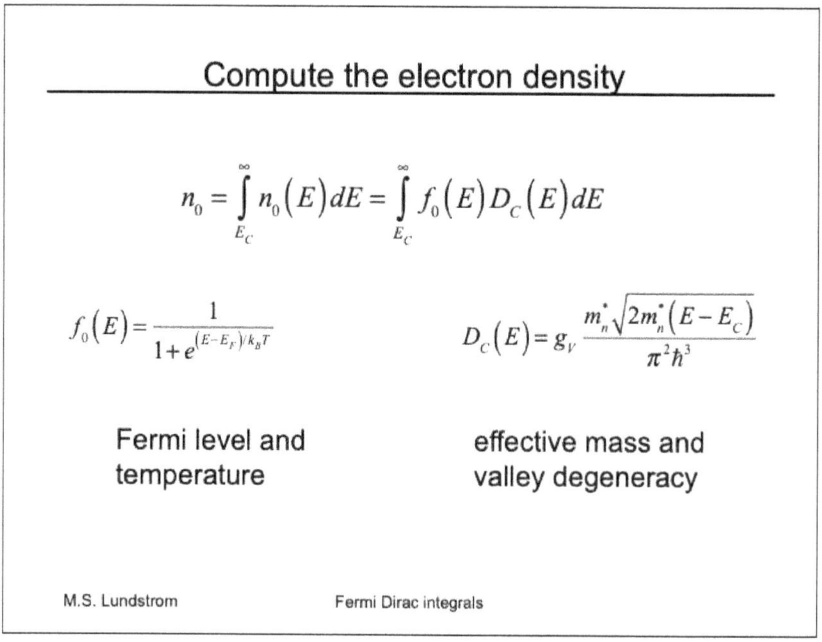

To compute the electron concentration, we integrate the number of electrons in an energy range, $n_0(E)dE$, over the conduction band. In this case, we **will not** make the nondegenerate approximation — we'll use the Fermi function and not its exponential approximation. As seen in the first equation above, we integrate from the bottom of the conduction band to infinity. You might worry about our density of states; it goes as $\sqrt{(E{-}E_c)}$, which only describes the DOS near the bottom of the conduction band. But the Fermi function helps out. There is essentially no probability that states well above E_c are occupied so it doesn't matter that the parabolic density of states is not accurate at high energies. We can also extend the integral to infinity and not worry about the fact that there is a top of the conduction band.

To compute $n_0(E_F, T)$, we multiply the number of states, times the probability that the states are occupied and integrate from the bottom of the conduction band to infinity. We can see that the result will depend on Fermi level and temperature, because those two parameters are in the Fermi function. The result will also depend on the effective mass of the semiconductor and the number of equivalent valleys because those quantities are in the density of states.

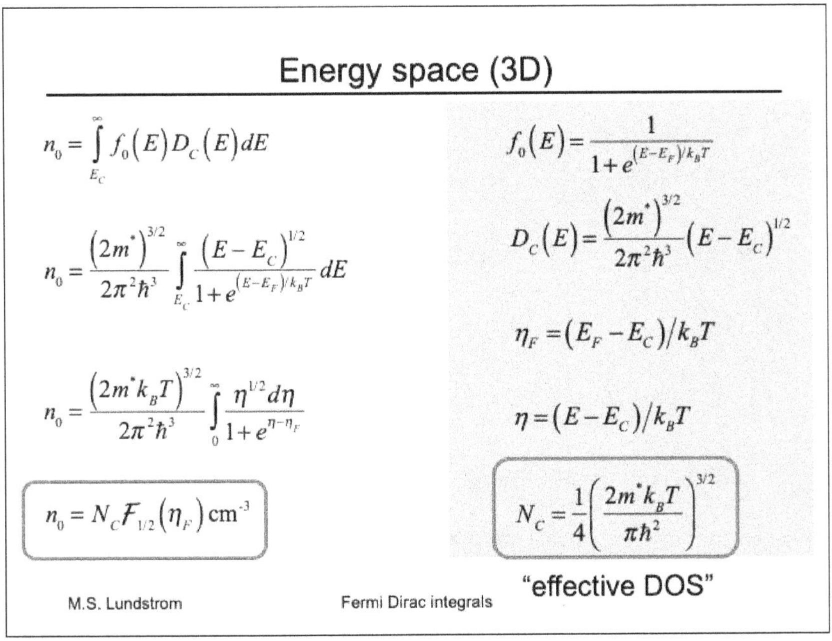

It is convenient now to make two definitions, First, we'll define a normalized Fermi level with respect to the bottom of the conduction band, $(E_F - E_c)/k_BT$, and call this dimensionless number η_F. We'll also define a normalized energy, $(E - E_c)/k_BT$ and call this dimensionless number η. With these definitions, we can rewrite the integral as shown in the third line on the left in the slide above.

Now it's simply a matter of evaluating the integral and then we have our answer. There's only one problem, the integral cannot be done analytically. It has to be done numerically.

We'll give our integral a name; it is called a **Fermi-Dirac integral of order 1/2**. The term that we have out front depend on the effective mass and the temperature, just as we expected it would. We give that quantity a name, N_C, the **conduction band effective density of states**. We have derived the expression that we're after. Next, let's talk a little bit more about this Fermi-Dirac integral.

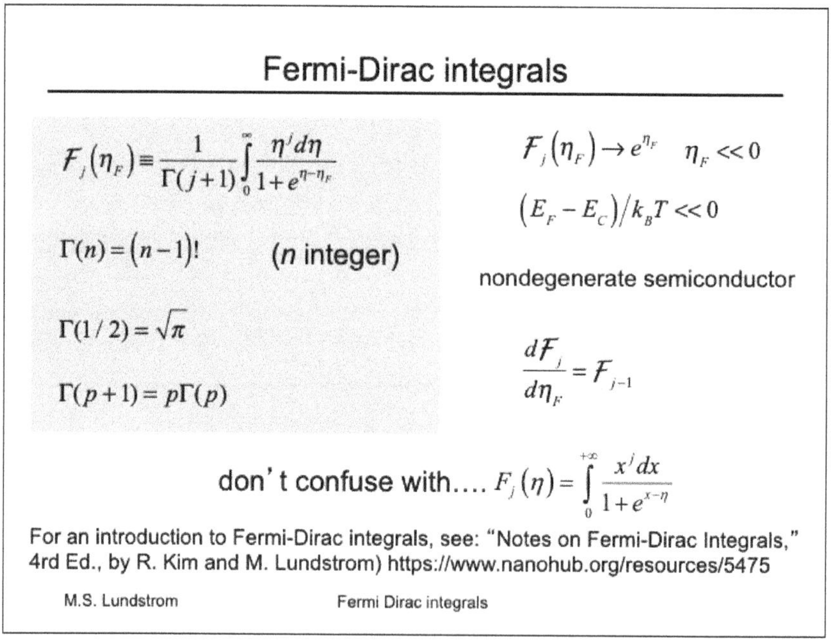

13.4 The Fermi-Dirac integral

The definition of a Fermi-Dirac integral of order j is shown in the shaded box above. For the 3D carrier density with parabolic bands, $j = 1/2$.

Fermi-Dirac integrals have a very nice property — if the normalized Fermi energy is much less than zero (which means that the Fermi energy is well below E_c) then the FD integral simplifies and becomes an exponential. This is a nondegenerate semiconductor. **In a nondegenerate semiconductor the Fermi-Dirac integral reduces to an exponential.** As shown above, there's also a nice property related to the derivative. If you should need to make use of FD integrals, the notes listed above contain all you need to know about using Fermi-Dirac integrals for most semiconductor problems.

Finally, when you are reading papers or books, be careful. The Fermi-Dirac integral is sometimes defined without the factor out front that involves a Γ function. Without that numerical factor, the FD integral is written as a roman F; with the factor, is it written as a script F, $\mathcal{F}$. Sometimes people aren't careful about which F they write, so be certain that you understand which version of the Fermi-Dirac integral is being used.

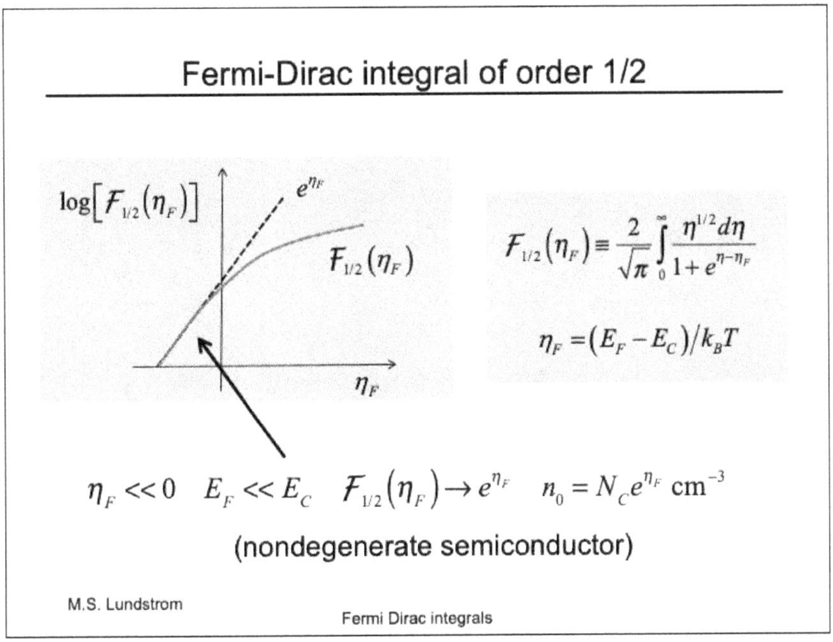

Let's take a closer look at the Fermi-Dirac integral of order 1/2, which occurs when we work out the carrier density in 3D. That function, $\mathcal{F}_{1/2}$, vs. η_F is plotted above. When $\eta_F < 0$, the Fermi level is below the conduction band, and the semiconductor is nondegenerate. When $\eta_F > 0$, the Fermi level is above the bottom of the conduction band and the semiconductor is degenerate.

As shown above in the plot above, for nondegenerate semiconductors ($\eta_F \ll 0$), the FD integral approaches the exponential, which simplifies matters greatly. For a given argument, η_F, **the FD integral is always equal to or less than the corresponding exponential.**

For a nondegenerate semiconductor, there is a very simple relation between the carrier concentration and the Fermi level. It's just the exponential relation shown above at the lower right. This is the expression that we will use in the remainder of these lectures.

A nondegenerate semiconductor is one for which the Fermi level lies within the forbidden gap and doesn't get too close to the bottom of the conduction band or too close to the top of the valence band. We'll sometimes use the exponential approximation to the FD integral outside of the safe region, but don't forget that the results will only be approximate.

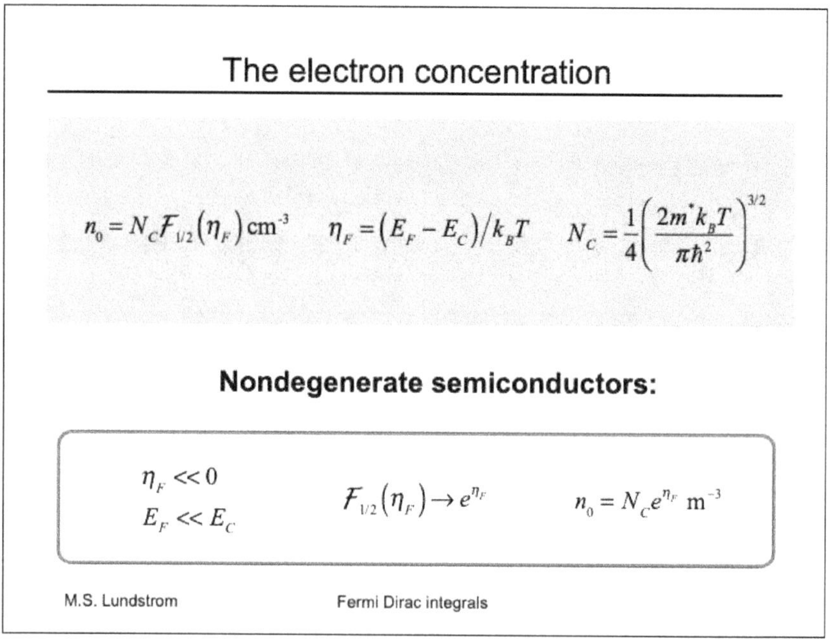

13.5 Electron and hole concentrations

We have succeeded in relating the electron density to the Fermi energy. In general, that relation involves Fermi-Dirac integrals, as shown in the shaded box above, but these are a little complicated because we need to do them numerically. For a nondegenerate semiconductor the relation becomes much simpler, the Fermi-Dirac integral reduces to an exponential and the electron density becomes $N_C \exp\left[(E_F - E_c)/k_B T\right]$.

Exponentials are much easier than Fermi-Dirac integrals to use. They illustrate all of the key points, and they generally get us close, or extremely close to the correct answer. When necessary, we can follow up with a calculation involving FD integrals. The key results for a nondegenerate semiconductor are:

$$\boxed{\begin{aligned} n_0 &= N_C e^{(E_F - E_c)/k_B T} \\ N_C &= \frac{1}{4}\left(\frac{2m_n^* k_B T}{\pi\hbar^2}\right)^{3/2} \end{aligned}}$$

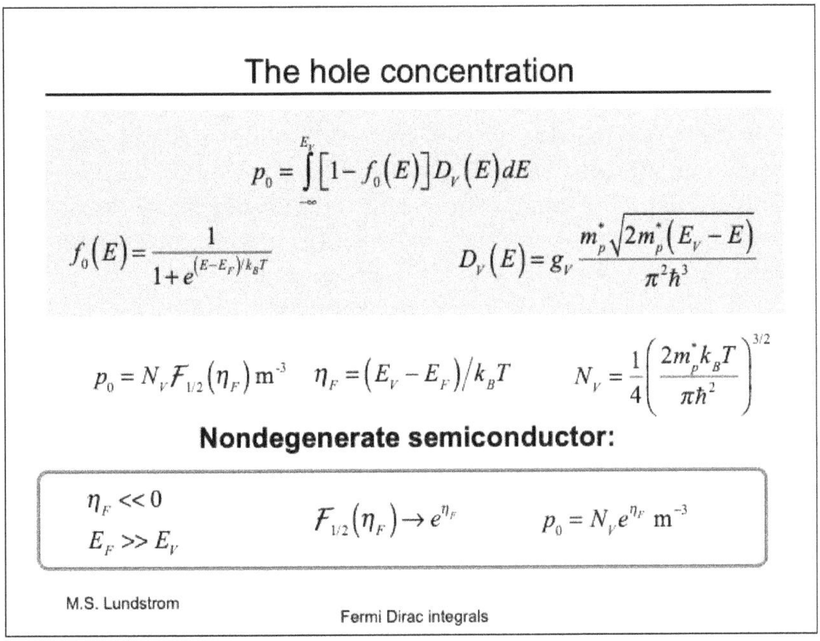

We could go through the same kind of calculation for holes, but in this case, we are interested in the probability that the state is empty. When we do the calculation, we get answers that look much like the electron density results. We get an effective density of states for the valence band times a Fermi-Dirac integral that depends on the normalized Fermi energy. In this case, the normalized Fermi energy is how far the Fermi energy is below the top of the valence band in units of k_BT. The effective density of states for the valence band, involves the effective mass for holes. For a nondegenerate P-type semiconductor, the hole concentration becomes $N_V \exp[(E_v - E_F)/k_BT$.

The key results are:

$$\boxed{\begin{aligned} p_0 &= N_V e^{(E_v - E_F)/k_BT} \\ N_V &= \frac{1}{4}\left(\frac{2m_p^* k_BT}{\pi\hbar^2}\right)^{3/2} \end{aligned}}$$

13.6 Summary

In this lecture, we discussed how to use the Fermi function and the density of states to relate the concentration of electrons in the conduction band and holes in the valence band to the location of the Fermi energy and to the temperature. For parabolic bands in 3D, these relations involve a Fermi-Dirac integral of order 1/2, but if the semiconductor is nondegenerate Fermi-Dirac integrals simplify to exponentials and we are left with simple expressions for the equilibrium electron and hole concentrations:

$$\boxed{\begin{aligned} n_0 &= N_C e^{(E_F - E_c)/k_B T} \\ p_0 &= N_V e^{(E_v - E_F)/k_B T} \\ N_C &= \frac{1}{4}\left(\frac{2m_n^* k_B T}{\pi\hbar^2}\right)^{3/2} \\ N_V &= \frac{1}{4}\left(\frac{2m_p^* k_B T}{\pi\hbar^2}\right)^{3/2}, \end{aligned}}$$

where N_C and N_V are the effective densities of states and m_n^* and m_p^* are the density of states effective masses for electrons and holes. It's worth remembering that the effective densities of states go as effective mass and temperature to the 3/2 power.

These important relations only apply to nondegenerate semiconductors, but they typically get us close. When necessary, we can follow up with more careful calculations that require Fermi-Dirac integrals. The important point, however, is that these simple, nondegenerate expressions illustrate all of the key principles that will be discussed in these lectures.

13.7 References

The book by Pierret has a good discussion of Fermi-Dirac integrals.

[1] R.F. Pierret, *Advanced Semiconductor Fundamentals*, (2nd Ed.), Pearson Education, Inc., 2003 (see Ch. 4, Sec. 4.1.1).

For information about how to compute Fermi-Dirac integrals, see.

[2] R. Kim, X. Wang, and M. Lundstrom, *Notes on Fermi-Dirac Integrals*, 4th Ed., Oct. 5, 2019. `https://arxiv.org/abs/0811.0116`.

Lecture 14

Carrier Concentration vs. Fermi Level

14.1 Introduction

In the previous lecture, we learned how to relate the equilibrium electron and hole concentrations, n_0 and p_0, to the position of the Fermi level. In this lecture, we continue that discussion.

For a nondegenerate semiconductor, the equilibrium electron concentration is related to the Fermi level according to

$$n_0 = N_C\, e^{(E_F - E_c)/k_B T} \quad \text{m}^{-3},$$

where the effective density of states for the conduction band is

$$N_C = \frac{1}{4}\left(\frac{2m_n^* k_B T}{\pi \hbar^2}\right)^{3/2} \quad \text{m}^{-3}.$$

The equilibrium hole concentration is related to the Fermi level according to a similar set of equations,

$$p_0 = N_V\, e^{(E_v - E_F)/k_B T} \quad \text{m}^{-3},$$

where the effective density of states for the valence band is

$$N_V = \frac{1}{4}\left(\frac{2m_p^* k_B T}{\pi \hbar^2}\right)^{3/2} \quad \text{m}^{-3}.$$

In this lecture, we'll do our calculations and examples for silicon. The density of states effective masses and the corresponding effective densities of states for silicon at $T = 300$ K are:

$$m_n^* = 1.182 m_0 \quad m_p^* = 0.81 m_0 \quad \text{(DOS effective masses)}$$

$$N_C = 3.23 \times 10^{19} \quad N_V = 1.83 \times 10^{19} \quad \text{cm}^{-3} \quad \text{(effective DOS)}.$$

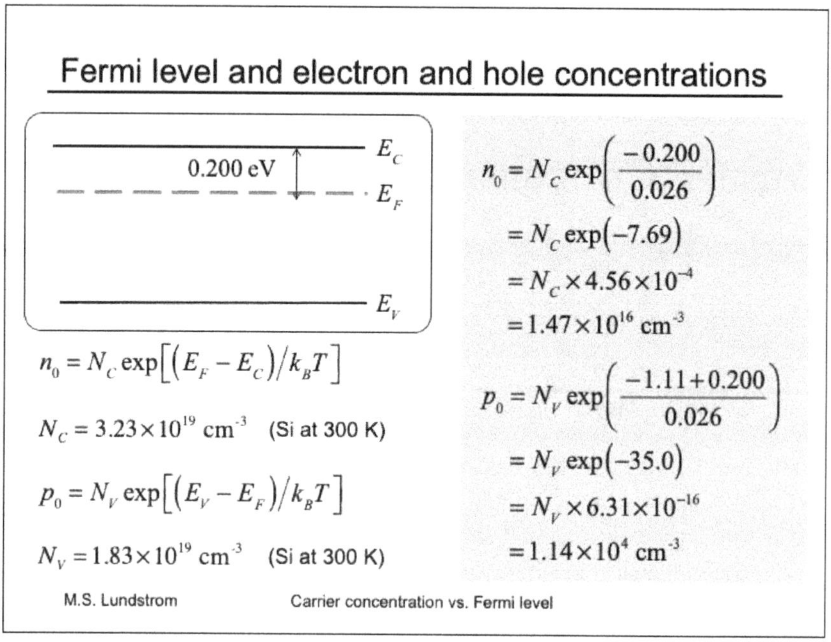

14.2 Fermi levels and carrier concentrations

Consider the example shown above at the upper left. The Fermi level is 0.2 eV below the bottom of the conduction band, so $E_F - E_c = -0.2$ eV. The thermal energy is $k_B T = 0.026$ eV at room temperature. As shown on the right above, to find the electron concentration, we multiply the exponential by the effective density of states. The result would be considered to be a moderate electron concentration.

The Fermi level is located well above the top of the valence band, so we don't expect there to be too many holes. We can use our expression for p_0 above recognizing that $E_v - E_F)$ is a large negative number.

$$\begin{aligned} E_v - E_F &= E_v - E_c + E_c - E_F \\ &= -E_G + (E_c - E_F) \quad . \\ &= -1.11 + 0.2 \quad \text{eV} \end{aligned}$$

Plugging numbers in, we get a very small hole concentration.

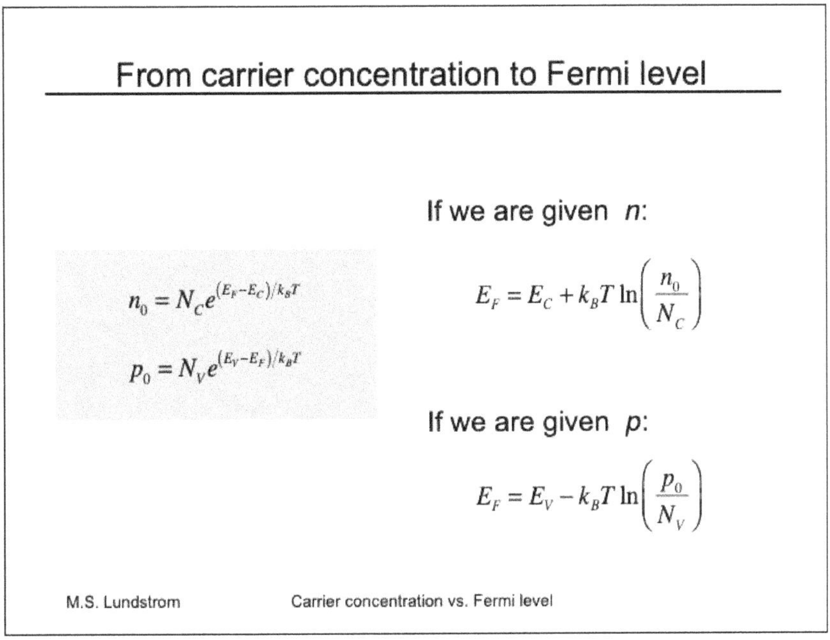

It is easy to go from the Fermi level to the carrier concentration, but sometimes, we may be given the carrier concentrations and asked for the Fermi level. That is straightforward too. We just solve the equations in the shaded box for the Fermi level.

For an N-type semiconductor, we might be given the electron density. We can then determine where the Fermi level is by using first expression shown above on the right.

For a P-type semiconductor, we might be given the hole density. We can then determine where the Fermi level is by using second expression shown above on the right.

Given n_0 or p_0, we can compute E_F, or given E_F, we can compute n_0 and p_0. We've seen how to compute the equilibrium electron and holes concentrations given the location of the Fermi level. As we will see later, the equilibrium product of the electron and hole concentrations is a **very important quantity** for a semiconductor. We'll discuss $n_0 p_0$ next, but first recall that for an intrinsic semiconductor, $n_0 = p_0 = n_i$, so for an intrinsic semiconductor, $n_0 p_0 = n_i^2$. As we discuss next, even for a doped semiconductor, $n_0 p_0 = n_i^2$ (as long as the semiconductor is nondegenerate).

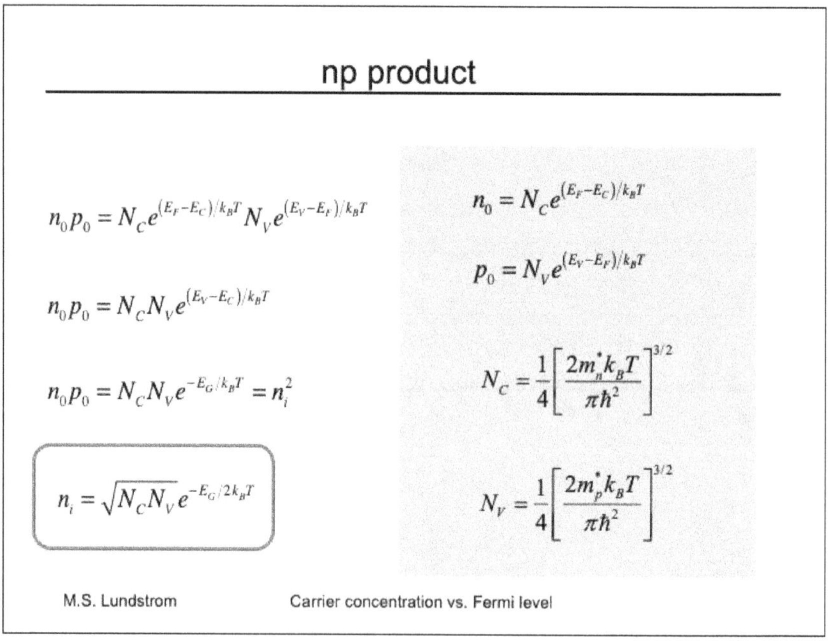

14.3 The equilibrium *np* product

We have expressions for n_0 and p_0, so let's multiply the two together. As shown above on the left, the Fermi level drops out of the product. In equilibrium, **the np product is independent of the location of the Fermi level.** (This is the case for the nondegenerate semiconductor that we are assuming.)

The equilibrium np product depends on the electron and hole effective masses (which are in the effective densities of states), the band gap of the semiconductor, and temperature. For an intrinsic semiconductor, $n_0 = p_0 = n_i$, so, $n_0 p_0 = n_i^2$. As shown above, $n_0 p_0$ doesn't depend on the Fermi level, so $n_0 p_0 = n_i^2$ for a nondegenerately doped semiconductor as well. This conclusion follows directly from the mathematical expressions we developed for n_0 and p_0. Later on, when we discuss generation and recombination, we will give a more physical explanation for why $n_0 p_0$ is always n_i^2.

When we discussed n_i earlier in the course, we pointed out that it depends exponentially on band gap and temperature. We now have an expression that verifies these exponential dependencies and that allows us to compute n_i from the material parameters of the semiconductor.

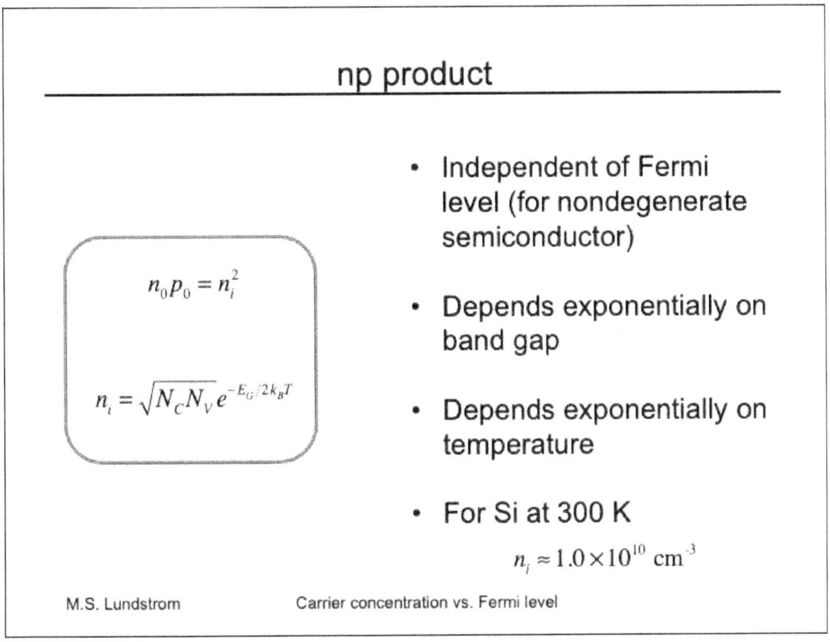

It is important to remember that the equilibrium np product is independent of where the Fermi level is located. That is not the case when we have a degenerate semiconductor; then we have to use Fermi-Dirac integrals.

We've learned that the equilibrium np product (n_i^2) depends exponentially on band gap. The wider the band gap, the harder it is to break covalent bonds and produce intrinsic carriers. It also depends exponentially on temperature. The higher the temperature, the easier it is for the thermal energy to break covalent bonds. For silicon, if we plug numbers in and are careful about how we do the calculation, we find that $n_i \approx 10^{10}$ per cubic centimeter, the number that we've been using in earlier lectures.

Re-visiting the hole concentration

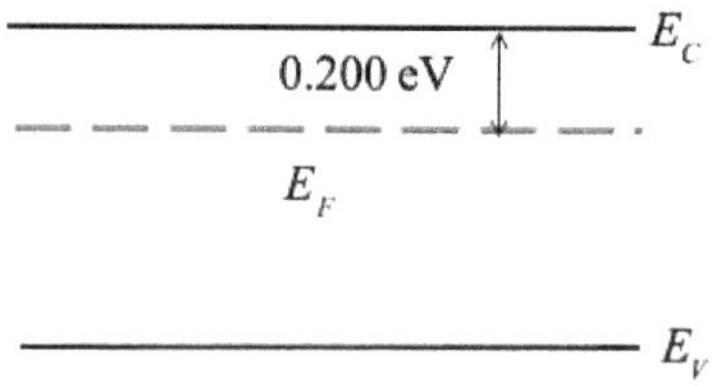

$$n_0 = N_C \exp\left(\frac{(E_F - E_C)}{k_B T}\right) = 1.47 \times 10^{16}\ \text{cm}^{-3}$$

$$p_0 = N_V \exp\left(\frac{(E_V - E_F)}{k_B T}\right) = 1.14 \times 10^{4}\ \text{cm}^{-3}$$

M.S. Lundstrom Carrier concentration vs. Fermi level

Let's re-do the example that we did earlier in Sec. 14.2, but do it a different way. Remember, for that example, we had an N-type semiconductor with a Fermi level near the conduction band. In this case, electrons are the **majority carriers** because most of the charge carriers are electrons, and holes are the **minority carriers** because there are fewer holes than electrons.

In Sec. 14.2, we first computed the electron density, and then we computed the hole density using the expression shown above, which relates the hole concentration to location of the Fermi level with respect to the top of the valence band. Doing the math, we found $p_0 = 1.14 \times 10^4$ holes per cubic centimeter.

There is another (usually easier) way to do this problem. We begin as we did earlier, by computing the electron concentration since it's an N-type semiconductor, but then we use the fact that $n_0 p_0 = n_i^2$, Since we know n_i^2, we find that $p_0 = n_i^2/n_0$, which gives us 0.68×10^4 holes per cubic centimeter.

The two approaches give results that are close, but in principle they should be the same. We discuss why we get different results next.

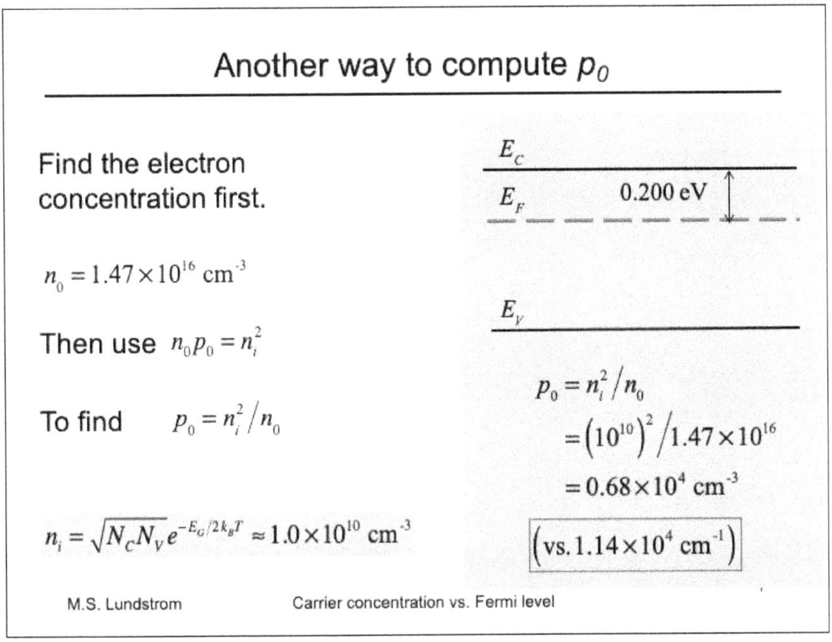

We've discussed two ways to compute the minority hole concentration, p_0, in an N-type semiconductor. The first way gave $p_0 = 1.14\times 10^4$, and the second way gave 0.68×10^4 holes per cubic centimeter. If we round both of them off, they're about 10^4 holes per cubic centimeter. That's 12 orders of magnitude less than the electron concentration, so that's not really too bad, but why do the two approaches give answers that are a little different?

To see why, plug in the numbers that we've been using for N_C, N_V, and E_G and compute n_i. You will see that it doesn't give us exactly 1×10^{10}. The intrinsic carrier concentration is **very sensitive** to these parameters. People who need an accurate intrinsic carrier concentration spend a lot of time worrying about temperature-dependent band gaps, temperature-dependent effective masses, etc., and they make small corrections for details, like an excitonic correction to the band gap. So the band gap and effective densities of states we used don't give us exactly the n_i we assumed.

When doing these kinds of calculations, ask yourself: "Do I have more confidence in the value of n_i or in the effective densities of states and band gap?" then choose the approach that you have more confidence in. Or, to get a quick approximate answer, just assume $n_i = 10^{10}\,\text{cm}^{-3}$ for Si at $T = 300$ K.

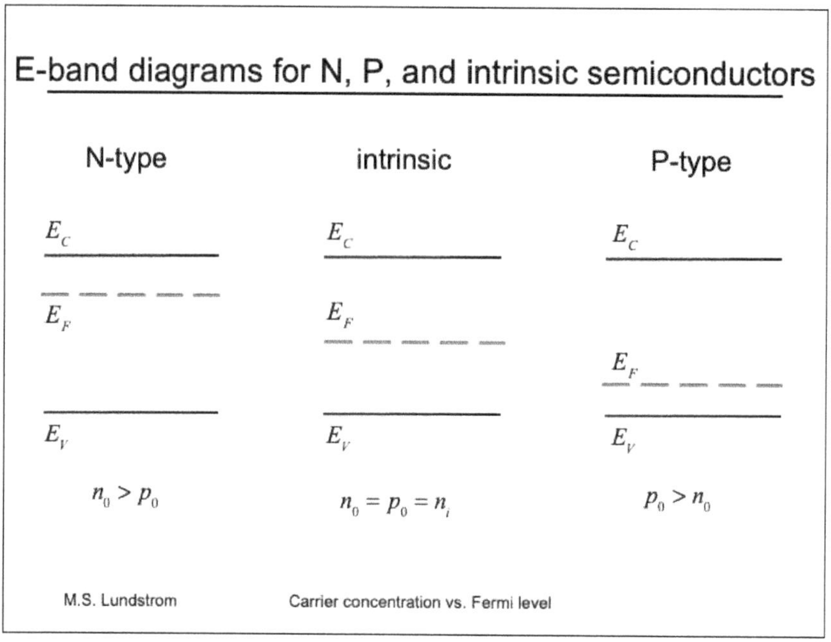

The np product is an important quantity to understand. Reading an energy band diagram and determining whether the semiconductor is N-type or P-type is also important. If we look at the energy band diagram shown above on the left, we see the Fermi level up near the bottom of the conduction band, so we know that there will be more electrons than holes; this is an N-type semiconductor, but n_0p_0 is still n_i^2.

If we look an energy band diagram shown above on the right, we see the Fermi level near the top of the valence band, so we know that there will be more holes than electrons; this is a P-type semiconductor, but n_0p_0 is still n_i^2.

Finally, for an intrinsic semiconductor, the Fermi level will be about in the middle of the gap as shown in the middle plot above. For an intrinsic semiconductor there are equal numbers of electrons and holes, but is the Fermi level *exactly* at the middle of the gap?

The intrinsic Fermi level

$$n_0 = N_C e^{(E_F - E_C)/k_B T}$$

$$p_0 = N_V e^{(E_V - E_F)/k_B T}$$

$$n_0 = p_0 = n_i \qquad E_F = E_i$$

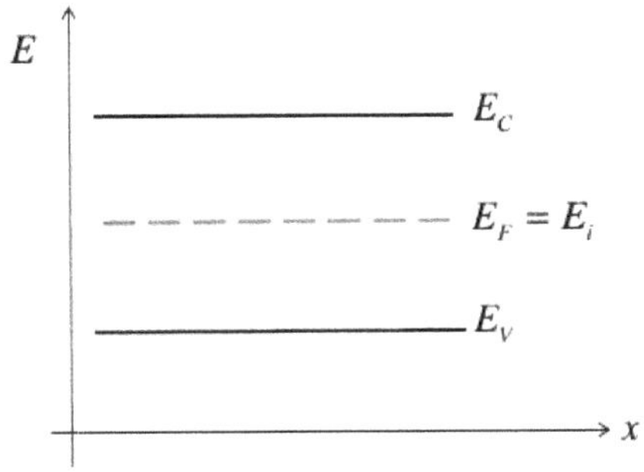

$$N_C e^{(E_i - E_C)/k_B T} = N_V e^{(E_V - E_i)/k_B T}$$

$$E_i = \frac{E_C + E_V}{2} + \frac{k_B T}{2} \ln\left(\frac{N_V}{N_C}\right)$$

M.S. Lundstrom Carrier concentration vs. Fermi level

We can easily calculate the location of the Fermi level for an intrinsic semiconductor as follows.

We know how the electron concentration is related to the Fermi level, and we know how the hole concentration is related to the Fermi level. We also know that in an intrinsic semiconductor, the electron concentration is equal to the hole concentration. If we set $n_0 = p_0$, we can solve the resulting expression for the Fermi level of an intrinsic semiconductor. By convention, we call the Fermi level under intrinsic conditions E_i instead of E_F; E_i is the **intrinsic Fermi level** or just the **intrinsic level**.

If we do this calculation, we get the expression shown above on the lower left. We find that E_i is offset from the middle of the band gap by a small factor that depends on how much different the effective density of states in the valence and conduction bands are. Remember that those two quantities depend on the effective masses, so the Fermi level is either a little above or a little below the middle of the band gap depending on the relative values of the effective masses for electrons and holes.

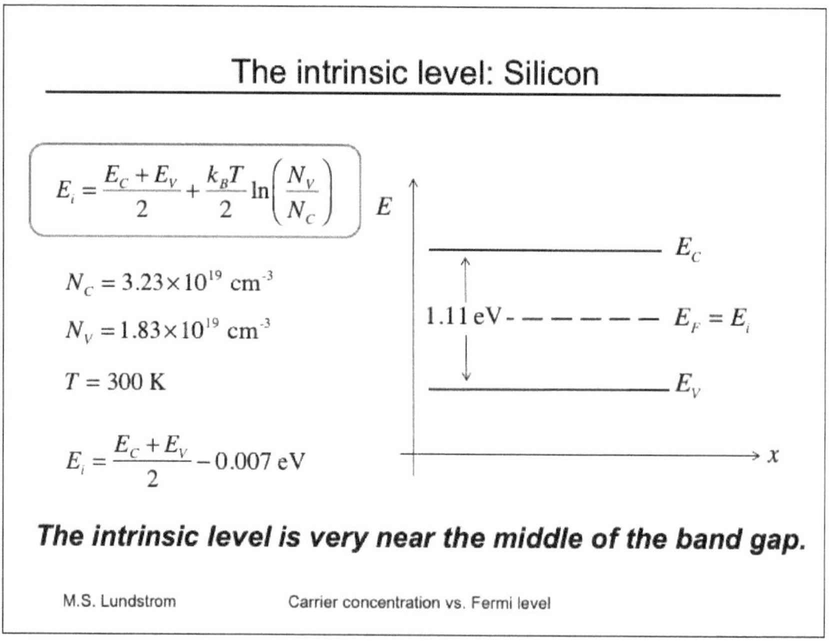

Let's compute the intrinsic level for silicon. We know what the effective densities of states are, so we can compute the correction term and see how far away from the middle of the band gap we are. Putting numbers in, we find that for Si, the intrinsic level is just 0.007 eV below the middle of the band gap. The intrinsic Fermi level is, indeed, located very close to the middle of the band gap. There are some semiconductors like gallium arsenide where there's more of an asymmetry between the conduction band and the valence band, but we take the logarithm of (N_V/N_C), so the intrinsic Fermi level never gets very far from the middle of the band gap.

Remember: The intrinsic level is very near the middle of the band gap. To see this quantitatively for GaAs, which has a large difference between the electron and hole effective masses, compute E_i for GaAs assuming that $E_G = 1.4$ eV, $N_C = 4.21 \times 10^{17}\ \text{cm}^{-3}$, and $N_V = 9.52 \times 10^{18}\ \text{cm}^{-3}$.

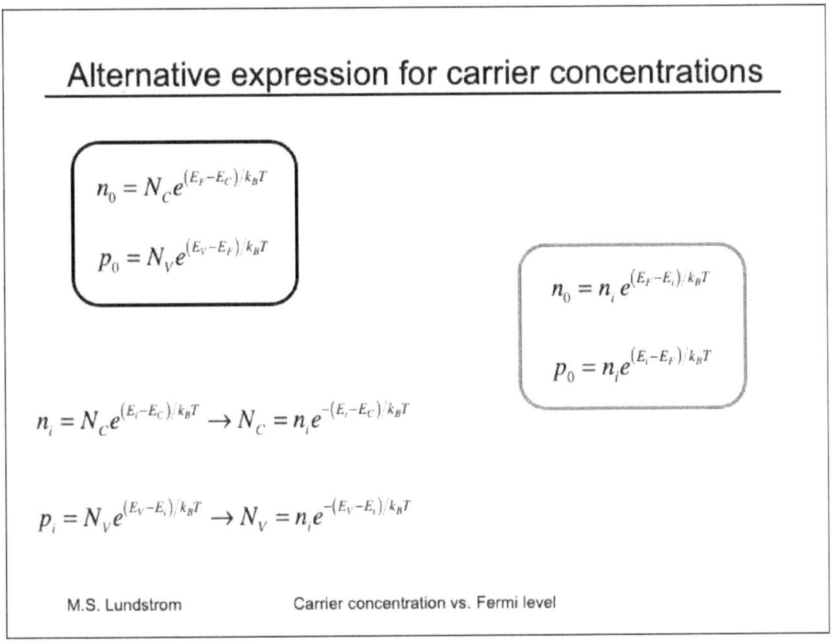

Knowing the intrinsic level, we can relate the carrier concentrations to the Fermi level in a different way. The two relations shown above on the upper left relate the carrier concentrations to the Fermi level and they're perfectly adequate, but there's another way to do this that's commonly done and is also useful.

We know that n_i is related to the Fermi level. By setting $E_F = E_i$ in the first two equations on the left, and by setting n_0 and p_0 equal to n_i, we can solve these equations for N_C and N_V, and then eliminate N_C and N_V in the first two equations. The result is the two equations shown above at the right.

Frequently, we know n_i for a semiconductor, so expressing the carrier concentrations in terms of n_i and the Fermi level with respect to E_i is convenient. As shown by the equations at the right, if the Fermi level is above the intrinsic level, there are more electrons than holes, and if the Fermi level is below the intrinsic level, there are more holes than electrons.

The two expressions at the upper left and the two on the right are equivalent. We'll use one or the other depending on which one is more convenient for the problem at hand. You should get used to being able to work back and forth and use either one of these approaches.

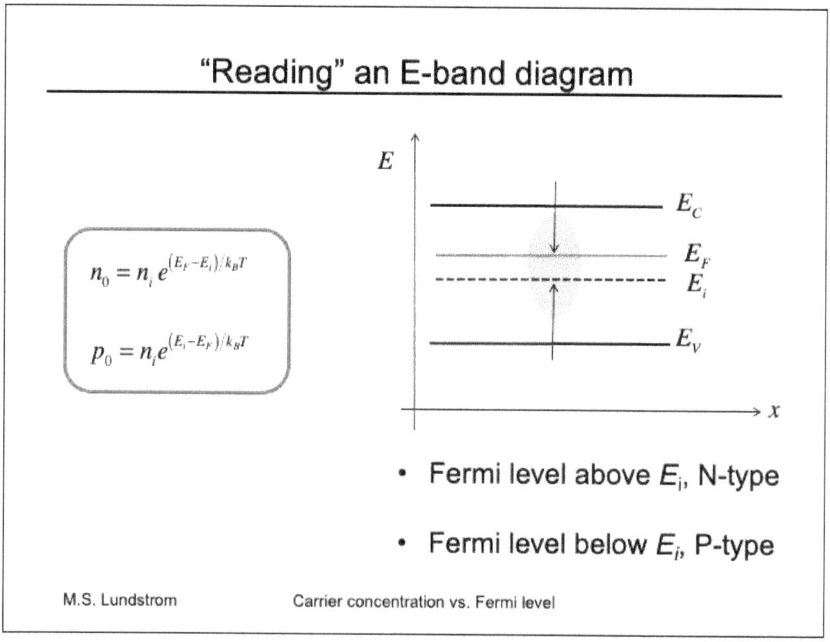

As discussed earlier, it is easy to read an energy band diagram and determine whether we have an N-type or a P-type semiconductor. If the Fermi level is closer to the conduction band than to the valence band, it's N-type. If it's closer to the valence band than to the conduction band, it's P-type. We can also do this from the relative position of the Fermi level and the intrinsic level.

If the Fermi level is above the intrinsic level (approximately the middle of the bandgap), it's N-type. If the Fermi level is below the intrinsic level, it's P-type. So very easy to read an energy band diagram quickly and determine whether we're dealing with an N-type or a P-type semiconductor. Finally, as shown above, E_i is typically shown on an energy band diagram as a dashed line midway between E_c and E_v.

14.4 Summary

Let's summarize the key equations. First, we have developed two equivalent ways to relate the equilibrium electron and hole concentrations to the location of the Fermi level in a nondegenerate semiconductor.

$$n_0 = N_C\, e^{(E_F - E_c)/k_B T} = n_i\, e^{(E_F - E_i)/k_B T}$$

$$p_0 = N_V\, e^{(E_v - E_F)/k_B T} = n_i\, e^{(E_i - E_F)/k_B T}\,.$$

The effective densities of states are

$$N_C = \frac{1}{4}\left(\frac{2m_n^* k_B T}{\pi\hbar^2}\right)^{3/2}$$

$$N_V = \frac{1}{4}\left(\frac{2m_p^* k_B T}{\pi\hbar^2}\right)^{3/2},$$

where m_n^* and m_p^* are the density of states effective masses for electrons and holes. We also have an expression that relates n_i to the band gap, temperature, and effective masses

$$n_i = \sqrt{N_C N_V}\, e^{-E_G/2k_B T}\,,$$

and we should remember that

$$n_0 p_0 = n_i^2$$

for a nondegenerate semiconductor in equilibrium. These are important relations that we will rely on. They only apply to nondegenerate semiconductors, but they typically get us close, and when necessary, we can follow up with more careful calculations using Fermi-Dirac integrals. To test your understanding, write the corresponding expressions for n_0, p_0, and $n_0 p_0$ without assuming that the semconductor is nondegenerate.

We continue this discussion of carrier densities in the next lecture, which is about how the carrier concentrations vary with doping density.

14.5 Reference

The book by Pierret provides plots of $n_i(T)$ vs. T for Si, Ge, and GaAs.

[1] R.F. Pierret, *Advanced Semiconductor Fundamentals*, (2nd Ed.), Pearson Education, Inc., 2003 (see Ch. 4, Sec. 4.4.3).

Lecture 15

Carrier Concentration vs. Doping Density

15.1 Introduction

Let's continue our discussion about carrier concentrations in semiconductors under equilibrium conditions. The topic for this lecture is how the carrier concentration depends on doping.

We have developed expressions that relate the Fermi level to the electron and hole concentrations. For a nondegenerate semiconductor, the equilibrium electron concentration is related to E_F according to

$$n_0 = N_C\, e^{(E_F - E_c)/k_B T} = n_i\, e^{(E_F - E_i)/k_B T} \quad \text{m}^{-3}\,,$$

where the effective density of states for the conduction band is

$$N_C = \frac{1}{4}\left(\frac{2m_n^* k_B T}{\pi \hbar^2}\right)^{3/2} \quad \text{m}^{-3}\,.$$

The equilibrium hole concentration is related to the position of the Fermi level according to a similar set of equations.

$$p_0 = N_V\, e^{(E_v - E_F)/k_B T} = n_i\, e^{(E_i - E_F)/k_B T} \quad \text{m}^{-3}\,,$$

where the effective density of states for the valence band is

$$N_V = \frac{1}{4}\left(\frac{2m_p^* k_B T}{\pi \hbar^2}\right)^{3/2} \quad \text{m}^{-3}\,.$$

Because the important principles are most clearly illustrated with the non-degenerate approximation to the Fermi Dirac integral, we will make that assumption in this lecture — realizing that from time to time when we're solving real problems, we may have to do the extra work of evaluating Fermi-Dirac integrals.

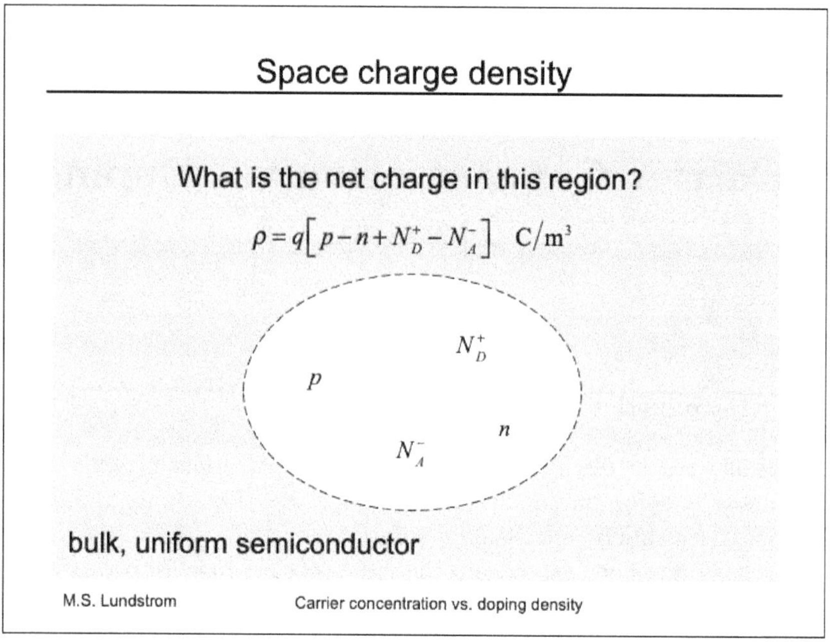

15.2 Space charge neutrality

Consider a specific region shown in white above in a uniformly doped semiconductor. The ionized donors have a positive charge, electrons have a negative charge, ionized acceptors have a negative charge, and holes have a positive charge. Electrons and holes are mobile, but dopants are fixed in the lattice.

The net charge density, ρ, is q times the concentration of positively charged holes, p, minus the concentration of negatively charged electrons, n, plus the concentration of positive charges due to ionized donors, N_D^+ minus the concentration of negative charges due to ionized acceptors, N_A^-.

You may have heard the phrase: "nature abhors a vacuum." Nature also doesn't like a charge imbalance, so the mobile charges move around, positive charges attract negative charges, and in the end for this uniform semiconductor the net charge is zero. When the doping density varies with position, however, a non-zero space charge results. When the doping changes abruptly, as in a PN junction, very large space charge densities result. **Space charge neutrality**, $\rho = 0$, will be our starting point for calculations with uniformly doped semiconductors.

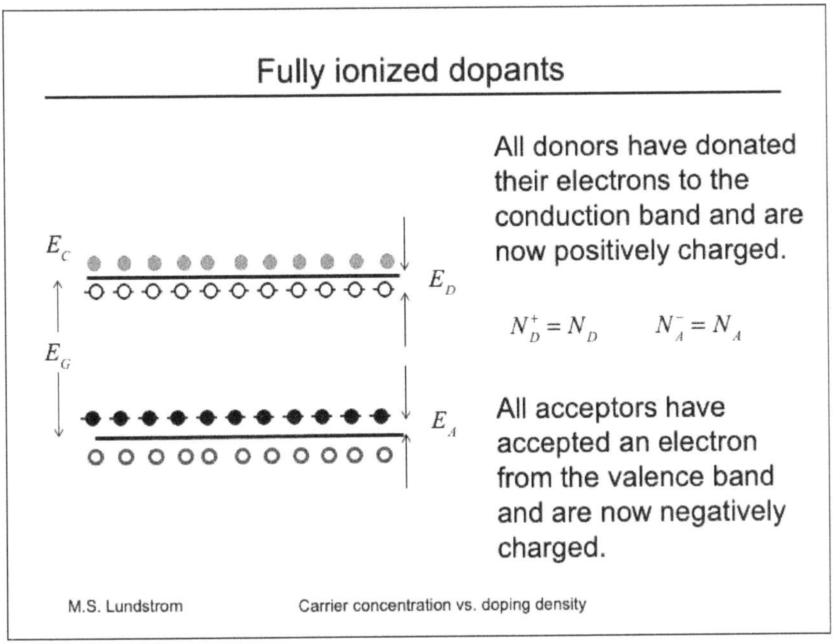

In this lecture, we will assume space charge neutrality ($\rho = 0$); we will also assume that the dopants are fully ionized, every donor has donated its electron to the conduction band, and every acceptor is negatively charged because an electron from the valence band has occupied each acceptor. We will assume that $N_D^+ = N_D$ and $N_A^- = N_A$. With this assumption,

$$\rho = q\left[p - n + N_D^+ - N_A^-\right] = 0 \rightarrow \rho = q\left[p - n + N_D - N_A\right] = 0\,.$$

In this lecture, we will also assume equilibrium conditions, so

$$\rho = q\left[p_0 - n_0 + N_D - N_A\right] = 0\,.$$

The subscripts, '0', indicate that we are dealing with a semiconductor in equilibrium. Because equilibrium conditions have been assumed, we also know that

$$n_0 p_0 = n_i^2\,.$$

Assuming that we know the doping density, the above two equations are two equations for the two unknowns, n_0 and p_0. We just need to solve these equations to relate the carrier concentrations to the doping density.

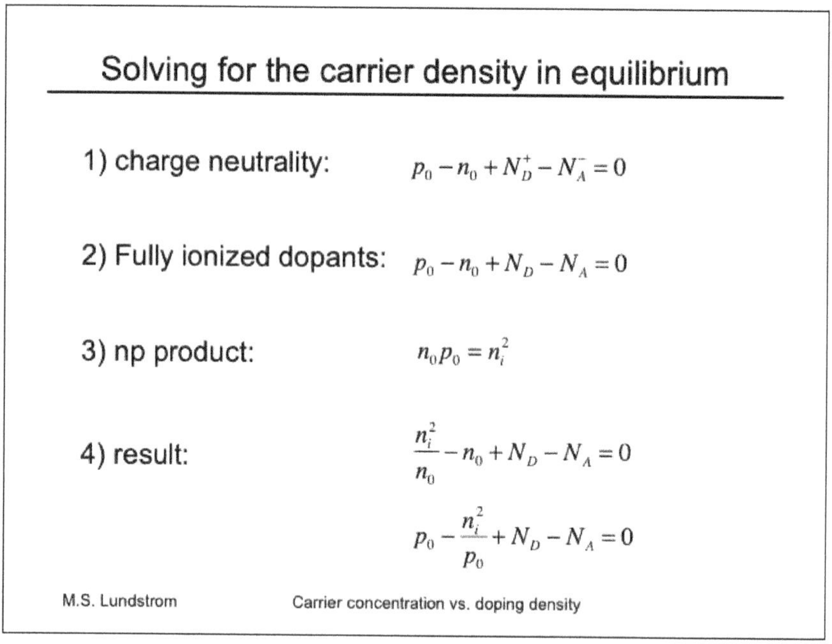

15.3 General expressions for carrier concentrations

The calculation proceeds as shown above.

We begin with the statement of charge neutrality (step 1 above). Next, we assume fully ionized dopants (step 2). Then we recall the fact that the equilibrium np product is n_i^2 (step 3). The result is that we can either eliminate p_0 and get an expression in terms of just n_0, or we can eliminate n_0 and get an expression in terms of just p_0 (step 4).

The final result is two quadratic equations that can be solved for n_0 or p_0. Generally, we would solve for p_0 in a P-type semiconductor and for n_0 in an N-type semiconductor, and then find the corresponding minority carrier concentrations from $n_0 p_0 = n_i^2$.

Results

$N_D > N_A$

$$n_0 = \frac{N_D - N_A}{2} + \left[\left(\frac{N_D - N_A}{2}\right)^2 + n_i^2\right]^{1/2}$$

$$p_0 = \frac{n_i^2}{n_0}$$

N-type

$N_A - N_D$

$$p_0 = \frac{N_A - N_D}{2} + \left[\left(\frac{N_A - N_D}{2}\right)^2 + n_i^2\right]^{1/2}$$

$$n_0 = \frac{n_i^2}{p_0}$$

P-type

M.S. Lundstrom Carrier concentration vs. doping density

The final results of solving the quadratic equations are shown above. For an N-type semiconductor, we solve a quadratic equation for n_0 and then use $n_0 p_0 = n_i^2$ to find p_0. For a P-type semiconductor, we solve a quadratic equation for p_0 and then use $n_0 p_0 = n_i^2$ to find n_0. Let's see how this works by doing some examples.

Example 1

Consider Si doped with phosphorus at $N_D = 2.00 \times 10^{15}$ cm^{-3} The temperature is 300 K. What are n_0 and p_0?

Recall that at 300 K in Si, $n_i = 1.00 \times 10^{10}$ cm^{-3}

Assume that the donors are fully ionized.

$$n_0 = \frac{N_D - N_A}{2} + \left[\left(\frac{N_D - N_A}{2}\right)^2 + n_i^2\right]^{1/2} = \frac{N_D}{2} + \left[\left(\frac{N_D}{2}\right)^2 + n_i^2\right]^{1/2}$$

$N_D >> n_i$ $\qquad n_0 = N_D = 2.00 \times 10^{15}$ cm^{-3}

$p_0 = n_i^2 / n_0$ $\qquad p_0 = \left(10^{10}\right)^2 / 2 \times 10^{15} = 5 \times 10^4$ cm^{-3}

M.S. Lundstrom Carrier concentration vs. doping density

The first example is silicon doped with moderate concentration of phosphorous or arsenic. We'll also assume that the temperature is 300 K, so it's safe to assume that the dopants are fully ionized. What are the electron and hole concentrations?

Remember that at 300 K, $n_i \approx 10^{10}$ per cm^3. We can evaluate the expression, but you can see that for $N_D \gg n_i$ and $N_A = 0$, the expression reduces to $n_0 = N_D$.

Knowing the electron concentration, the hole concentration is readily found from n_i^2/n_0.

Finally, note that this is an **extrinsic semiconductor**, $N_D \gg n_i$. For problems like this, we really don't need to take the time to solve the quadratic equation because it is clear the $n_0 = N_D$.

Example 2

Consider Si doped with phosphorus at $N_D = 2.00 \times 10^{15}\ \text{cm}^{-3}$ and Boron at $N_A = 1.00 \times 10^{15}\ \text{cm}^{-15}$.
The temperature is 300 K. What are n_0 and p_0?

$$n_0 = \frac{N_D - N_A}{2} + \left[\left(\frac{N_D - N_A}{2}\right)^2 + n_i^2\right]^{1/2}$$

$$N_D - N_A >> n_i \qquad n_0 = N_D - N_A$$

$$n_0 = 1.00 \times 10^{15}\ \text{cm}^{-3}$$

$$p_0 = n_i^2 / n_0 \qquad p_0 = \left(10^{10}\right)^2 / 1 \times 10^{15} = 1 \times 10^{5}\ \text{cm}^{-3}$$

M.S. Lundstrom Carrier concentration vs. doping density

The second example has the same concentration of N-type dopants, but now we have added some P-type dopants. Again, we want to determine the electron and hole concentrations. We could evaluate the quadratic expression, but you can see that for $(N_D - N_A) \gg n_i$, the equation simplifies to $n_0 = N_D - N_A$. To find the hole concentration, it's just n_i^2/n_0 again.

This is also an extrinsic semiconductor because $N_D - N_A \gg n_i$. We don't need to solve the quadratic equation for problems like these two examples.

For extrinsic semiconductors like these two examples, the majority carrier concentration is just the net doping density and the minority carrier concentration is obtained from $n_0 p_0 = n_i^2$.

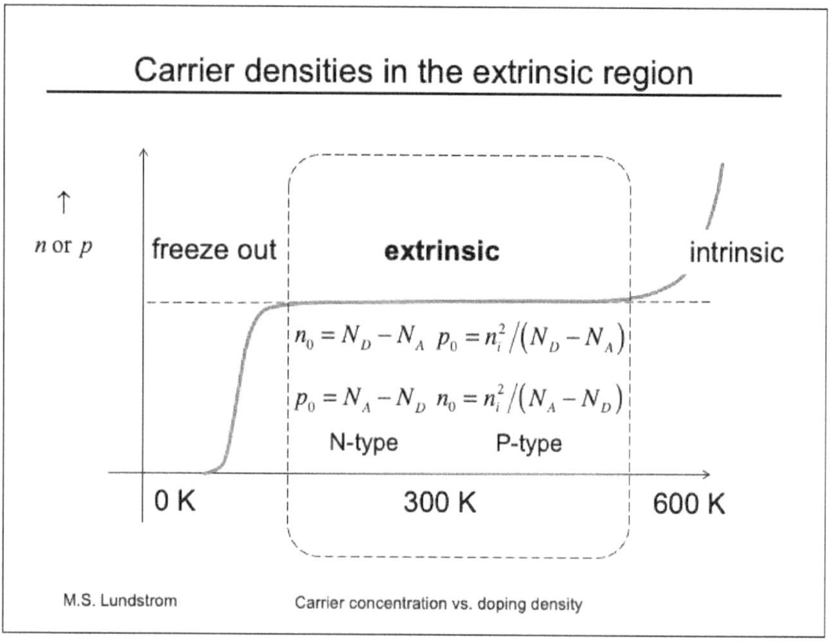

15.4 The extrinsic region

The carrier concentration vs. temperature is shown above. In this lecture, we have been working in the **extrinsic region**, where the temperature is moderate, the dopants fully ionized, and where the dopants overwhelm the intrinsic carriers. It is very easy to calculate the carrier concentrations in the extrinsic region.

For silicon, the net doping density is typically much greater than the intrinsic carrier concentration and at room temperature, the dopants are fully ionized. If we are dealing with an N-type sample, then the electron concentration is equal to the net N-type doping, $N_D - N_A$, and we determine the minority hole concentration from the n_0p_0 product. If we are dealing with a P-type sample, the hole density is equal to the net P-type doping, $N_A - N_D$, and we get the minority electron concentration from the n_0p_0 product.

In these lectures, we will be dealing with semiconductors in the extrinsic region, but semiconductor engineers sometimes encounter low temperatures and sometimes high temperatures. How to compute the carrier concentrations at low and high temperatures is the subject of the next lecture.

15.5 Summary

In this lecture, we developed expressions for the equilibrium carrier concentrations when the dopants are fully ionized (i.e. the temperature is not too low). We found two quadratic equations. For an N-type sample ($N_D > N_A$),

$$n_0 = \frac{N_D - N_A}{2} + \left[\left(\frac{N_D - N_A}{2}\right)^2 + n_i^2\right]^{1/2}$$

and for a P-type sample ($N_A > N_D$),

$$p_0 = \frac{N_A - N_D}{2} + \left[\left(\frac{N_A - N_D}{2}\right)^2 + n_i^2\right]^{1/2} .$$

In both cases, we determine the minority carrier concentration from

$$n_0 p_0 = n_i^2$$

In the **extrinsic region**, things are simple. In this region, the dopants are fully ionized (i.e. the temperature is not too low), and n_i is negligible compared to the net doping density. In this case, the quadratic equations simplify, and we find for an N-type sample ($N_D > N_A$),

$$n_0 = N_D - N_A$$

and for a P-type sample ($N_A > N_D$),

$$p_0 = N_A - N_D .$$

Again, in both cases, we determine the minority carrier concentration from

$$n_0 p_0 = n_i^2$$

Sometimes, however, the temperature is too low to fully ionize the dopants or it is too high to make n_i negligible. In that case, things get a little more complicated as discussed in the next lecture.

15.6 Reference

For more discussion about carrier concentrations vs. doping density, see:

[1] R.F. Pierret, *Advanced Semiconductor Fundamentals*, (2nd Ed.), Pearson Education, Inc., 2003 (see Ch. 4, Sec. 4.4.3).

Lecture 16

Carrier Concentration vs. Temperature

16.1 Introduction

In Lecture 6, we discussed qualitatively how the carrier concentration varies with temperature. At low temperatures, the dopants and intrinsic carriers **freeze out**, and there are no charge carriers, so the semiconductor becomes an insulator. In the previous lecture, we discussed the moderate temperature, **extrinsic region**. At high temperatures, there is an **intrinsic region**, where the intrinsic carrier concentration becomes comparable or even greater than the doping density. This lecture is a quantitative discussion of the freeze-out and intrinsic regions, but first, let's quickly review the extrinsic region.

In the extrinsic region, the net ionized doping density overwhelms the intrinsic carriers, and we can just write down the carrier concentrations. Assuming that the dopants are fully ionized,

$$n_0 = N_D^+ - N_A^- = N_D - N_A \qquad \text{(N-type)}$$
$$p_0 = N_A^- - N_D^+ = N_A - N_D \qquad \text{(P-type)}\,.$$

Typically, we would compute the majority carrier concentration first, and then obtain the minority carrier concentration from

$$n_0 p_0 = n_i^2\,.$$

When the temperature is too low to fully ionize the dopants, or when it is too high to ignore the intrinsic carriers, then we need to return to the quadratic equations that we obtained in Lecture 15 by assuming space charge neutrality.

$$n_0 = \frac{N_D^+ - N_A^-}{2} + \left[\left(\frac{N_D^+ - N_A^-}{2}\right)^2 + n_i^2\right]^{1/2} \qquad \text{(N-type)}$$

$$p_0 = \frac{N_A^- - N_D^+}{2} + \left[\left(\frac{N_A^- - N_D^+}{2}\right)^2 + n_i^2\right]^{1/2} \qquad \text{(P-type)}\,.$$

For low temperatures, n_i can be neglected and these equations reduce to

$$n_0 = N_D^+ - N_A^- \qquad \text{(N-type)}$$
$$p_0 = N_A^- - N_D^+ \qquad \text{(P-type)}\,.$$

When the temperature is too high to ignore the intrinsic carriers, we can safely assume that the dopants are fully ionized and

$$n_0 = \frac{N_D - N_A}{2} + \left[\left(\frac{N_D - N_A}{2}\right)^2 + n_i^2\right]^{1/2} \qquad \text{(N-type)}$$

$$p_0 = \frac{N_A - N_D}{2} + \left[\left(\frac{N_A - N_D}{2}\right)^2 + n_i^2\right]^{1/2} \qquad \text{(P-type)}\,.$$

In this lecture, we'll discuss both the low and high temperature cases.

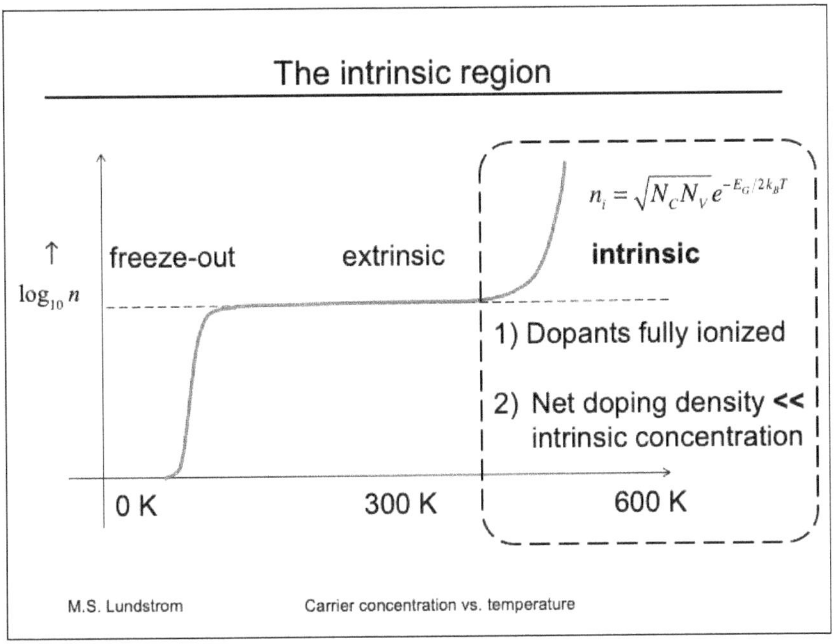

16.2 The intrinsic region

Let's shift our attention now from the moderate temperatures of the extrinsic region to higher temperatures, the **intrinsic region**. We saw in Lecture 14 that the intrinsic carrier concentration increases exponentially with temperature, so as we increase the temperature, we'll get to a point where the concentration of intrinsic carriers, created by breaking covalent bonds and exciting electrons across the bandgap, becomes comparable to the doping density. The dopants continue to be fully ionized, but the net doping density is no longer much larger the intrinsic carrier concentration.

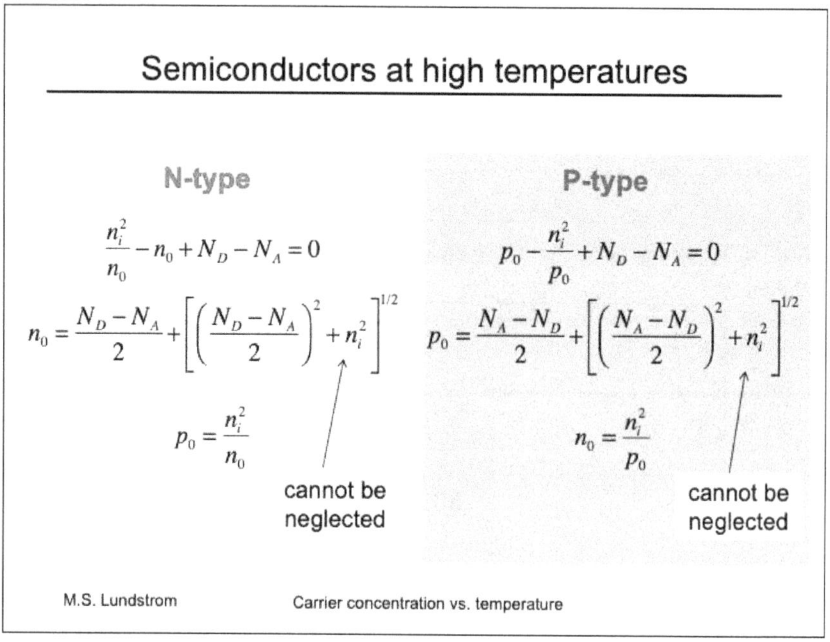

Let's see how this works. In the previous lecture, Example 1 was for Si at $T = 300$ K doped at $N_D = 2 \times 10^{15}$ donors per cc, but for Example 3, let's make the temperature 600 K. What are n_0 and p_0?

At 300 K, $n_i = 10^{10}$ and could be ignored, but at 600 K, it will be significantly higher. We could compute it from the intrinsic carrier concentration formula shown on the previous slide, but computing it actually requires some care. The formula probably would not give us the correct answer because we have to account for things like the temperature dependence of the bandgap, the temperature dependence of effective masses, and other details. For this example, we'll assume $n_i = 4 \times 10^{15}$ per cc at $T = 600$ K (see Fig. 4.17 of ref. [1]). We solve our quadratic equation, plug numbers in for the doping density and for intrinsic carrier concentration squared, and see that we can't neglect the n_i^2 term,

$$n_0 = 1.0 \times 10^{15} + \left[\left(1.0 \times 10^{15}\right)^2 + \left(4.0 \times 10^{15}\right)^2\right]^{1/2} = 5.1 \times 10^{15}\,\text{cm}^{-3}$$

The electron concentration is significantly higher than the N-type doping density. The minority hole concentration is $p_0 = n_i^2/n_0 = 3.3 \times 10^{15}\,\text{cm}^{-3}$, which is also significantly higher than the doping density.

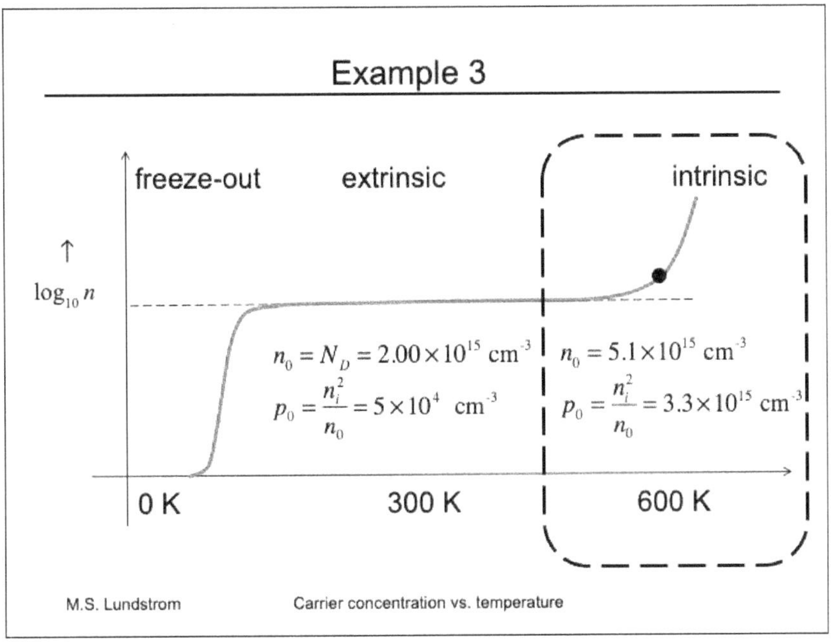

The results of the example are shown above. The extrinsic region is simple, but when the temperature is high enough, we need to include the intrinsic carriers. When we do that, we get a significantly higher electron concentration that exceeds N_D, and we also get a high hole concentration. The calculations are reasonably straight-forward using the quadratic equations that we developed from the assumption of space charge neutrality.

As we go to even higher temperatures, and the semiconductor becomes fully intrinsic, we lose the effects of the doping because the intrinsic carriers overwhelm the dopants, $n_0 \approx p_0 \approx n_i$. Selectively doping various regions of a semiconductor is how we create devices, but at high temperatures, we lose the effect of doping and devices fail. "Going intrinsic" determines the upper limit operating temperature of a semiconductor device. If we need semiconductor devices to operate at high temperatures, we would look for semiconductors such as silicon carbide or gallium nitride that have larger bandgaps than silicon because they will have smaller intrinsic carrier concentrations, and these effects will take place at higher temperatures than they do in silicon.

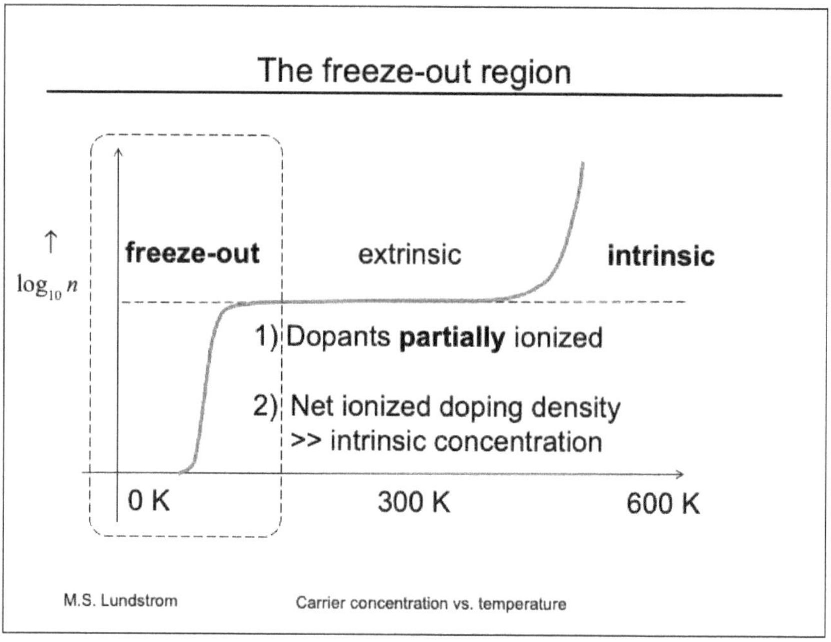

16.3 The freeze-out region

Let's shift our attention now to the low temperature, the freeze-out region. At low temperatures, there is not enough thermal energy to completely ionize the dopants, they're only partially ionized, or maybe not ionized at all. The thermal energy is also too low to create intrinsic carriers. Even though they are only partially ionized, the ionized doping density can still be much greater than the intrinsic carrier concentration. How would we solve for the carrier concentration in the freeze out region? We begin with space charge neutrality, but since n_i^2 is so small, we can neglect the minority carriers. If this is an N-type sample, we will also neglect the holes. The result is that the electron concentration is just equal to the net N-type ionized doping concentration. All we have to do is to compute the concentration of dopants that are ionized, and we have our answer.

At low temperatures

$$n_0 = N_D^+ - N_A^- \quad \text{(N-type)}$$
$$p_0 = N_A^- - N_D^+ \quad \text{(P-type)}\,.$$

At low temperatures, the minority carrier density is so low that there would rarely be any reason to compute it.

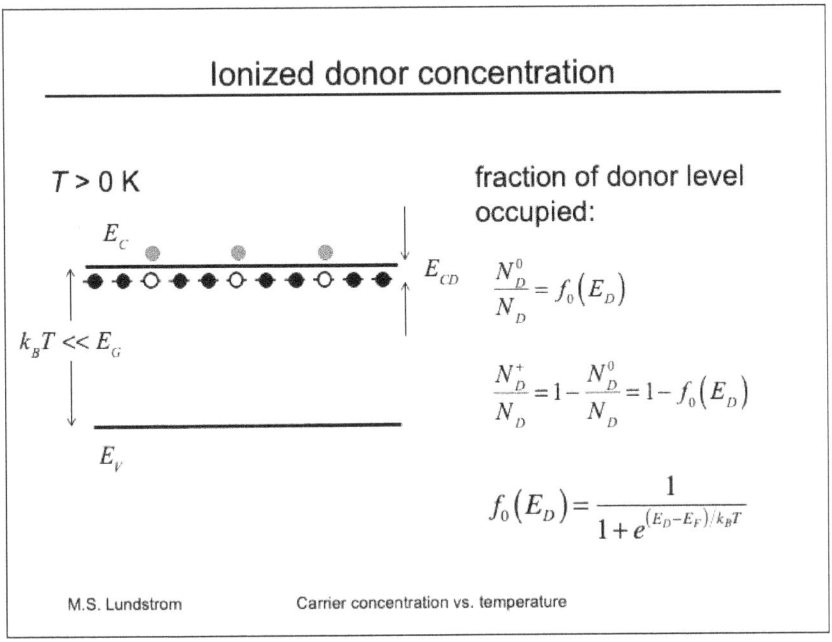

A low temperatures, only a few of the bonds that hold the fifth electron to the donor are been broken. The temperature is so low that there are essentially no intrinsic carriers; there is one electron in the conduction band for each donor that is ionized. Let's compute the fraction of the dopants that are ionized, N_D^+/N_D.

To compute N_D^+/N_D, we might expect that the fraction of these states that are occupied is given by a Fermi function. (This will get us close, but we'll see on the next slide that it is not quite correct.)

If we assume that the probability that an electron is on one of the donor sites is given by the equilibrium Fermi function, then the probability that the donors are ionized is given by $1 - f_0(E_D)$.

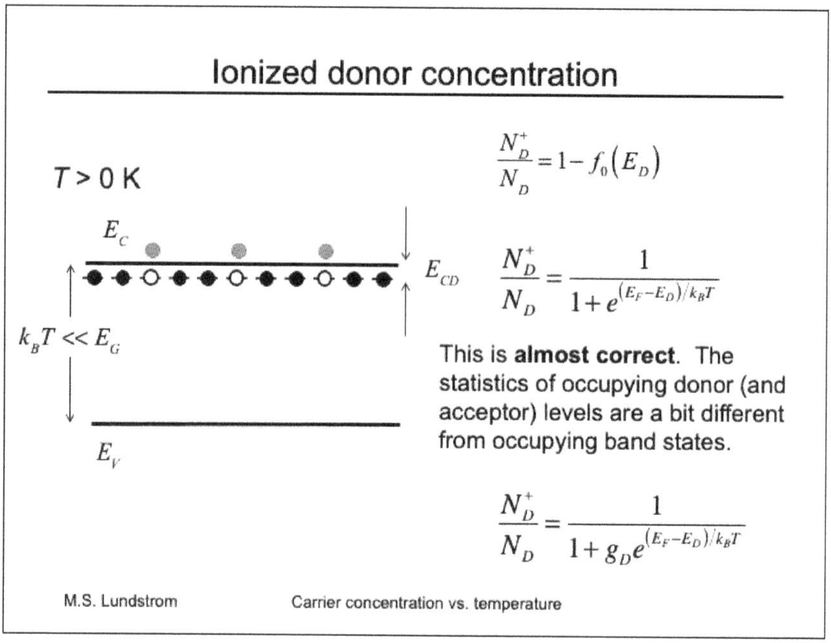

Assuming $N_D^+/N_D = 1 - f_0(E_D)$ is close, but it's not exactly right; the actual answer is a little bit different. The correct answer includes a donor **degeneracy factor**, g_D, as shown above. This occurs because the localized donor states are different from the de-localized, extended states in the conduction band. Each donor can accept a fifth electron that has spin up or spin down, but once it accepts an electron of either spin, the donor atom is neutral and cannot accept another electron. This doesn't occur for the distributed conduction band states, so the occupation statistics for donor states and conduction band states are different as shown above.

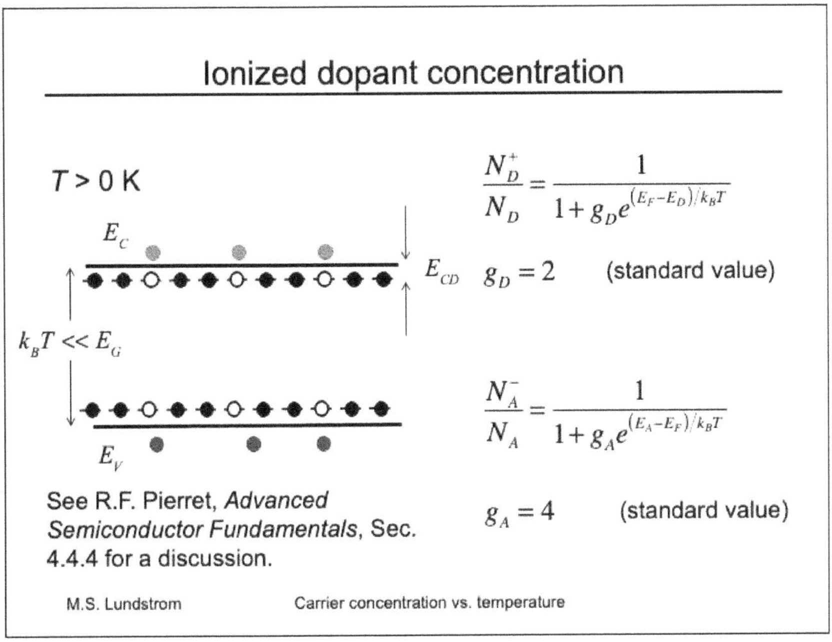

To summarize, the way we describe the fraction of donor or acceptor states that are ionized is by an expression that looks almost like a Fermi function, but with a **donor or acceptor degeneracy factor**. The standard factor used for the conduction band is two, and for the valence band, it is four. If you'd like to learn more about these factors and where they come from, see reference [1].

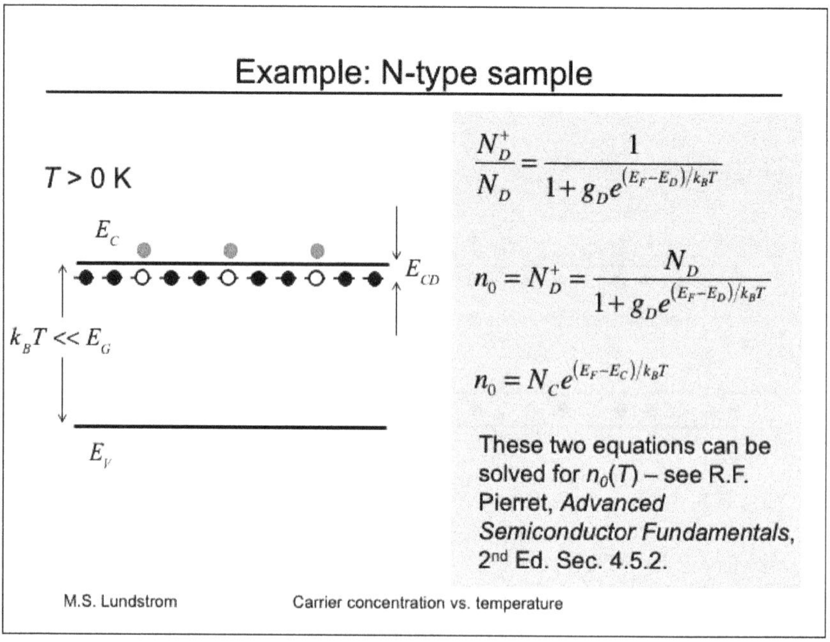

The slide above summarizes how the calculation is done for an N-type semiconductor at a low temperature. We'll just summarize the approach, for the details, see reference [1].

As shown in the first equation at the right, the fraction of donors that are ionized (empty donor states) is given by expression that is similar to $(1-f_0)$. As shown in the second equation on the right, the electron concentration is equal to the density of ionized donors. Finally, the third equation relates the electron concentration to the Fermi level.

If we equate the second equation to the third, we get an equation that can be solved for the Fermi level. Once we know the Fermi level, the third equation gives us the electron concentration.

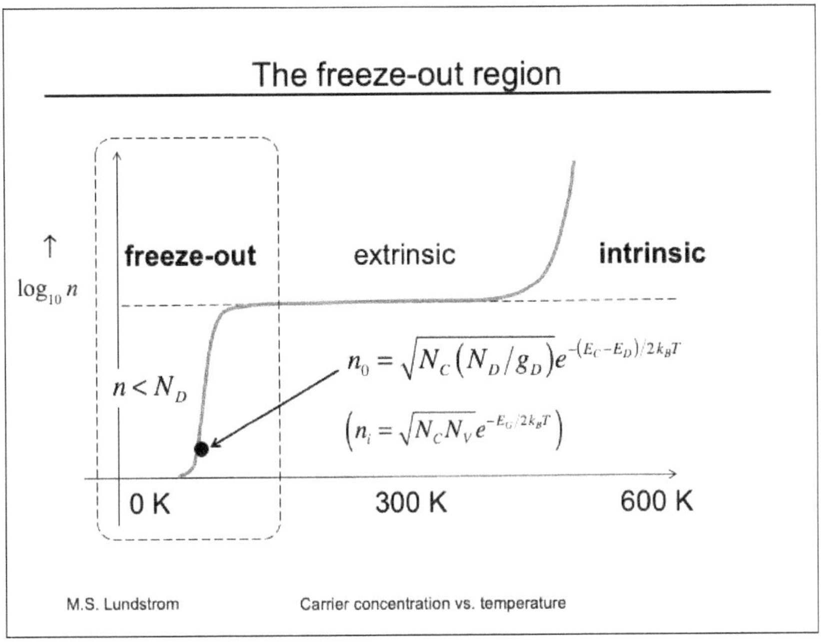

The result turns out to be a little bit complicated [1],

$$n_0 = \frac{N_\zeta}{2}\left[\left(1 + \frac{4N_D}{N_\zeta}\right)^{1/2} - 1\right]$$

$$N_\zeta = \frac{N_C}{g_D}\, e^{-(E_c - E_D)/k_BT} .$$

This equation describes the decrease in the free carrier concentration as the temperature decreases and fewer and fewer dopants are ionized.

At very low temperatures, the expression simplifies as shown on the slide above; n_0 depends exponentially on the binding energy of the fifth electron, $(E_c - E_D)$ and on temperature. The low temperature expression looks like the expression for n_i; there is a square root of effective density of states for the conduction band, but instead of an effective density of states for the valence band, we have the density of donor states divided by the donor degeneracy factor. The thermal energy is so small that there is essentially no probability that an electron from the valence band will be excited to the conduction band, so the valence band is out of the picture. The donor levels play the role of the valence band in this case. For n_i, we had $n_i \propto \exp[-(E_c - E_v)/k_BT]$, and in this case we have $n_0 \propto \exp[-(E_c - E_D)/k_BT]$. The result is intuitively what we might have expected.

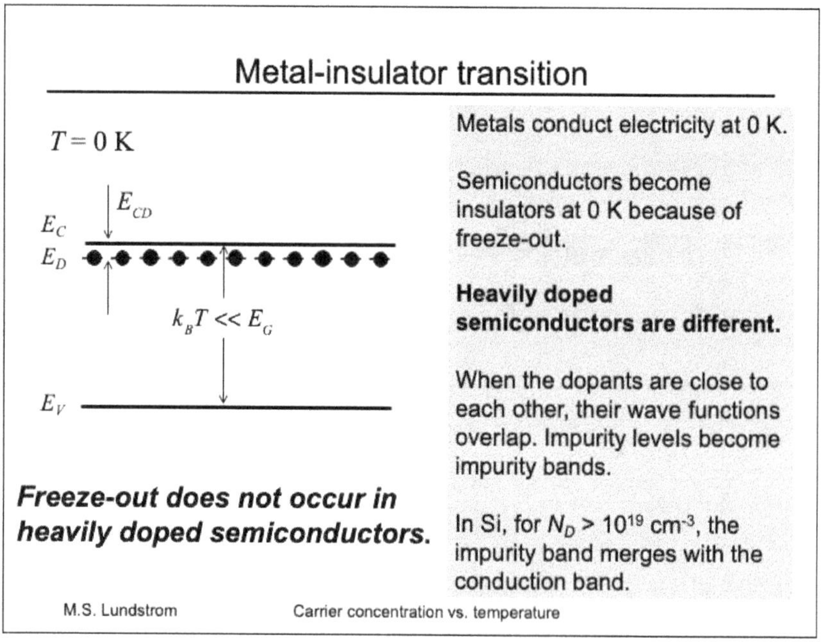

16.4 The metal-insulator transistion

We should briefly mention the **metal-insulator transition**. Recall that a metal conducts electricity at $T = 0$ K because the Fermi level is right in the middle of the band, so there are plenty of states for conduction. Semiconductors, however, become insulators at $T = 0$ K, because the dopants freeze out and there are no intrinsic carriers, so there are no carriers to conduct electricity, but heavily doped semiconductors are different. Recall that band states arise when the wave functions of adjacent atoms overlap, the states smear out and became delocalized bands that allow electrons to travel throughout the crystal. When we dope heavily, the wave functions of the adjacent dopants begin to overlap and delocalize. Instead of electrons sitting on specific dopants, these levels smear out into an **impurity band**, and electrons are free to move around and conduct.

When heavily doped semiconductors are cooled down to low temperatures, they do not become insulators; they continue to conduct electricity and behave as a metal because there is no freeze out to localized impurity states. This is known as the metal-insulator transition.

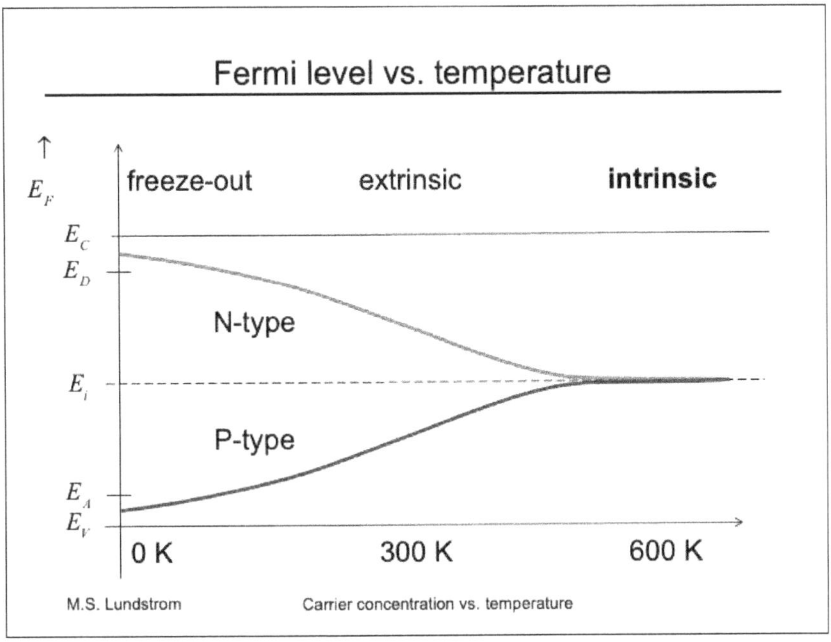

16.5 The Fermi level vs. temperature

Now that we understand how the carrier concentration varies with temperature, can we also understand how the Fermi level varies with temperature? Let's consider first an N-type semiconductor and ask how the Fermi level varies as we sweep the temperature from a low to a high temperature. The expected Fermi level vs. temperature characteristic is sketched above. At low temperatures, all of the donor states are occupied, so the Fermi level needs to be above the donor level. At very high temperatures, the semiconductor is intrinsic, so the Fermi level needs to be near the middle of the gap. As the temperature increases, the Fermi level smoothly transitions from a high energy near E_c to mid-gap as shown above.

Something similar happens for a P-type semiconductor. At low temperatures, all of the acceptor states are empty, so the Fermi level has to be below the acceptor level. At very high temperatures, the P-type semiconductor also becomes intrinsic, so the Fermi level is near the middle of the gap. As the temperature increases, the Fermi level smoothly transitions from a low energy near E_v to mid-gap as shown above.

16.6 Summary

In this lecture we have discussed how the carrier concentration varies with temperature. In the extrinsic region, where the donors are fully ionized and where the dopants overwhelm the intrinsic carriers, things are simple. In an N-type semiconductor, the electron concentration is just equal to the net donor density. In a P-type semiconductor, the hole concentration is just equal to the net acceptor density. When the semiconductor becomes hot, the intrinsic carriers become important, and the carrier concentration increases. Finally, as the semiconductor is cooled down, the dopants cannot be ionized, so the carrier density decreases, unless the semiconductor is doped very heavily. This qualitative understanding of how the carrier density varies with temperature is something that you now have a quantitative understanding of.

This wraps up Part 3. You should now be able to:

- Explain what the Fermi function is, how it is used, and the effect of the location of the Fermi level and temperature on the Fermi function.
- Understand the relation between the Fermi level, E_F, and the equilibrium carrier densities for degenerate and nondegenerate semiconductors.
- Relate the intrinsic carrier density to the effective densities of states and to the band gap and understand what the intrinsic Fermi level, E_i, is.
- Relate the equilibrium carrier densities to the location of the Fermi level relative to the conduction and valence band edges and to the intrinsic level.
- Compute the equilibrium carrier densities given the donor and acceptor concentrations and the temperature.

16.7 Reference

For more discussion about carrier concentration calculations and degeneracy factors for dopants, see:

[1] R.F. Pierret, *Advanced Semiconductor Fundamentals*, (2nd Ed.), Pearson Education, Inc., 2003 (see Ch. 4, Secs. 4.4.3, 4.4.4, and 4.5).

PART 4

Carrier Transport, Recombination, and Generation

Lecture 17

Current Equation

17.1 Introduction

So far in these lectures we have discussed semiconductors in equilibrium, focusing in Part 3 on equilibrium carrier concentrations. Semiconductor devices like transistors operate by applying voltages to the contacts. Solar cells function when light shines on them. These semiconductors are out of equilibrium. Current is an out of equilibrium phenomenon, and so are carrier recombination and generation. Upon completion of Part 4, you will:

- Be familiar with the general current equation and the terms transmission, channels, Fermi window, ballistic and diffusive transport.
- Understand how the current in a nano-device is related to the Fermi levels in the two contacts and how the current in the bulk is related to the quasi-Fermi level.
- Be able to compute the drift velocity from the electric field in a uniform semiconductor and understand conductivity and resistivity.
- Be able to relate the mobility to effective mass and momentum relaxation time and compute the diffusion coefficient given the mobility.
- Understand the drift-diffusion equation, its relation to the general current equation, and what is meant by high-field transport.
- Be familiar with important recombination processes, what low-level injection is, and how to determine the minority carrier lifetime when several processes occur simultaneously.
- Understand thermal generation, impact ionization, and optical absorption in direct and indirect gap semiconductors.

We begin Part 4 with the question: "What causes current to flow?" and develop a general expression that describes the current in a semiconductor.

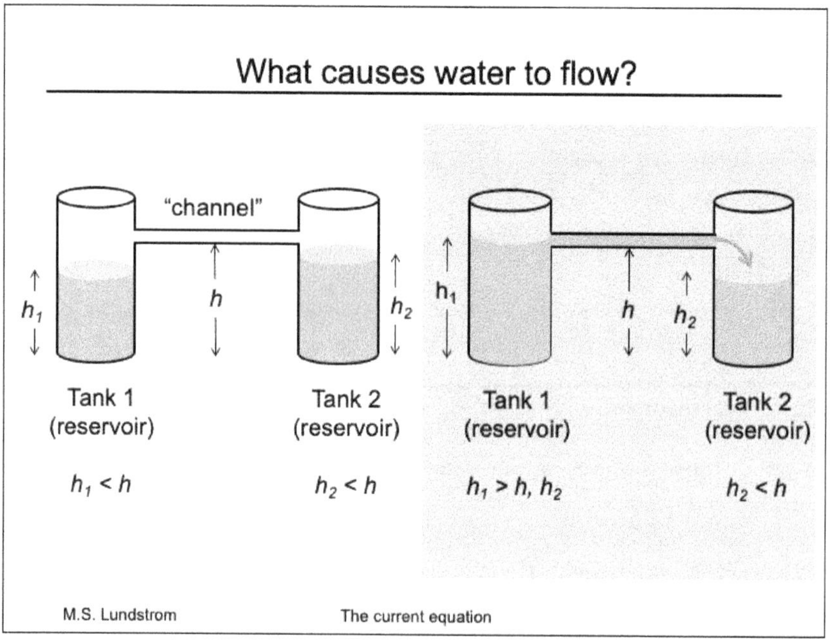

17.2 A hydrodynamic analogy

What causes current to flow? The general concepts can be illustrated with a hydrodynamic analogy, i.e. what causes water to flow between the two tanks shown above?

Consider the case shown above on the left. Both tanks are partially filled, tank 1 to a height, h_1 and tank 2 to a height, h_2 . These tanks are reservoirs of water. There is a pipe, at a height, h, which we will call a "channel" for water to flow between the two tanks when the height of the water in the two tanks allows water to flow. In the case shown above on the left, the height of the water in neither tank reaches the height of the pipe that connects the tanks. Obviously no water will flow between the two tanks. We have water in the tanks, but no flow of water between the tanks.

Now consider the case shown above on the right. In this case, tank 1 is filled to a height that is higher than the connecting pipe (the channel). Water will flow from tank 1 to tank 2. Water will continue to flow only so long as $h_1 > h$ and $h_2 < h$.

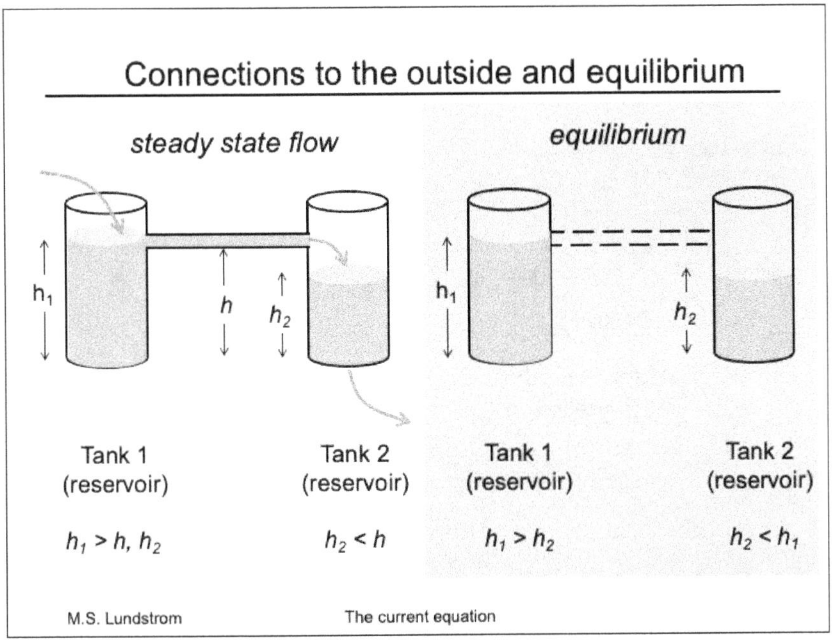

Consider again the case shown on the left above. It is important to note that to maintain the water levels in tank 1 and tank 2 as water flows between them, we have to connect the tanks to the outside world. In the case shown above, we would have to continually put water in the tank 1, and we would have to continually drain water out of tank 2. If the in-flow to tank 1 and the out-flow from tank 2 are identical, there will be a constant height in tank 1, a constant height in tank 2, and a constant, steady state flow of water from tank 1 to tank 2.

Finally, consider the case shown on the right in the slide above where the channel (pipe) connecting the two tanks has been removed and the tanks are not connected to the outside world. In this case, each tank is in **equilibrium** with no flow of what between the tanks and two different heights of water. If the channel is there, but small, the flow of water will be small, and the tanks will be **near equilibrium**. What does this simple hydrodynamic model tell us about electron flow?

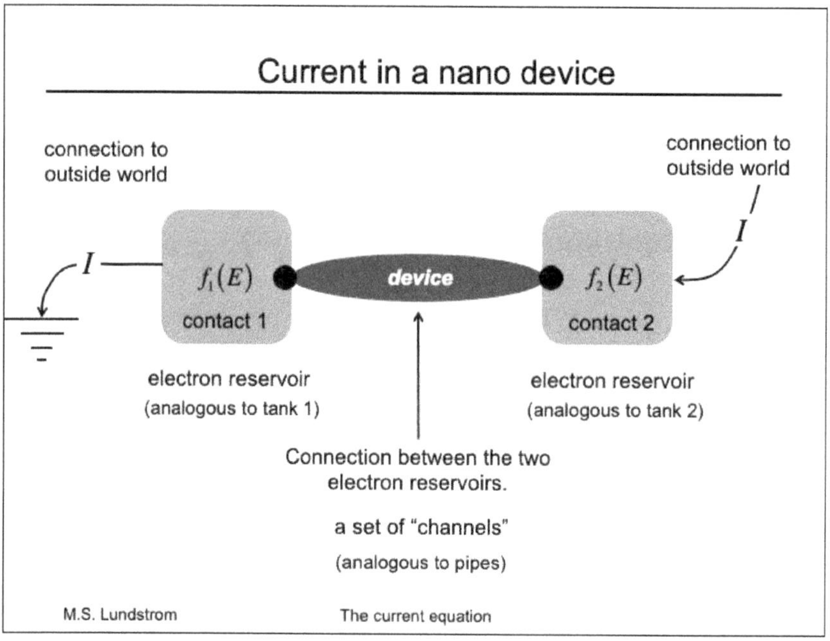

17.3 Current in a nano device

The flow of electrons produces electrical currents. Our goal is to understand the flow of electrons in semiconductors out of equlibrium. In learning chemistry, we begin with the simplest atom — the hydrogen atom and then to proceed to more complex atoms, molecules, compounds, etc. We will take a similar approach; we begin by understanding the current in a very small nanodevice, and then extend that understanding to large, bulk semiconductors. Shown above is a model nanoelectronic device. The device is a set of channels (i.e. the pipes) connected to contacts (i.e. the tanks), which inject electrons into the channel and collect electrons from the channel. The contacts are reservoirs of electrons that are large and metallic and are very nearly in thermal equilibrium. Each contact is characterized by a Fermi function, which describes how the states are filled. The Fermi levels in two the contacts are analogous to the height of the water in the two tanks. The Fermi levels may be different in the two contacts, and there are outside connections to a power supply or battery to maintain steady-state conditions as current flows into contact 2 and out of contact 1.

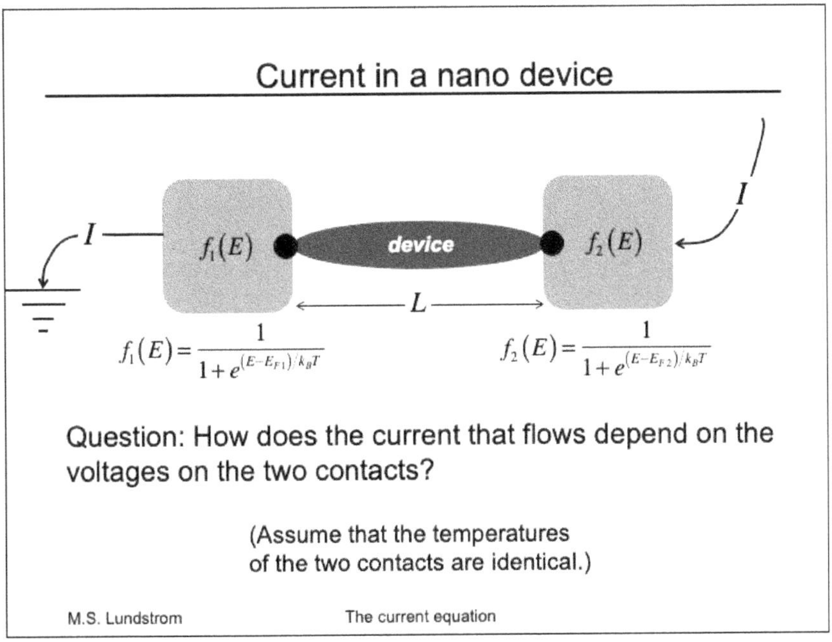

The device we're considering is shown again in the slide above, The length of the device is L. Each of the two contacts to the device is assumed to be very near thermal equilibrium, so each contact can be characterized by an equilibrium Fermi function. The Fermi level in the second contact might be different from the Fermi level in the first contact. The question is: "How does the current that flows depend on the voltages applied to the contacts?"

You might wonder how we can say that the contacts are in equilibrium when current is flowing. The answer is that we assume that the contacts are very large so that the small current is only a small perturbation to equilibrium in the contacts. The device itself, however, is significantly out of equilibrium.

We will assume in these lectures that the temperatures of the two contacts are identical. Temperature differences can also cause current to flow — even with the Fermi levels in the two contacts are the same. This is a **thermoelectric effect**, a slightly more advanced topic that we won't be discussing here. Our goal is to understand **isothermal current flow** in this device.

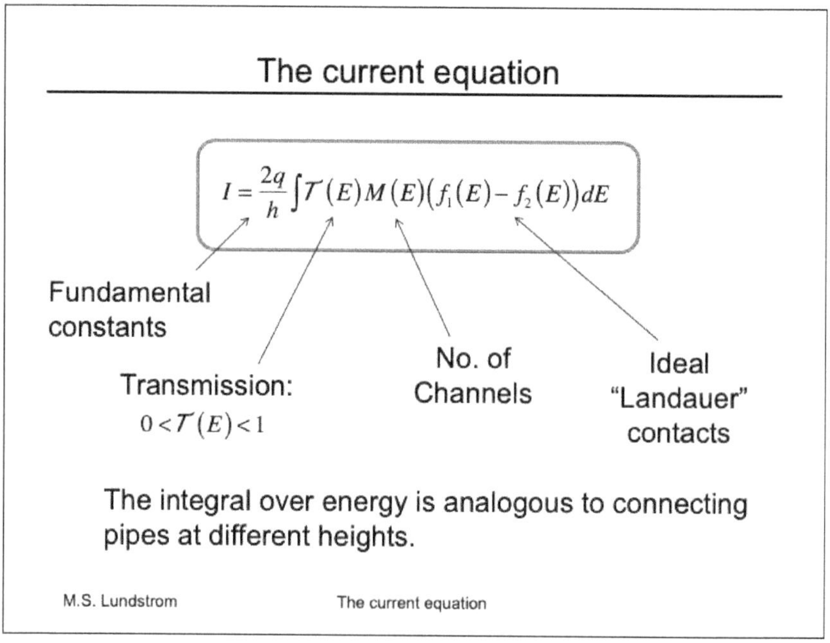

The equation shown above gives the current in the device. We won't derive this expression (it can be done relatively easily from something called the **Boltzmann transport equation**), but it's a simple, intuitive equation that we can readily understand and use. As seen in the slide above, out front we have some fundamental constants, $2q/h$, where q is the magnitude of the charge on an electron, and h is Planck's constant. There is an integral over energy, which is analogous to adding the contributions of a set of pipes at different heights in our hydrodynamic analogy.

Looking inside the integral, there is a parameter, $\mathcal{T}(E)$, called the **transmission**; it's simply a number between zero and one. There is also a parameter, $M(E)$, which describes how the channels (the connecting pipes in the hydrodynamic analogy) are distributed in energy. Finally, there's the difference between the Fermi functions of the two contacts.

This approach to current flow is called the **Landauer approach**, named after the physicist who introduced this way of thinking about current in terms of transmission.

The equation shown above is our starting point for understanding electrical currents in semiconductors. Next, we'll discuss the parameters inside the integral.

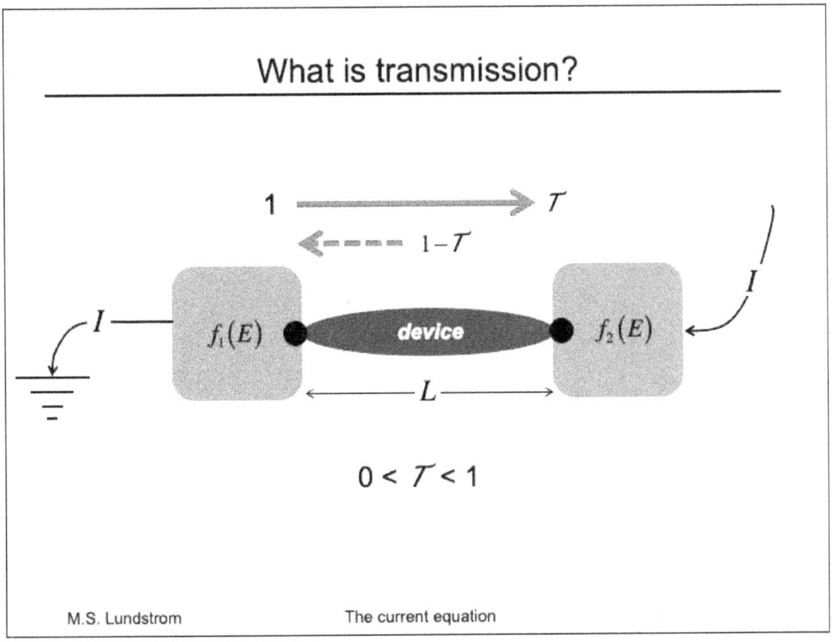

17.4 Transmission

The first parameter inside the integral is the **transmission**. Let's assume, as shown above, that there is a steady state electron flux of magnitude one that enters that enters the device from contact 1. Some fraction of that flux transmits across and enters contact 2. The fraction of the injected flux that crosses the device and enters the other contact is the transmission, $\mathcal{T}$. We're going to assume that electrons don't disappear (recombine) and that they aren't generated in the device, so conservation of electrons tells us that the fraction that doesn't emerge into contact 2 must have back-scattered and returned to contact 1. That fraction is $1-\mathcal{T}$. The transmission is a number between zero and one that tells us the fraction of the injected flux that emerges at the other end of the device.

We should mention that we are assuming ideal (so-called "absorbing") contacts. If an electron is injected from contact 1 and enters contact 2, it cannot reflect and re-enter the device. This is a good assumption for good electrical contacts.

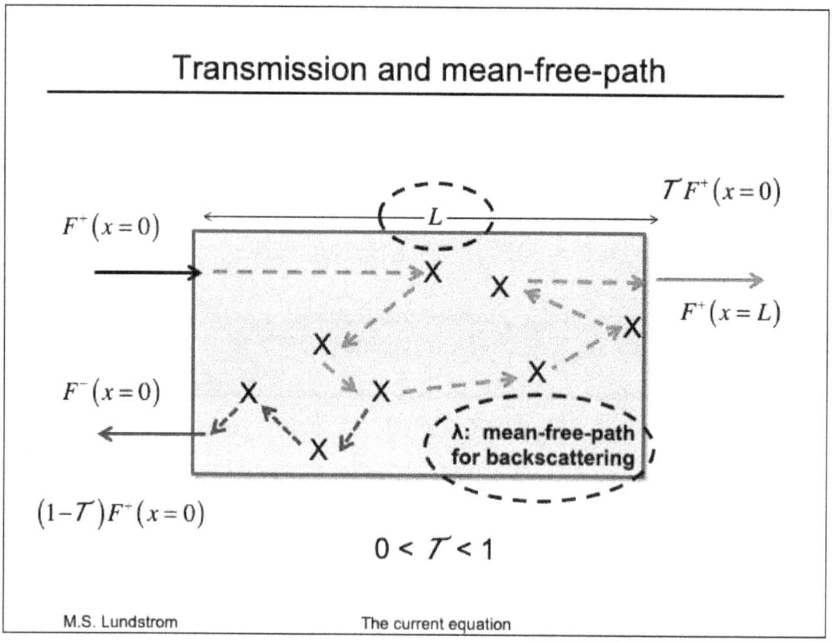

Transmission is a key parameter that we need to understand. Consider the slab of semiconductor slab of length, L, shown above with a flux of electrons, $F^+(x = 0)$, injected at $x = 0$. The average distance between scattering events, which knock the electrons around in random directions, is called the **mean-free-path** (MFP), λ. (Do not confuse this λ with wavelength; we are talking now about semiclassical particles, not waves.)

Electrons injected at $x = 0$ undergo a random walk, scattering off of defects, impurities, lattice vibrations, etc., and the fraction of the injected flux that emerges at $x = L$, is the transmission. It can also happen that the random scattering events might turn the electrons around and they could return to the first contact. That fraction is $1 - \mathcal{T}$.

The key parameters in determining whether the transmission is close to zero, close to one, or in between are the length of the slab, L, and the average distance between backscattering events, λ. The shorter the MFP, the stronger the scattering is, and the lower the probability that an electron will transmit across. For a given MFP, a longer slab will have a lower transmission. Similarly, the longer the MFP, the weaker the scattering is, and the higher the probability that an electron will transmit across. For a given MFP, a shorter slab will have a higher transmission.

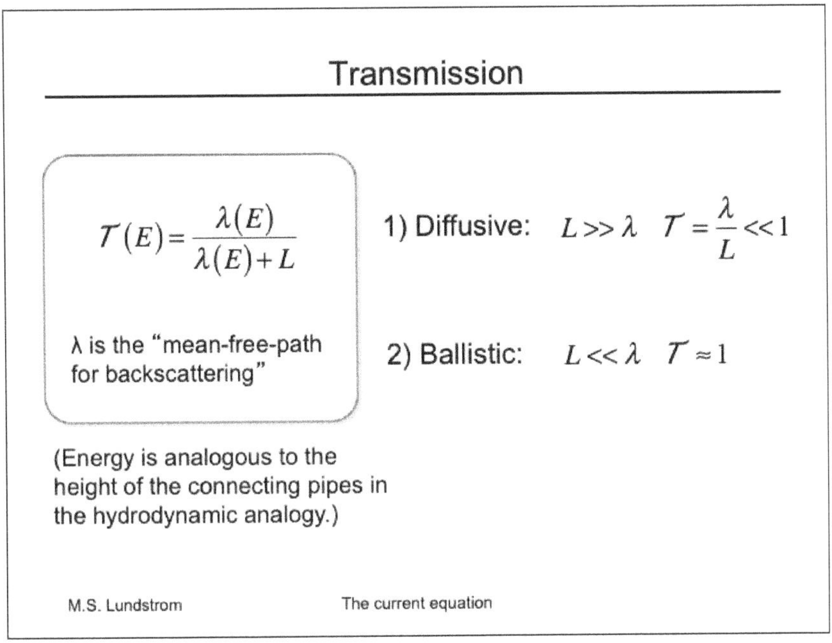

The simple expression shown above relates the transmission to the MFP, λ, and to the length, L. It can be derived from a simplified Boltzmann transport equation, and it generally works pretty well in practice. There are two limits, the **diffusive limit** and the **ballistic limit**.

The diffusive limit occurs when the device is many MFPs long. Under these conditions, the transmission just reduces to MFP divided by the length of the device — that's a small number. The diffusive limit, which applies to long devices, is the limit that will be most interested in.

There is also a ballistic limit. If the device is very short, or the MFP is very long, an electron injected at $x = 0$ will most likely zip right across and enter contact 2. In this limit, you can see from the expression shown above that the transmission is one. This is called **ballistic transport**. Modern day devices often operate between the ballistic and diffusive limits.

The transmission is generally a function of energy. Channels at different energies are analogous to pipes at different heights, and we are assuming that electrons don't change energies (i.e. don't change channels) within the device. This is true for ballistic transport. It is also true for elastic scattering, which doesn't change the energy of electrons, and it is approximately true when small voltages are applied.

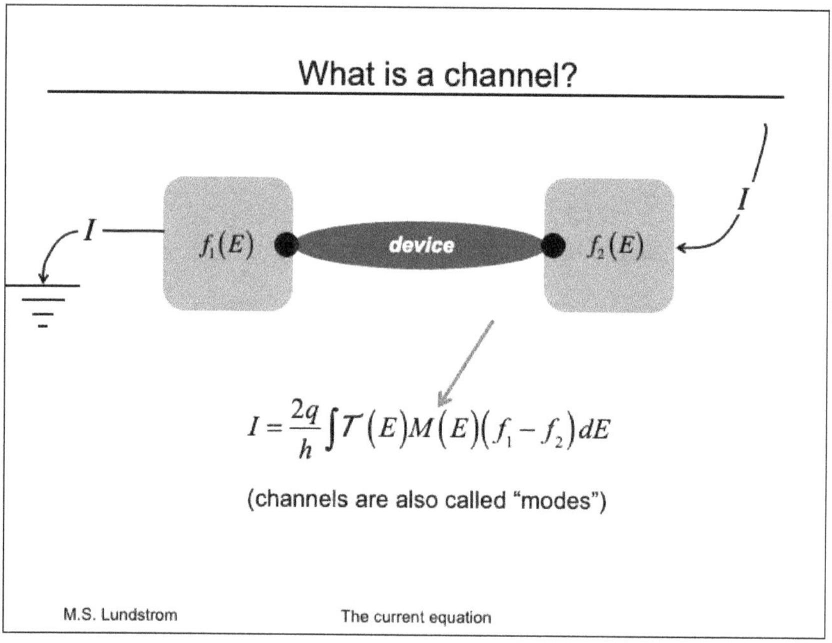

17.5 Channels

Another parameter in our current equation is $M(E)$, the number of channels. Channels are sometimes called **modes**, because they are analogous to the confined modes in an electromagnetic waveguide, each one is a channel. Let's discuss $M(E)$ briefly.

Channels for electron flow are like the pipes connecting the two tanks in our hydrodynamic analogy. From a more fundamental semiconductor perspective, a channel is a state that can hold an electron, but it has to be a state with a velocity, so that electrons can move. A channel is a state with a velocity,

$$M(E) \propto \upsilon^+(E)D(E)\,,$$

where $D(E)$ is the density of states, (the number of states per Joule), and υ^+ is the velocity (in m/s) in the forward direction (contact 1 to contact 2) for the channel at energy, E. We will assume that the velocity in the reverse direction for electrons that backscatter is the same.

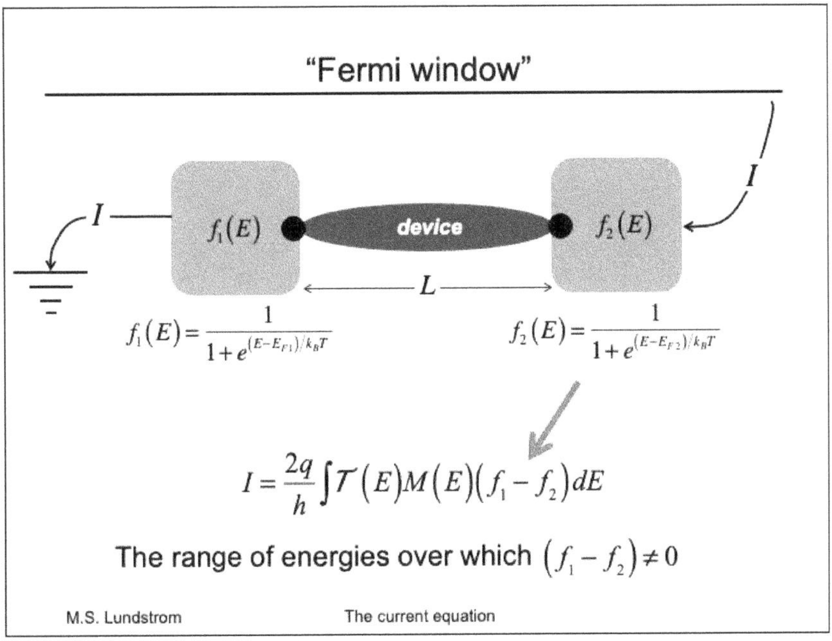

17.6 Fermi window

Finally, let's look at the last parameter inside the integral of our current expression, $(f_1(E) - f_2(E))$ — this is the difference in the Fermi functions of the two contacts. If current is flowing, then for some range of energies, $(f_1(E) - f_2(E))$ is non zero; this is the range of energies over which current flows. This range of energies is called the **Fermi window**, and it plays an important role in our understanding of why currents flow.

We haven't derived our current equation, but it's easy to understand the parameters inside the integral. The integral over energy means that we are simply adding the current contributions from channels at different energies (i.e. energies in the conduction band, or in the valence band). In our hydrodynamic analogy, this is analogous to adding the flows due to separate pipes at different heights. This approximation is exact in the ballistic limit, and for elastic scattering, which does not change the electron's energy. It is approximately true at low biases, even when there is inelastic scattering.

We can now understand, at a simple level, how current would flow in this nano device, and that will give us a deeper understanding of the Fermi window.

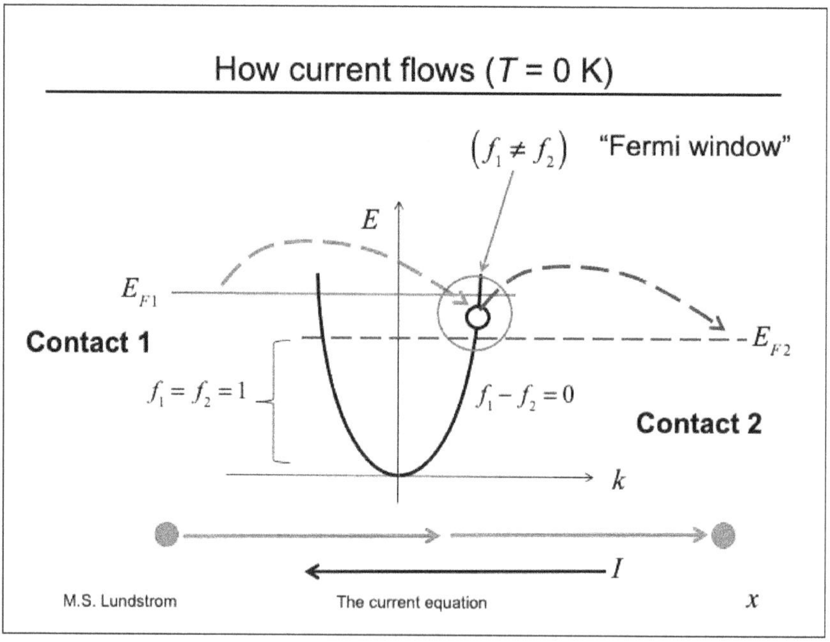

Assume that $T = 0$ K, and that the device is ballistic ($\mathcal{T} = 1$). At $T = 0$ K, the Fermi function is one below the E_F and zero above E_F. Assume that E_{F2} is a little lower in energy than E_{F1}. How does current flow?

Consider an energy below the lowest Fermi level, E_{F2}; the Fermi function is one for both cases, so $(f_1 - f_2) = 0$, and no current flows. Consider any energy above the highest Fermi level, E_{F1}; f_1 and f_2 are both zero, so $(f_1 - f_2) = 0$, and no current flows. The only energies that we need to consider are between the two Fermi levels where $(f_1 - f_2)$ is nonzero.

Consider a state located between E_{F1} and E_{F2} (the circle shown above). This state is below E_{F1}, so contact 1 "says" that it should be filled. An electron comes in from contact 1 and occupies the state, but this state is above E_{F2}, so contact 2 "says" that this state should be empty. The electron flows out of the state into contact 2, and we're back where we started from. This is a steady state process — electrons flow in from contract 1 and fill the state, then electrons flow out of that state to contact 2, and that produces a steady current in the opposite direction because electrons carry a negative charge.

This is how current flows in a small nano-device. **Differences in Fermi levels produce currents.**

Fermi window under small bias

$$I = \frac{2q}{h}\int T(E)M(E)\left(f_1(E) - f_2(E)\right)dE$$

$$f_1(E) = \frac{1}{1+e^{(E-E_{F1})/k_BT}} \approx f_0(E)$$

$$\frac{f_2(E)-f_1(E)}{\delta E_F} \approx \frac{\partial f_0}{\partial E_F} = -\frac{\partial f_0}{\partial E}$$

$$f_2(E) = \frac{1}{1+e^{(E-E_{F2})/k_BT}} \approx f_0(E)$$

$$f_1(E) - f_2(E) \approx -\left(-\frac{\partial f_0}{\partial E}\right)\delta E_F$$

$$E_{F2} = E_{F1} + \delta E_F$$

M.S. Lundstrom The current equation

17.7 Small bias, $T > 0$ K

Consider now the more general case in which $T > 0$ K. We apply a small voltage between the two contacts and find the Fermi window, $[f_1(E) - f_2(E)]$. As shown above, f_1 is an equilibrium Fermi function with its own Fermi level, E_{F1}, and f_2 is also an equilibrium Fermi function, but E_{F2} might be different. We assume that there is only a small difference in the Fermi levels, so $f_1 \approx f_2 \approx f_0$.

Now, let's assume that $E_{F2} = E_{F1} + \delta E_F$. As shown above, $(f_2 - f_1)/(E_{F2} - E_{F1})$ is an approximation to df_0/dE_F. In the Fermi function, E and E_F both appear in the argument of the exponential — the only difference is a negative sign for E_F, so $df_0/dE_F = -df_0/dE$. Finally, as shown in the last equation on the right, we find that $[f_1(E) - f_2(E)]$ is proportional to the difference in the two Fermi levels, δE_F.

The two Fermi levels are different because there are different voltages on the contacts. A positive voltage on a metal plate creates a potential well that pulls electrons down in energy. Similarly, applying a positive voltage to contact 2 lowers E_{F2}.

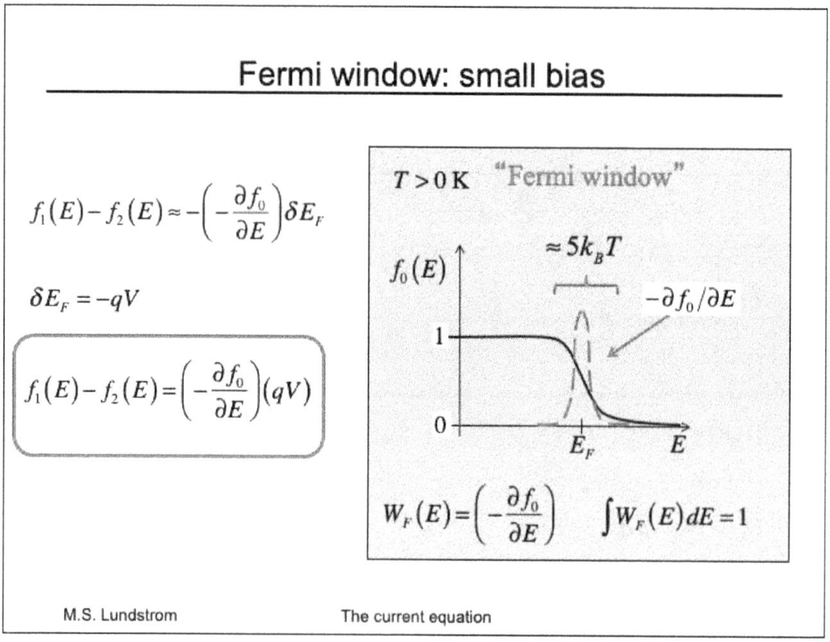

We now have an expression for the **Fermi window** when $(f_1 - f_2)$ is small. We've seen that we can write it as shown above in the first equation on the left. The difference in the Fermi levels in this equation, δE_F, comes from the voltage. A positive voltage to contact 2 lowers the Fermi level in contact 2. We conclude that $(f_1 - f_2)$ is proportional to the voltage that we apply to contact 2.

We know what the Fermi function looks like, so we can easily take the minus derivative of the Fermi function. As shown by the dashed line in the figure above, the derivative is non-zero only around the Fermi level. This is the Fermi window, the range of energies that is important for current flow. We know that the transition from 1 to 0 in the Fermi function is a few k_BT wide, so the width of this derivative is also a few k_BT wide.

The derivative shown above is also called a **window function**, $W_F(E)$, because it is the energy "window" over which current flows. We can show that the area under $W_F(E)$ is 1. As temperature decreases, $W_F(E)$ gets sharper, but the area under $W_F(E)$ is one, so $W_F(E)$ approaches a delta function at the Fermi energy as $T \to 0$ K. Knowing this will be useful to us later. Most semiconductor work is done near room temperature, but there'll be some cases where we'd like to go to $T = 0$ K and see what happens.

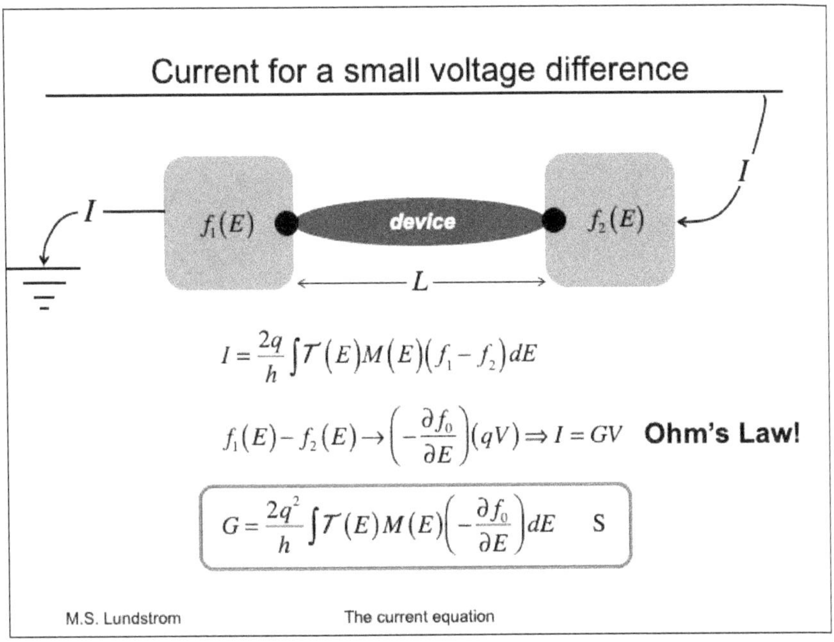

As shown above, the final result is that current is conductance times voltage — something we already knew, but now we have an expression for the conductance. The expression involves two fundamental constants, $2q^2/h$, the energy-dependent transmission $\mathcal{T}(E)$, the number of channels at the energy, E, $M(E)$, and the window function, $W_F(E) = -df_0/dE$, which tells us which energies contribute to the current. The units of conductance are Siemens (amperes/volt). The expressions above for the current at small bias and for the conductance are quite general; they apply to semiconductors, to metals, and to novel materials like graphene. Let's take a closer look now and apply these results to a large bulk semiconductor.

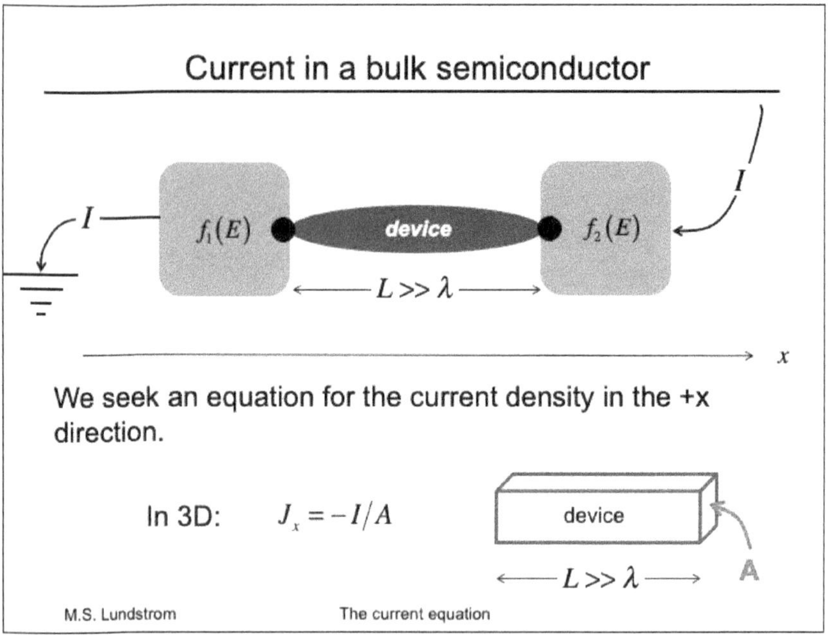

17.8 Current in a bulk semiconductor

Consider a device that is very long, many mean free paths long. In this case, the current is limited by the scattering of electrons within the device; the contacts drop out of the picture. We seek an equation for the electron current in the $+x$ direction, which is opposite to the direction of the current, I flowing into contact 2, the $-x$ direction because electrons carry a negative charge.

Our device is a large, bulk semiconductor with a large cross sectional area, A. (Because the cross-sectional area is large, there are many channels, too many to count.) Since the cross sectional area, A, is arbitrary, we'll be most interested in the current density, the current per unit area in the $+x$ direction, which is minus the current flowing into contact 2 divided by the cross-sectional area, A. With just a little work, we can find the current density in the $+x$ direction.

Current equation in the bulk

$$G = \frac{2q^2}{h}\int \mathcal{T}(E)M(E)\left(-\frac{\partial f_0}{\partial E}\right)dE \qquad \mathcal{T}(E) = \frac{\lambda(E)}{\lambda(E)+L} \rightarrow \frac{\lambda(E)}{L}$$

"diffusive"

$$J_x = -I/A = -GV/A = -\left\{\frac{2q^2}{h}\int \frac{\lambda(E)}{L}\frac{M(E)}{A}\left(-\frac{\partial f_0}{\partial E}\right)dE\right\}V = \sigma(-V/L)$$

$$\sigma \equiv \frac{2q^2}{h}\int \lambda(E)\frac{M(E)}{A}\left(-\frac{\partial f_0}{\partial E}\right)dE \quad \text{(conductivity)}$$

$$\frac{qV}{L} = \frac{E_{F1}-E_{F_2}}{L} = -\frac{dE_F}{dx}\ ?$$

M.S. Lundstrom The current equation

As shown in the gray box above, we begin with our general expression for conductance. The device is long, so the transmission in the diffusive limit is $\mathcal{T} = \lambda/L$. The current density is the current, $-GV/A$; the minus sign is because we want the current density in the $+x$ direction. As shown above, we can write, $J_x = \sigma(-V/L)$, where σ is the **conductivity**. The term, $-V/L$ looks like minus the gradient of the electrostatic potential, the electric field, but we should be more careful.

Remember that the voltage produces a difference in Fermi levels, so we can replace qV/L by $(E_{F1}-E_{F2})/L$, which looks like minus the gradient of the Fermi level. Recall, however, that the Fermi level is an equilibrium concept, and the Fermi level is constant in equilibrium. There is, however, something analogous to a Fermi level that applies out of equilibrium and that quantity can have a slope. That quantity is the **quasi-Fermi level.** In this case, an applied voltage produced a difference in Fermi levels, but differences in carrier density can also produce differences in Fermi levels. The most general way to express the current density is as being proportional to the gradient of the quasi-Fermi level.

Current equation in the bulk

$$J_x = -I/A = -\left\{ \frac{2q^2}{h} \int \frac{\lambda(E)}{L} \frac{M(E)}{A} \left(-\frac{\partial f_0}{\partial E} \right) dE \right\} V$$

$$\frac{qV}{L} = -\frac{dF_n}{dx}$$

$$J_{nx} = \sigma_n \frac{d(F_n/q)}{dx} \qquad \sigma_n = \frac{2q^2}{h} \int \lambda(E)\left(M(E)/A\right)\left(-\frac{\partial f_0}{\partial E} \right) dE$$

σ_n: Conductivity (S/m)

M.S. Lundstrom The current equation

As shown above, we replace the gradient in the voltage along this long semiconductor with the gradient of the quasi-Fermi level, and we end up with a current expression that is simple, general, and powerful.

Assuming the integral over energy is over channels in the conduction band, we use the electron quasi-Fermi level and get expressions for the electron current density, J_{nx}, and the electron conductivity, σ_n.

As shown above, the electron current density is the conductivity times the gradient of the quasi-Fermi level divided by q. The quantity, F_n/q, is the **quasi-Fermi potential**; its gradient has units of volts/m, just like the electric field. So current is proportional to the gradient of something like the Fermi level, but it's the gradient of an analogous quantity that is valid out of equilibrium.

We also now have a general expression for the conductivity that we can evaluate for any material. The units of conductivity are S/m, although semiconductor people would generally quote it in S/cm.

When the integral over energy is over channels in the valence band, we use the hole quasi-Fermi level, F_p, and get expressions for the hole current density, J_{px} and the hole conductivity, σ_p. So what exactly is the quasi-Fermi level?

The quasi-Fermi level or electrochemical potential

"quasi-Fermi level" $J_x = \sigma_n \frac{d(F_n/q)}{dx}$ "electrochemical potential"

The quantity, F_n, is analogous to the Fermi level, but it is defined out of equilibrium and reduces to the Fermi level in equilibrium.

$$n_0 = N_C e^{(E_F - E_C)/k_B T} \rightarrow n = N_C e^{(F_n - E_C)/k_B T}$$

In equilibrium, the current is zero: $J_x = 0 = \sigma_n \frac{d(F_n/q)}{dx} \rightarrow \frac{dE_F}{dx} = 0$

Very important point:
The Fermi level is constant in equilibrium.

17.9 Quasi-Fermi levels

The quasi-Fermi level is analogous to the Fermi level; it is the corresponding quantity out of equilibrium. Sometimes it's called the **electrochemical potential**. The electrochemical potential in equilibrium is the Fermi level and out of equilibrium, it's the quasi-Fermi level.

What, exactly is the quasi-Fermi level? You'll recall that when we evaluated equilibrium electron densities, the important quantity was the location of the Fermi level with respect to the bottom of the band of the conduction band. Out of equilibrium, we have an analogous expression, but the quasi-Fermi level replaces the Fermi level. The corresponding expression for the hole concentration is $p = N_V \exp[(E_v - F_p)/k_B T)]$, where F_p is the hole quasi-Fermi level.

In equilibrium, the electron current is zero. In equilibrium the two quasi-Fermi levels become the one Fermi level, so our current equation tells us that the Fermi level must be constant in equilibrium in order to have zero current. We conclude that **the Fermi level is spatially uniform in equilibrium**.

17.10 Summary

We now have very general and powerful expressions for the near-equilibrium electron and hole currents densities in a bulk semiconductor under low bias.

$$\boxed{\begin{aligned} J_{nx} &= \sigma_n \frac{dF_n/q}{dx} \\ J_{px} &= \sigma_p \frac{dF_p/q}{dx} \,. \end{aligned}}$$

The currents are proportional to gradients of the corresponding quasi-Fermi level. The constants of proportionality are the electron and hole **conductivities**, σ_n and σ_p, and we have a very general expression for the conductivity that we can evaluate for the material of interest,

$$\sigma = \frac{2q^2}{h} \int \lambda(E)(M(E)/A) \left(-\frac{\partial f_0}{\partial E}\right) dE \,.$$

Note that this expression gives the electron conductivity, σ_n, if the integral is over channels in the conduction band and the hole conductivity, σ_p, if the integral is over channels in the valence band. There are some approximations. First, we assume that the cross-sectional area, A, is large, so that $M(E) \propto A$, and, therefore, $M(E)/A$ is independent of area. We assume that the semiconductor is long, so that $L \gg \lambda$. Finally, we assume that the applied bias is low, so that the gradients of the electrochemical potentials are small. These are the assumptions that we will make as we develop the conventional theory of semiconductors, but for high voltages, the equations will be modified.

To summarize, for bulk semiconductor **gradients in the quasi-Fermi level cause currents to flow**. For short devices ($L < \lambda$), **differences in the Fermi levels of the two contacts cause current flow**.

If you've encountered semiconductors before, you may have heard of **drift currents** and **diffusion currents**. As we will discuss in the next two lectures, these are special cases of the more general expression.

17.11 Reference

For a deeper understanding of the transmission approach to current (also called the Landauer approach), see:

[1] S. Datta, *Lessons from NanoElectronics: A. Basic Concepts*, 2nd Ed., World Scientific, 2017.

Lecture 18

Drift Current

18.1 Introduction

In the previous lecture, we developed general expressions for the electron and hole currents in terms of the gradients of the quasi-Fermi levels. Semiconductor engineers frequently talk about **drift currents** and **diffusion currents**. The point of this lecture and the next is to discuss the drift and diffusion currents and relate them to the more general approach that we discussed in the previous lecture.

This lecture is about the drift current, which is the current in the presence of an electric field when the carrier density is uniform. There will be some important parameters that we will introduce. One is called the **mobility** and another is the **conductivity**. We actually saw the conductivity in the previous lecture where we developed a general expression for current in terms of the material parameters of the semiconductor. We'll present a simpler expression for the conductivity in this lecture. We'll also introduce a third quantity called the **resistivity**, which is simply one over the conductivity.

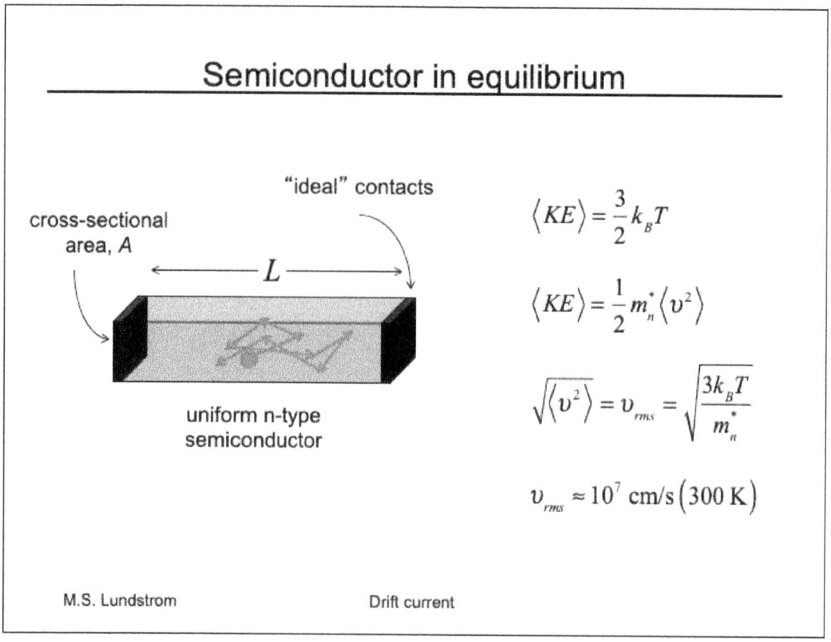

18.2 Near-equilibrium, diffusive transport

In equilibrium, electrons bounce around in random thermal motion. They're knocked here and there when they encounter an impurity, a lattice vibration, a crystal defect, etc. Electrons zip around in random directions with no average velocity in any given direction. No current is flowing.

Electrons in the conduction band of a nondegenerate semiconductor may be considered to be an ideal gas of particles for which, you may recall, particles are in random thermal motion with an average kinetic energy of $3k_BT/2$, where k_B is Boltzmann's constant and T is the temperature in Kelvin. The average kinetic energy is also one-half mass times the average velocity squared, $m^* \langle \upsilon^2 \rangle /2$. If we equate those two expressions for average kinetic energy, we find the **root mean square thermal velocity**. An expression for the rms thermal velocity is shown above. The lighter the effective mass is, the higher the velocity. The higher the temperature is the higher the velocity. For typical semiconductors, we find that this velocity is roughly 10^7 cm/s, which is quite high. Equilibrium electrons zip around in random thermal motion at a very high average speed, but there is no current flow because there's no average velocity in any direction.

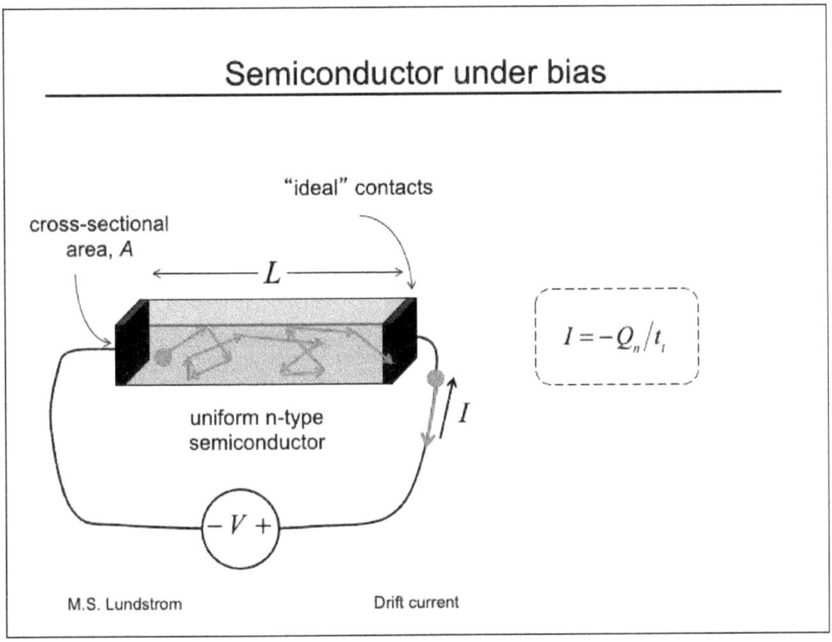

Now what happens if we apply a small voltage across the semiconductor? The positive voltage on the right contact will attract electrons, and electrons will move from the left to the right across the semiconductor. They're still bouncing around in random thermal motion, but now there's a small additional component superimposed upon the large and random thermal velocity. The ensemble of electrons slowly moves from the left to the right. We say that the electrons **drift** in the electric field, and the average velocity at which they drift from the left to the right is called the **drift velocity**.

To find the current, we remember that the definition of current is charge divided by time. How much mobile charge, Q_n, is in the semiconductor? How much time, t_t, does it take for this charge to flow out of the semiconductor? The minus sign in the equation above is due to the fact that we're defining current to be positive when it flows *into* contact 2. Note that Q_n is not the total charge (which includes the charge of the ionized donors), but, rather, only the charge due to the mobile electrons that carry the current. The sign of Q_n is negative.

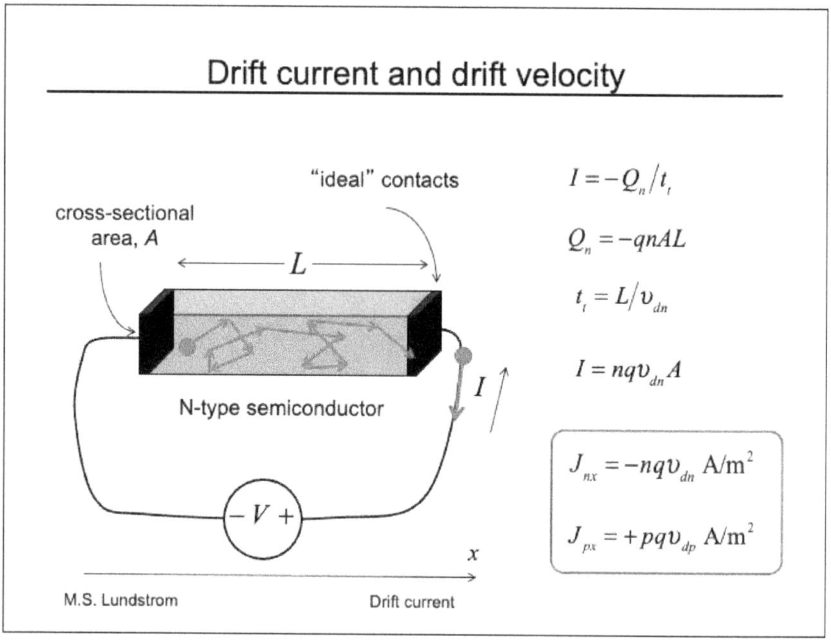

18.3 Drift current and drift velocities

The charge in the semiconductor is $-q$, times n, the electron density, times the volume of the semiconductor, AL. The charge is negative because we are dealing with negatively charged electrons in this N-type example.

How long does it take for the mobile charge to move out of the semiconductor? The **transit time**, t_t, is the length of the slab divided by the average velocity at which the electrons are moving. We end up with a very simple expression for the current. The more electrons in the slab, the larger the current, the faster they're moving, the larger the current, and the bigger the cross sectional area, the larger the current.

The cross sectional area in this example is arbitrary, so we divide by A to find a simple expression for the electron current density. If we had a P-type semiconductor with holes, everything would be the same except there would a positive sign instead of a negative sign, and the average velocity at which the holes drift replaces the average velocity at which electrons drift.

As shown in the red box above, we end up with two simple expressions for the drift current densities. Next, we need to relate the average velocities to the electric field or to the voltage across the device.

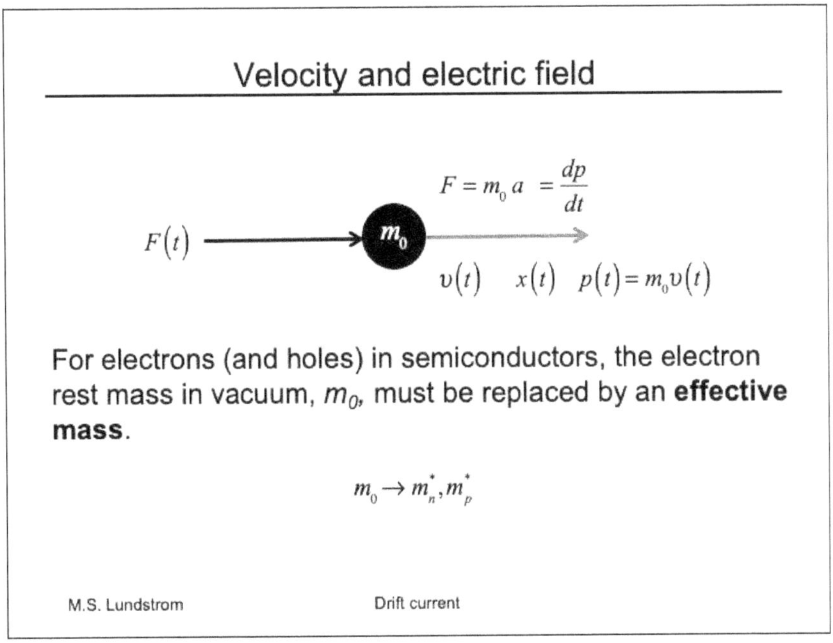

To relate the drift velocity to the electric field, recall that according to Newton's law, force is mass times acceleration; we can also write force as dp/dt, where p is the momentum of the particle.

If we apply a force to the particle, we can find the velocity, momentum, and position as a function of time. Since we're in a semiconductor, we simply replace the rest mass of an electron in vacuum, m_0, by the effective mass, m^*, and then we can use Newton's laws.

In equilibrium, electrons are in random thermal motion with a distribution of velocities (and, therefore, momenta) that can be found from the Fermi function in the non-degenerate (Maxwellian) approximation:

$$\begin{aligned} f_0 = e^{(E_F - E)/k_BT} &= e^{(E_F - E_c)/k_BT} \times e^{-p^2/2m^* k_BT} \\ &= e^{(E_F - E_c)/k_BT} \times e^{-m^* \upsilon^2/2k_BT}, \end{aligned}$$

where we have assumed a parabolic band semiconductor, so $E = E_c + p^2/2m^*$ and $p = m^*\upsilon$. When we apply a force (electric field), there is a small, average velocity (momentum) in the direction of the force. It is this drift velocity, υ_d, (or drift momentum, p_d) that we need to calculate.

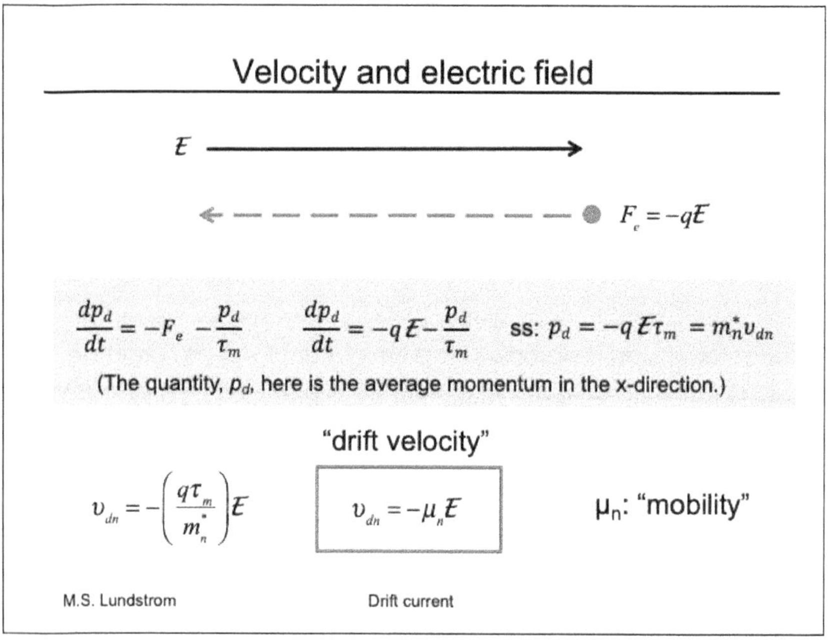

Let's assume an electric field pointing in the $+x$ direction, from the left to the right. The force on electrons, $-q$ times the electric field, is in the opposite direction because the charge on an electron is negative. As shown above in the gray box, Newton's law says that the time rate of change of momentum is equal to the force on the electrons. We write this equation for an electron with the average momentum in the direction of the force, p_d. The additional term on the right hand side is like friction, it slows the electron down. We write this frictional term as -p_d/τ_m The quantity, τ_m, the **momentum relaxation time**; p_d/τ_m is the rate at which the momentum along $+x$, which was acquired from the electric field, is lost due to scattering processes that randomize momentum. After the initial acceleration, which is very quick, what is the steady-state electron velocity?

To find the steady-state average momentum (and velocity), we set $dp_d/dt = 0$. The important result shown in the red box is that **the drift velocity is proportional to the electric field**. We'll lump all the constants together into a very important parameter called the **electron mobility**. So the velocity is minus mobility times electric field. The minus sign comes from the negative charge on the electron. As seen above, the electron mobility depends on the scattering time and the effective mass.

Drift velocities in Si

$$\upsilon_{dn} = -\mu_n \mathcal{E} \qquad \mu_n = 1360 \frac{\text{cm}^2}{\text{V-s}}$$

$$\upsilon_{dp} = +\mu_p \mathcal{E} \qquad \mu_p = 480 \frac{\text{cm}^2}{\text{V-s}}$$

pure (undoped) Si at 300 K

Assume a modest strength electric field $\mathcal{E} = 100 \frac{\text{V}}{\text{cm}}$

$$\upsilon_{dn} = -1.4 \times 10^5 \frac{\text{cm}}{\text{s}} \qquad \upsilon_{dp} = +0.5 \times 10^5 \frac{\text{cm}}{\text{s}} \qquad \upsilon_{rms} \approx 1 \times 10^7 \frac{\text{cm}}{\text{s}}$$

Drift velocities are much lower than the thermal velocity.

M.S. Lundstrom Drift current

The electron drift velocity is proportional to the electric field, and so is the hole drift velocity, but holes drift in the opposite direction because they have a positive charge. The mobilities are well known and have been measured carefully in silicon and in other important semiconductors. For intrinsic Si, the electron mobility is almost three times higher than the hole mobility. Let's assume a modest electric field of 100 volts per centimeter and see what the drift velocities are. If we plug numbers in, we find that the electron drift velocity is about 1.4×10^5 cm/s, and the hole drift velocity is about 0.5×10^5 cm/s. These velocities are about 100 times smaller than the rms thermal velocity, which is about 10^7 cm/s.

To summarize, the charge carriers in semiconductors zip around in random thermal motion with a thermal velocity that is about 10^7 cm/s. If there is a voltage applied across the semiconductor, then superimposed on the very high random thermal velocity is a small average velocity that causes the charge carriers to drift at an average velocity we call the **drift velocity**. The drift velocity is proportional to the electric field. For small or modest electric fields, the drift velocity is much smaller than the thermal velocity.

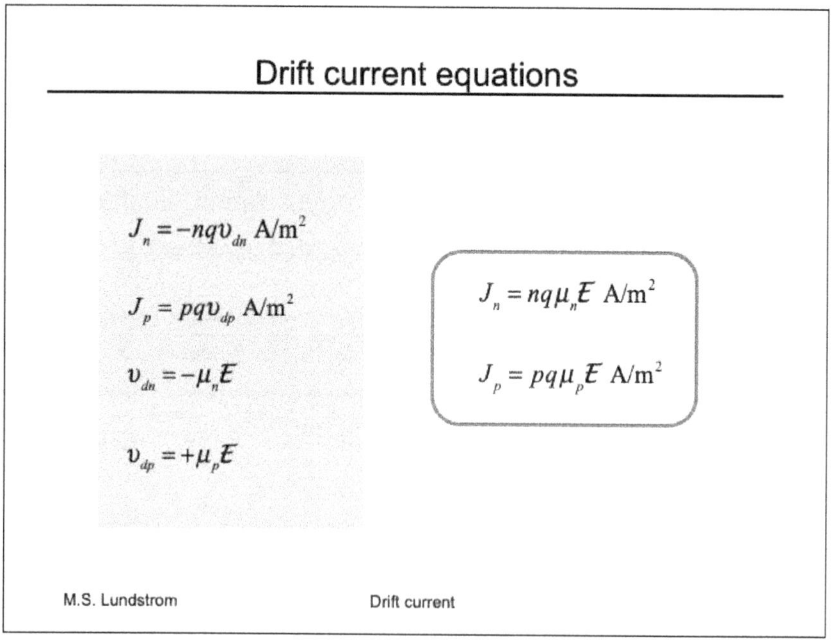

As shown above in the gray box, we have two drift current equations, one for electrons and one for holes. Remember that we deduced the current equations and found that the electron and hole currents were proportional to the electron and hole densities and to their drift velocities. As also seen in the gray box, the drift velocities are proportional to the electric field. The constant of proportionality is the mobility for holes and electrons. If we put these equations together, we get the very important equations shown above on the right. These equations describe the drift currents in a semiconductor, the current due to the presence of an electric field. The mobility is a very important semiconductor parameter. The higher the mobility, the higher the drift velocity, which gives a higher current.

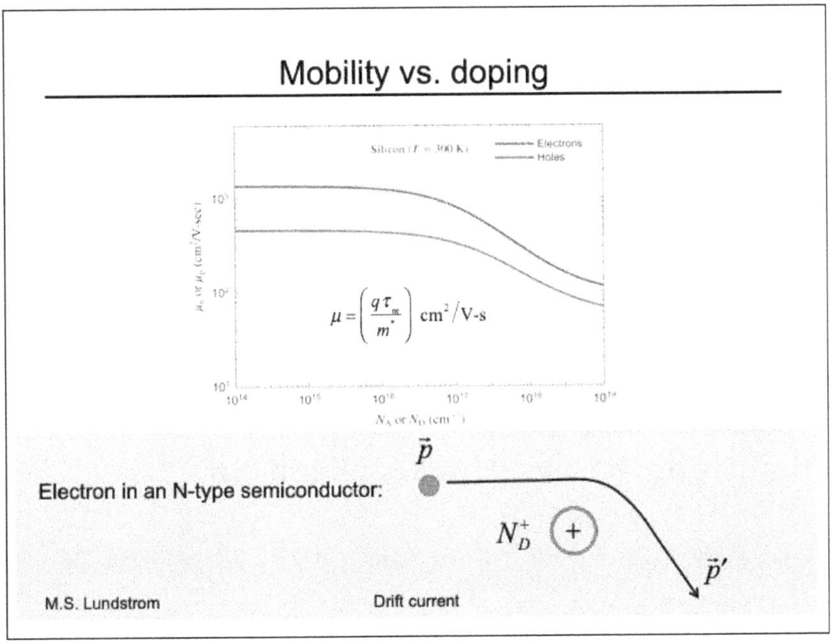

18.4 Mobility

Mobility is a key parameter. The plots above show the mobility of electrons and holes in silicon at room temperature as a function of the doping density (after [1]). For lightly doped Si at the left of the plot, the electron mobility is about 1360 $\text{cm}^2/\text{V}-\text{s}$, and the hole mobility is about 460 $\text{cm}^2/\text{V}-\text{s}$. You can see, however, that as we dope the semiconductor more and more, we don't just get more and more carriers, we also lower their mobility. This happens because of something called **ionized impurity scattering**.

Dopants do more than simply donate an electron to the conduction band or a hole to the valence band. When ionized, dopants introduce a charged defect into the semiconductor. As shown on the bottom in the slide above, when an electron moves near an ionized (charged) impurity, it will be deflected. The velocity in the x direction can be lowered by this deflection. This is called ionized impurity scattering. The more impurities we put in, the more chances that electrons will encounter a charged impurity and be deflected. Increased concentrations of ionized dopants lower the scattering time, and that lowers the mobility.

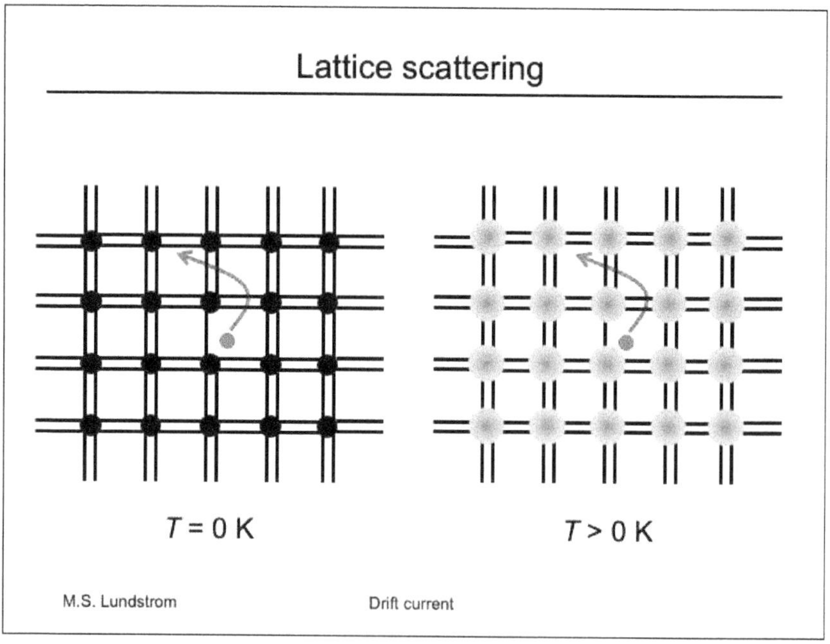

There are other ways that the electron can be scattered; one of them is **lattice scattering**. Shown above at the left is our cartoon of a silicon lattice at $T = 0$ K. Ignoring zero-point energy, each atom is in a particular location in the crystal lattice. At finite temperatures, atoms in the crystal have a kinetic energy of $3k_BT/2$. They're vibrating in random thermal motion, so their locations are not precise. As shown above on the right, the spacing between atoms changes as the atoms vibrate. The changing lattice spacing changes the band gap a little, and that interaction with the electrons (the deformation potential) knocks them in random directions. As the temperature increases, the magnitude of the lattice vibration increases, which increases scattering, decreases the scattering time, and lowers the mobility.

Electrons can also scatter off of other electrons, but since these processes just transfer momentum from one electron to another, the overall momentum of the ensemble of electrons is conserved. As a result, electron-electron scattering tends to be less important.

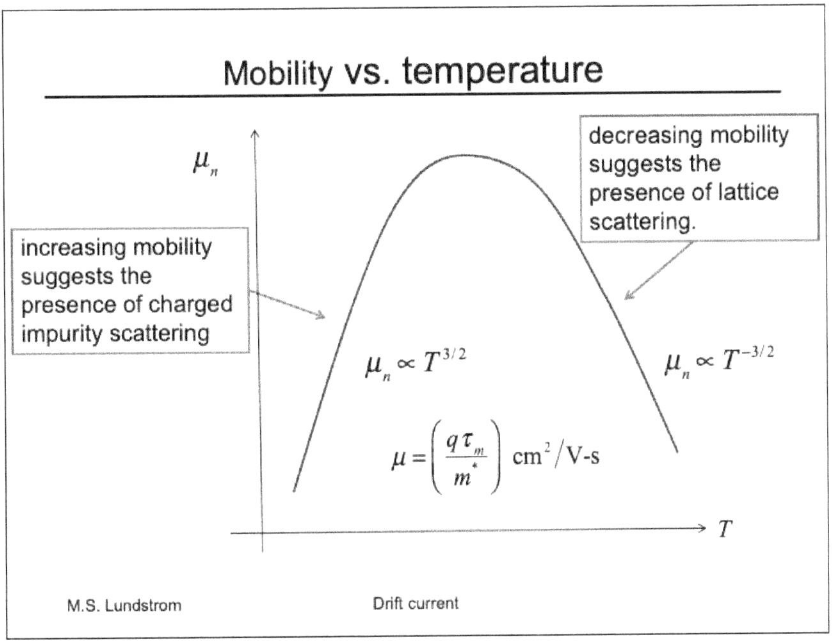

The measured mobility vs. temperature for a semiconductor generally looks as sketched above. At low temperatures, the lattice vibrations are not important, but what is important is ionized impurity scattering. As we increase the temperature, the thermal velocity increases. Electrons zip past charged impurities more rapidly, they're deflected less so the mobility increases as we increase the temperature. In simple semiconductors, the mobility increases as $T^{3/2}$ when ionized impurity scattering dominates.

As the temperatures increases, the mobility reaches a maximum and then decreases. It starts to drop because the lattice vibrations begin to dominate. The stronger the lattice vibrations are, the more they knock electrons around, and the lower the mobility is. In simple semiconductors, the mobility decreases as $T^{-3/2}$ when lattice scattering dominates.

If measurements show a mobility that increases with temperature, it suggests that charged impurity scattering dominates. Conversely, a mobility that decreases with increasing temperature suggests that lattice scattering dominates.

Measuring the mobility vs. temperature tells semiconductor technologists a lot about a semiconductor.

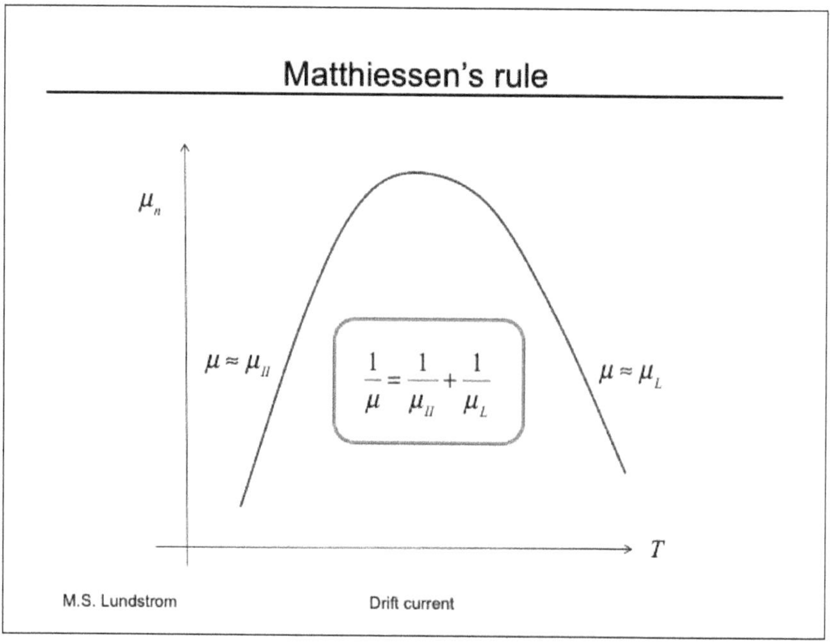

In general, a semiconductor will be at a temperature where both scattering mechanisms are important. How do we account for both of them? The answer, as shown above, is that the inverse of the total mobility is the sum of the inverses of the individual mobilities as if they were acting alone. You can see that if one of them acting alone has a very high mobility, it drops out, and the lower of the two mobilities dominates the overall mobility. This prescription for combining mobilities acting alone is called **Matthiessen's rule** and can be expanded to more than two scattering mechanisms. Another way to understand Matthiessen's rule is to think of scattering probabilities. The probability per second of scattering is one over the scattering time. So if we have two independent scattering mechanisms operating at the same time, we should add scattering probabilities. Mobility is proportional to scattering time, which means that we should add inverse mobilities.

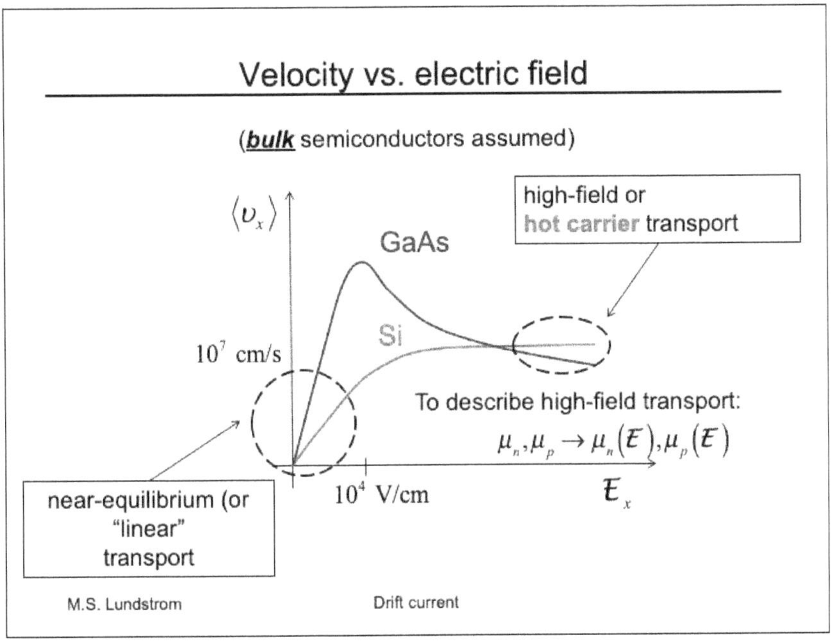

18.5 High-field transport

The average velocity vs. electric field is sketched above for electrons in Si and GaAs. In **near-equilibrium** or **linear transport**, the electric field is small, and the average velocity is proportional to the electric field. As electric fields increase, electrons gain kinetic energy and find more ways to scatter, so the mobility decreases. For large electric fields, the velocity saturates at about 10^7 cm/s for silicon.

Other semiconductors can display distinctly different velocity vs. electric field characteristics. For example, in GaAs, the effective mass of electrons is very light, so the mobility is much higher than for Si. For very high electric fields, the velocity in GaAs saturates at roughly 10^7 cm/s, as it does in Si, but the velocity vs. field characteristics for GaAs has a complex shape that will be described in the next slide.

Under high electric fields, electrons have a lot of kinetic energy, and the velocity vs. electric field characteristic becomes nonlinear. If we write the kinetic energy as $3k_BT_e/2$, where T_e is the **electron temperature**, then the electron temperature is "hot." We call this **hot carrier transport**.

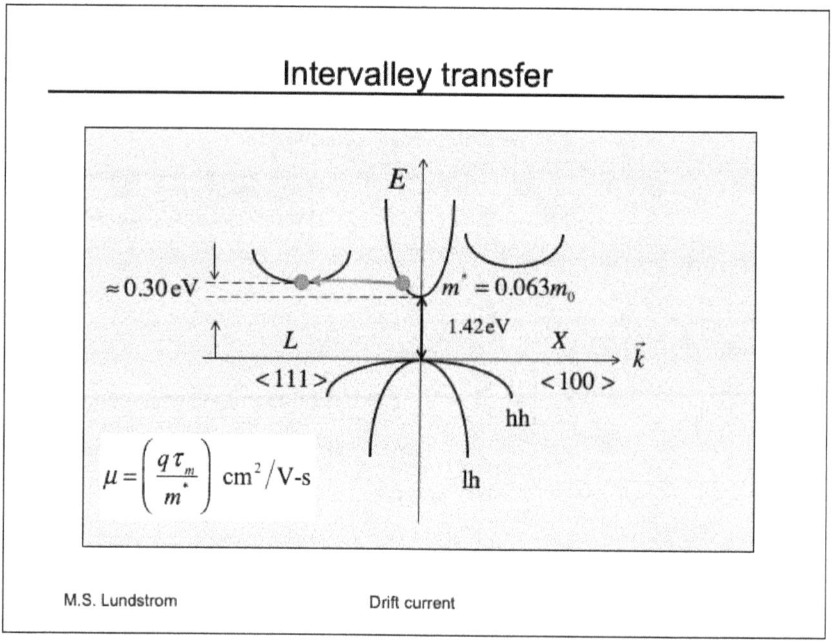

For Si, increasing electric fields increase the electron energy, which increases the scattering rate. When the field exceeds about 10^4 V/cm in Si, the drift velocity saturates because increases in velocity due to the electric field are balanced by decreases in velocity due to increased scattering.

The unusual velocity vs. electric field characteristic for electrons in GaAs is explained by the band structure. As shown above, GaAs is a direct band gap semiconductor with a very light effective mass for electrons, but there are also **satellite valleys** at higher energies that have larger effective masses. Because the effective mass of the lowest valley is small, the mobility of electrons in GaAs is very high, but when the electric field is strong enough, electrons gain energy and are well above the bottom of the conduction band and also above the minimum energy of the satellite valleys. A scattering event can then transfer electrons to one of the satellite valleys where the effective mass is much larger. There are also more ways for an electron in one of these satellite valleys to scatter among the different upper valleys, so the mobility plummets when this **intervalley transfer** occurs. Intervalley transfer explains the unusual velocity vs. electric field characteristic of GaAs, which can be exploited to build microwave oscillators.

Current, conductivity, resistivity

$$J_{nx} = nq\mu_n \mathcal{E}_x \ \text{A/cm}^2 \qquad J_{nx} = \sigma_n \mathcal{E}_x \ \text{A/cm}^2 \qquad \sigma_n = nq\mu_n \ (\text{units?})$$

$$J_{px} = pq\mu_p \mathcal{E}_x \ \text{A/cm}^2 \qquad J_{px} = \sigma_p \mathcal{E}_x \ \text{A/cm}^2 \qquad \sigma_p = pq\mu_p$$

$$J_x = J_{nx} + J_{px} = \left(\sigma_n + \sigma_p\right)\mathcal{E}_x = \sigma \mathcal{E}_x \ \text{A/cm}^2$$

$$J_x = \sigma \mathcal{E}_x \ \text{A/cm}^2 \qquad \mathcal{E}_x = \frac{1}{\sigma} J_x = \rho J_x \ \text{V/cm}$$

$$\rho = \frac{1}{\sigma} = \frac{1}{\sigma_n + \sigma_p} = \frac{1}{nq\mu_n + pq\mu_p} \ \Omega\text{-cm}$$

M.S. Lundstrom Drift current

18.6 Current, conductivity, and resistivity

Shown above is the expression for the electron drift current. We can lump n, q, and μ_n together and recognize that as the conductivity, σ_n, of the semiconductor. You can work out the units and convince yourself that the units are Siemens per meter. We'll usually quote results in Siemens/cm. In general, there will be both electron and hole currents, and the total current is simply the sum of the two. Alternatively, we could add the electron and hole conductivities and write the total current as total conductivity times electric field.

There is also an electromagnetic displacement current that should be added. The displacement current is proportional to the time rate of change of the electric field. For very high-speed devices, this term should be included, but for most semiconductor problems, the displacement current can be neglected, as we have done here.

Instead of expressing the current as conductivity times electric field, we can solve for the electric field and find that it is one over conductivity times the current density. One over conductivity has a name, it's called the **resistivity**. The conductivity and resistivity are inverses of each other.

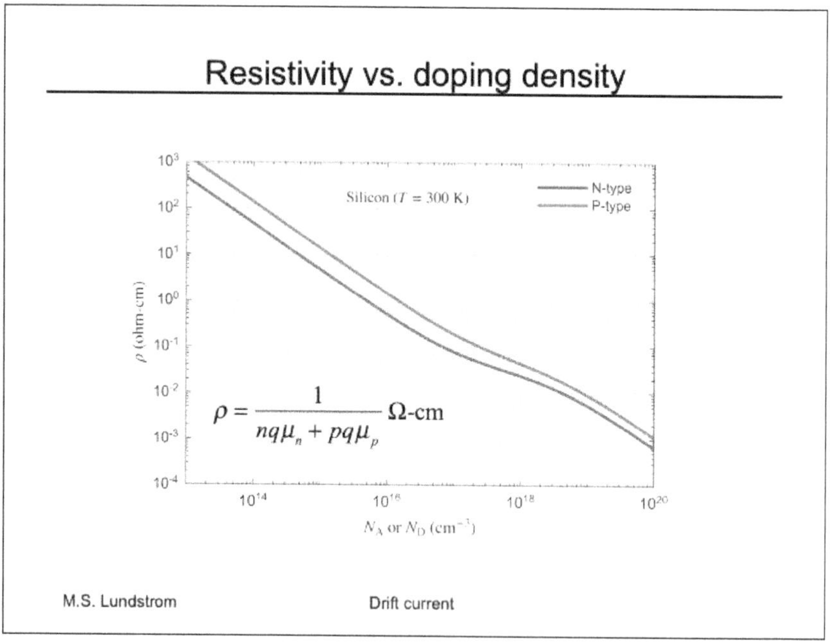

Shown above is a plot of the resistivity vs. doping density for electrons and holes in silicon (after [1]). As the plot above shows, the resistivity does not drop linearly as we increase the doping density. At a given doping density, the resistivity is higher than might have expected because, as we have seen, when we dope the semiconductor more heavily, the mobility decreases. For well characterized materials like Si, if we need the resistivity at a particular doping density, we can easily look it up.

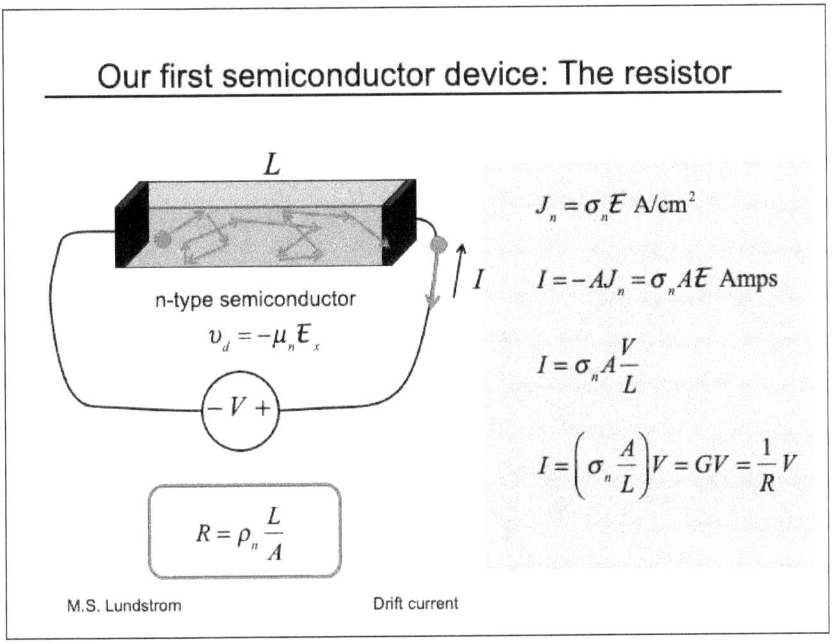

Let's discuss a semiconductor device, a simple resistor. What is the current when we apply a voltage across the resistor? We begin with our current equation. Since we want the current itself not the current density, we multiply by the cross-sectional area. We also recognize that the electric field is voltage divided by the distance between the two contacts, L. When we lump these parameters together, we find that the current is proportional to voltage. The quantity in the parentheses in the last equation in the gray box above is the **conductance**, or one over the **resistance**. As shown in the equation on the lower left, the resistance of this resistor is resistivity (a material parameter) times length, L, divided by cross-sectional area, A. This result makes a lot of sense. The more resistive the material is, the higher its resistivity and the higher the resistance. The longer the resistor is, the harder it is for current to flow through it, so the higher the resistance is. The fatter the resistor is (the bigger the cross sectional area) the more parallel paths for the current flow there are, so the resistance is lower. The IV characteristic of our simple semiconductor device is readily understood.

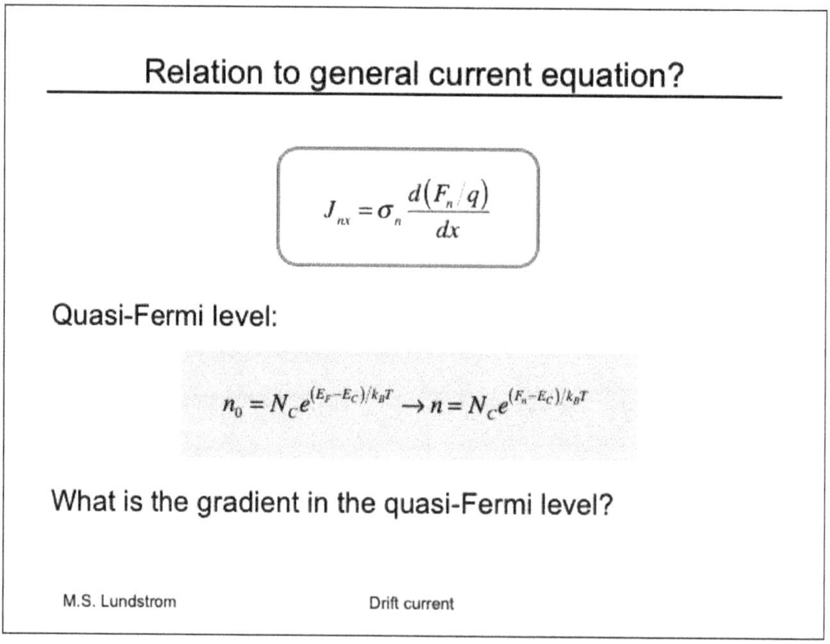

18.7 Relation to the general current equation

Before we wrap up this lecture, we should ask: "What does this lecture have to do with the previous lecture?" In the previous lecture, we presented a very general expression for current, and now we seem to be talking about a different way that current can flow. Well, not really...

Let's begin with our general expression for the near-equilibrium current in a bulk semiconductor from Lecture 17,

$$J_{nx} = \sigma_n \frac{d(F_n/q)}{dx},$$

and see how it is related to the drift current. What's important in the general expression is the gradient of the quasi-Fermi level. Remember that the quasi-Fermi level is analogous to the Fermi level, so we can write it in terms of the electron density, as shown in the gray box. We can solve that equation for the quasi-Fermi level, take its gradient, and see what happens.

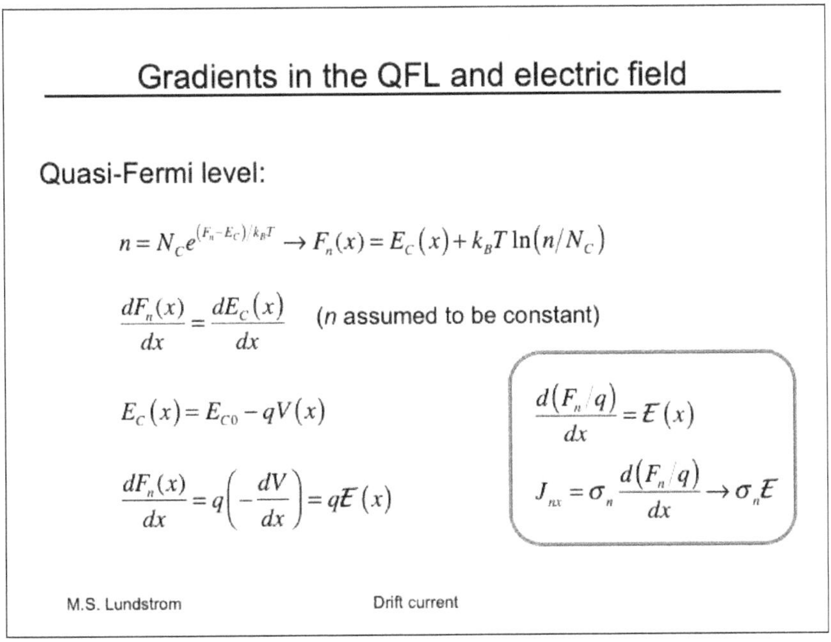

We begin with the equation that relates the electron density to the quasi-Fermi level (QFL), and solve for the QFL. The result is shown on the first line above. Next, take the gradient of the QFL. (We assume here that the electron density is spatially uniform.) We see that the derivative, dF_n/dx, is just the derivative of the bottom of the conduction band.

The bottom of the conduction band moves up and down as the electrostatic potential increases and decreases. A positive voltage lowers the bottom of the conduction band, so when we take the slope, we're just taking minus the gradient of the electrostatic potential, but this is the definition of the electric field. We conclude that the gradient of the quasi-Fermi potential is q times the electric field. Returning to our general current expression for bulk semiconductors, we see that

$$J_{nx} = \sigma_n \frac{d(F_n/q)}{dx} \rightarrow J_{nx} = \sigma_n \mathcal{E}_x \, .$$

This important result shows us that when the general expression for current flow is examined under conditions where the electron density is constant, then it reduces to the drift current, which is proportional to the electric field. **The drift current is a special case of the more general current equation.**

18.8 Summary

The key take away from this lecture is that the general current equation for bulk semiconductors under near-equilibrium conditions simplifies when the carrier density is uniform,

$$J_{nx} = \sigma_n \frac{d(F_n/q)}{dx} \rightarrow J_{nx} = \sigma_n \mathcal{E}_x \,.$$

Instead of the current being proportional to the gradient of the quasi-Fermi level, it is proportional to the electric field. The constant of proportionality is the conductivity, which we related to the material parameters, mean-free-path and number of channels in the previous lecture as

$$\sigma = \frac{2q^2}{h} \int \lambda(E)(M(E)/A) \left(-\frac{\partial f_0}{\partial E} \right) dE \,.$$

Note that this expression gives the electron conductivity, σ_n, if the integral is over channels in the conduction band and the hole conductivity, σ_p, if the integral is over channels in the valence band.

In this lecture, we wrote the electron conductivity as

$$\sigma_n = nq\mu_n = 1/\rho_n \,,$$

where n is the electron concentration, μ_n is the electron mobility, and ρ the resistivity. This expression is commonly used in semiconductor work because the conductivity and carrier density are readily measured, so it is possible to deduce the mobility experimentally from $\mu_n = \sigma_n/(nq)$. The best way to compute the mobility is to equate these two expressions and solve for the mobility in terms of the material parameters. There are analogous equations for the hole current density and the hole conductivity. Both electrons and holes carry current, so in general, we'll have both electron and hole drift currents. The total current is the sum of the two currents, and the total conductivity is the sum of the electron and hole conductivities.

The drift current discussed in this lecture is a special case of the more general current equation, but gradients in the carrier density can also cause current the flow. That is the subject of the next lecture.

18.9 Reference

For much more discussion about the drift current in semiconductors, see:

[1] R.F. Pierret, *Advanced Semiconductor Fundamentals*, (2nd Ed.), Pearson Education, Inc., 2003 (see Ch. 6, Sec. 6.1).

Lecture 19

Diffusion Current

19.1 Introduction

Let's continue our discussion of electric currents in semiconductors. We've talked about the general current equation and about the drift current, which is a special case. In this lecture we'll discuss the diffusion current, another special case.

We saw in the last lecture that when the carrier densities are spatially uniform,

$$J_{nx} = \sigma_n \frac{d(F_n/q)}{dx} \rightarrow J_{nx} = \sigma_n \mathcal{E}_x$$
$$J_{px} = \sigma_p \frac{d(F_p/q)}{dx} \rightarrow J_{px} = \sigma_p \mathcal{E}_x \,.$$

Under these conditions, we have only a drift current, which is proportional to the electric field. The key take away from the previous lecture was that the general current equation for bulk semiconductors under near-equilibrium conditions simplifies to the drift current equation when the carrier density is uniform.

What if the carrier densities aren't constant? That's the subject of this lecture. We'll assume in this lecture that there is no electric field, but there is a gradient in the carrier density.

This lecture begins with Fick's Law of diffusion, which has been known for a long time. We'll show how Fick's Law leads to hole and electron diffusion currents. We'll also discuss the physics of diffusion, and then we'll relate the diffusion current to the general current equation.

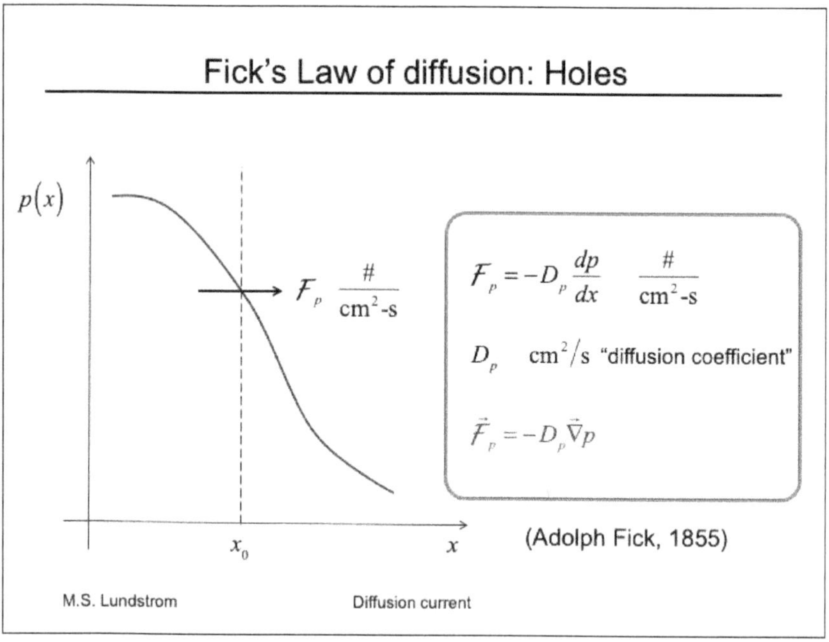

19.2 Diffusion fluxes and diffusion currents

Ficks' Law of diffusion states that if we have a non-uniform concentration of particles, in our case it might be the density of holes in the valence band, then there will be a flux of particles from high concentration to low concentration. The flux at $x = x_0$ is the number per square centimeter per second of particles that cross a plane at $x = x_0$. This flux of particles is proportional to the concentration gradient, and the constant of proportionality is the **diffusion coefficient**, which has the units of m^2/s. (Semiconductor people typically quote the diffusion coefficient in units of cm^2/s.) In three dimensions, the flux is minus the diffusion coefficient times the gradient of the particle concentration.

Although the figure above is for holes diffusing, the same thing happens for electrons. Electrons are particles, they move down their concentration gradient, and the flux of electrons is given by Fick's Law of diffusion with an electron diffusion coefficient instead of the hole diffusion coefficient. Diffusion has nothing to do with the charge of the particles, any type of particle in random thermal motion diffuses from regions of high concentration to regions of lower concentration.

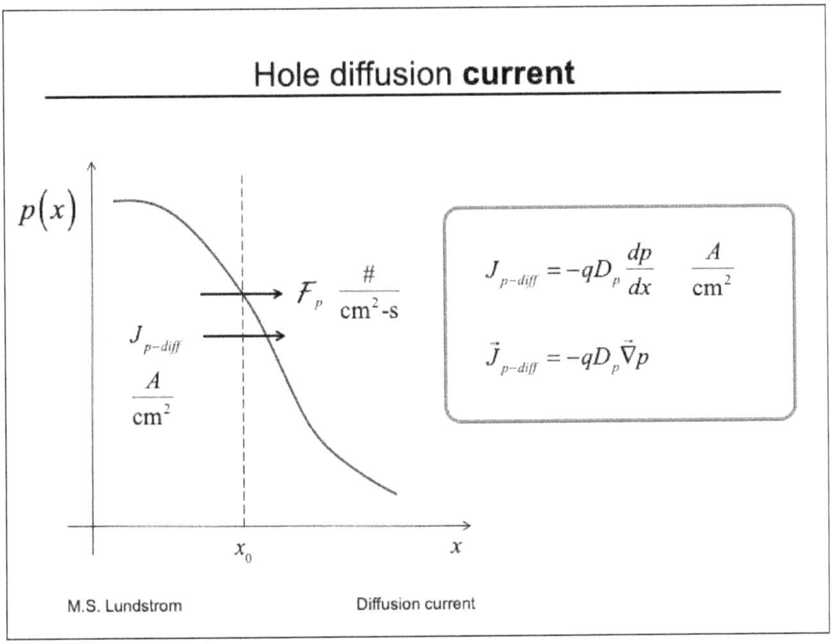

To find the currents, we simply multiply the flux of particles by the charge that each particle carries. For holes, the charge is $+q$, so there will be a hole current in the direction of the hole flux — in the $+x$ direction in the example above. Note that there is a minus sign because holes flow down the concentration gradient. The hole diffusion current has the SI units of amperes per square meter.

Electrons also diffuse from high concentration to low concentration, but electrons have a negative charge. The electron diffusion current is in the opposite direction of the electron flux. We simply multiply the electron flux by the charge, $-q$, and we get an expression for the electron diffusion current:

$$J_{n-diff} = (-q)\left(-D_n \frac{dn}{dx}\right) = +qD_n \frac{dn}{dx} .$$

Note that the electron diffusion current and the hole diffusion current have opposite signs, but the electron and hole drift currents had the same signs. You should be sure that you understand the reason.

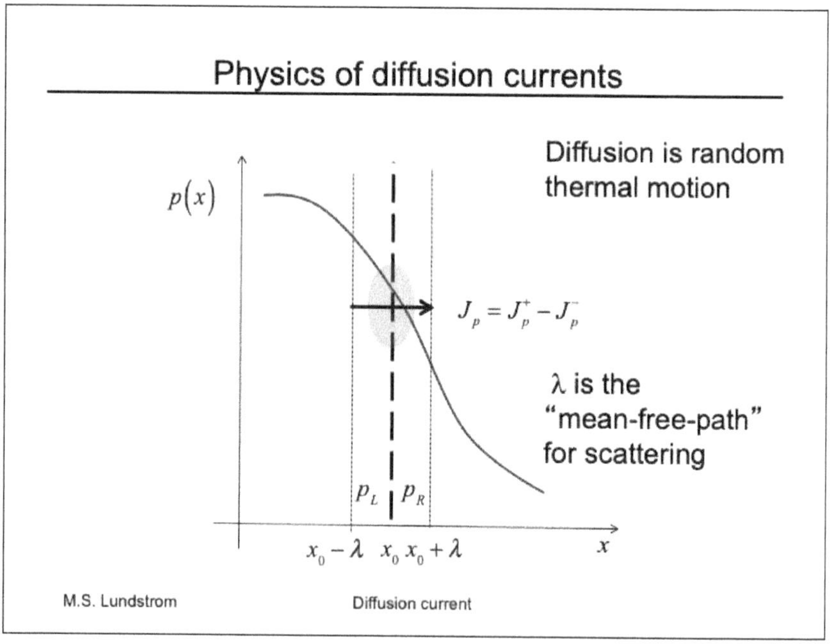

19.3 The physics of diffusion

It is observed that particles move from high concentration to low concentration. What is the underlying physics?

To understand the physics of diffusion, let's take a look at a specific location, x_0 as shown above, and remember that λ is the mean-free-path for particle backscattering. Let's look to the left one mean-free-path and to the right one mean-free-path. In these two regions, we will assume that no scattering occurs. The concentration of holes on the left of x_0 is p_L. The concentration of holes on the right is p_R. We're interested in the net current in the middle of this region at $x = x_0$. There will be holes moving from the left to the right across x_0 leading to a current, J_p^+. There will also be holes moving from the right to the left across x_0 leading to a current, J_p^-. We simply subtract the two directed currents and that's the net hole current in the positive x direction.

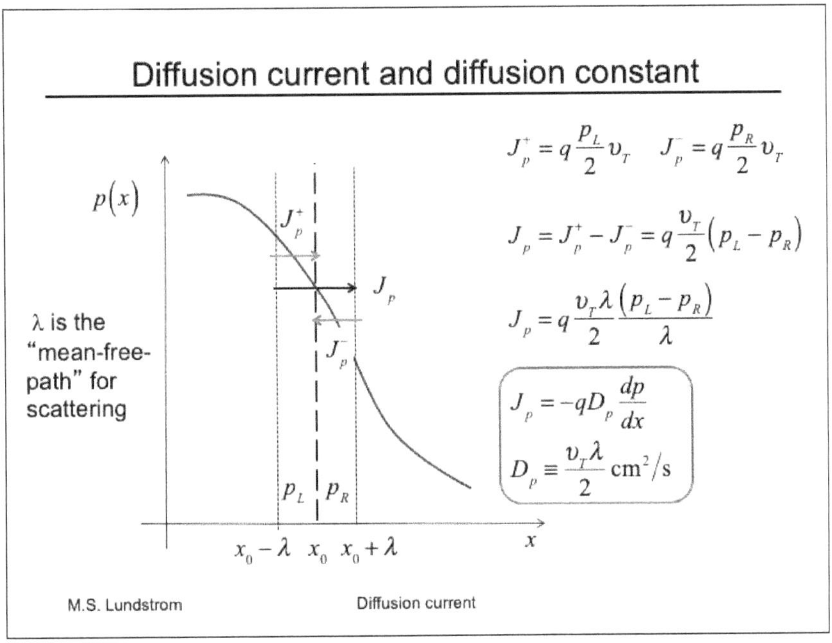

As shown above in the first line on the right, only half of the holes on the left of x_0, p_L, have a velocity in the $+x$ direction because they are randomly distributed in velocity. Similarly, only half of the holes on the right, p_R, have a velocity in the $-x$ direction. Assume that the holes all move at the same thermal velocity, υ_T, so we get the directed currents, J_p^+ and J_p^-, as shown above.

To find the net current at $x = x_0$, we subtract the two directed currents, then multiply and divide by the mean-free-path. You can see that result looks like a derivative of the hole density. If we lump some constants together, we've derived Fick's Law of diffusion.

You can also see that we have derived an expression for the hole diffusion coefficient, and you can see that it depends on the mean-free-path and on the thermal velocity (which makes sense because diffusion is driven by random thermal motion). We see that the higher the temperature, the higher the thermal velocity, so the higher the diffusion coefficient. The shorter the MFP, the lower the diffusion coefficient and the harder it is for particles to diffuse. Just as we had a simple expression for the mobility, we now have a simple expression for the diffusion coefficient.

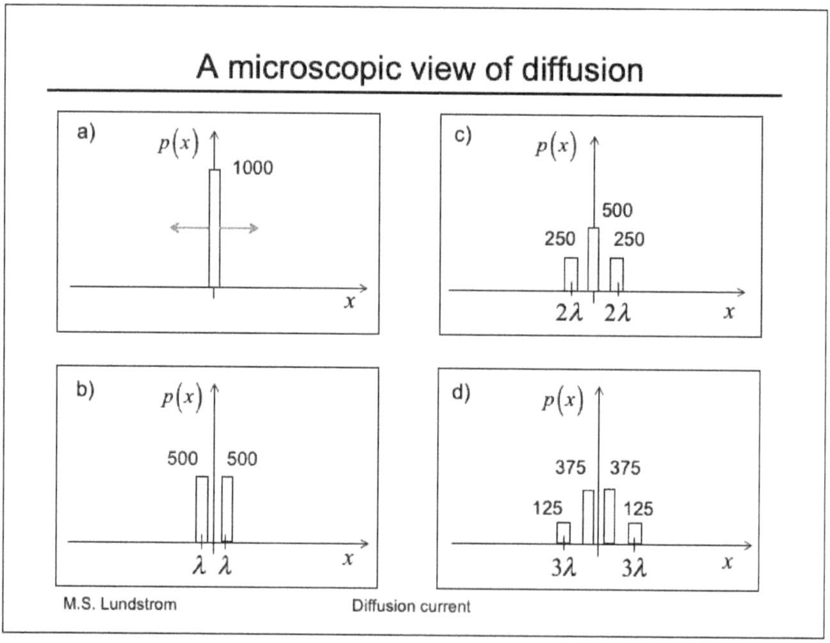

There is another way to discuss the physics of diffusion, a microscopic perspective. As shown in a) above, let's assume that at time, $t = 0$ we put 1000 particles bunched up in a very thin region right near $x = 0$. Assume that these particles are in random thermal motion, and let them go. Half of them will be going in the $+x$ direction and half in the $-x$ direction. Let's wait until one scattering time and see what happens.

Figure b) shows what happens after one scattering time; 500 have moved to the right by a distance of one MFP, and 500 have moved to the left. Then they scatter and their velocities are randomized; half of the 500 are moving to the left, half to the right. What happens after one more scattering time?

Figure c) shows what happens after one more scattering time. Again, the particles are still in random thermal motion. Of the particles that are at $x = \pm 2\lambda$, half of them will move to the left and half will move to the right. Of the particles that are in the middle 500 bar, 250 will move to the left and 250 will move to the right.

Figure d) shows where the particles are after one more scattering time. We see now what is happening. The high concentration at one point is spreading out, and particles are diffusing down the concentration gradient.

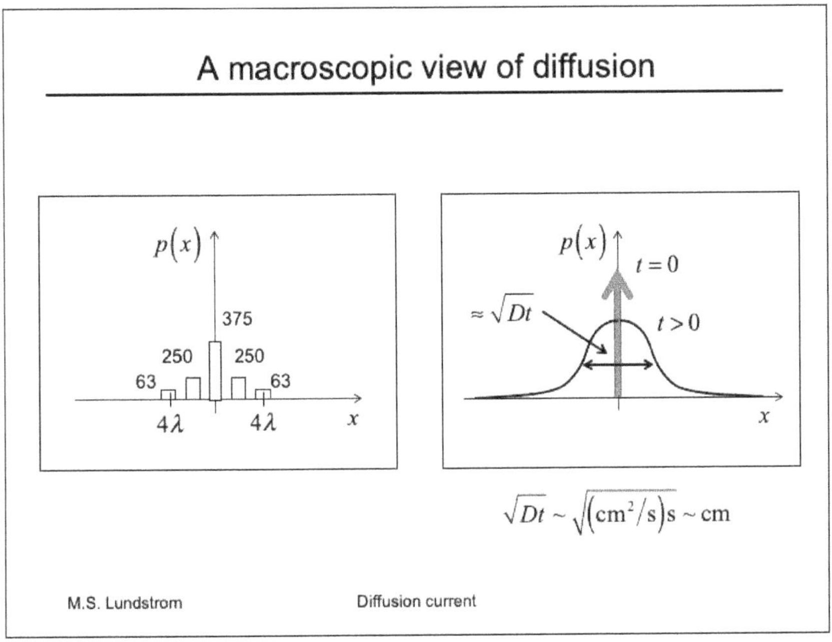

The figure on the left above shows what happens after one more scattering time. The distribution of particles is beginning to resemble a Gaussian. In fact, if we solve the diffusion problem mathematically we'll find out that if we started with a delta function at $t = 0$ and let these particles diffuse for a time t, we'll get a Gaussian distribution that looks like the one shown above on the right. It has a standard deviation that's proportional to the square root of diffusion coefficient times time, $\sqrt{Dt}$ That's actually useful to remember — **the distance that particles diffuse in a time, t, is on the order of $\sqrt{\mathbf{Dt}}$.** Depending on the specifics of the problem, there may be a numerical factor on the order of unity, but $\sqrt{Dt}$ is roughly the distance over which particles in random thermal motion diffuse.

You might wonder what force is pushes particles from high concentration to low concentration. For the drift current, there was a real force — the force due to the electric field that was pushing electrons or holes. In this case, is there is no force. Diffusion is a purely statistical effect. Random thermal motion in the presence of a concentration gradient causes particles to move from high concentration to low concentration. Sometimes people refer to this as a **statistical force**, but there is no actual physical force driving diffusion.

Diffusion currents and diffusion coefficients

$$J_{n-diff} = +qD_n \frac{dn}{dx} \quad \frac{A}{\text{cm}^2} \qquad D_n = \frac{\upsilon_{Tn}\lambda_n}{2} \ \text{cm}^2/\text{s}$$

$$J_{p-diff} = -qD_p \frac{dp}{dx} \quad \frac{A}{\text{cm}^2} \qquad D_p = \frac{\upsilon_{Tp}\lambda_p}{2} \ \text{cm}^2/\text{s}$$

M.S. Lundstrom — Diffusion current

19.4 Relation to the general current equation

We discussed drift currents and mobility in the last lecture, and now we have diffusion currents and diffusion coefficients. Just as we had a simple expression for the mobility in terms of material parameters, effective mass and scattering time, we now have a simple expression for the diffusion coefficient in terms of the thermal velocity and mean-free-path.

Now that we understand diffusion, let's go back to the results of Lecture 17 and connect the diffusion current to the general expression for the near-equilibrium current in a bulk semiconductor,

$$J_{nx} = \sigma_n \frac{d(F_n/q)}{dx} .$$

Recall that the quasi-Fermi level, F_n, is analogous to the equilibrium Fermi level, E_F. Just as E_F is related to the electron concentration in equilibrium, F_n is related to the electron concentration out of equilibrium:

$$n_0 = N_C e^{(E_F - E_c)/k_B T} \rightarrow n = N_C e^{(F_n - E_c)/k_B T} .$$

Let's solve for F_n, take its gradient, put it in the general current equation, and see what happens.

Gradients in the QFL

Quasi-Fermi level:

$$n = N_C e^{(F_n - E_C)/k_B T} \rightarrow F_n(x) = E_C(x) + k_B T \ln(n/N_C)$$

$$\frac{dF_n(x)}{dx} = k_B T\left(\frac{1}{n}\frac{dn(x)}{dx}\right) \qquad (E_C \text{ assumed to be constant})$$

$$J_{nx} = \sigma_n \frac{d(F_n/q)}{dx} = (nq\mu_n)\frac{k_B T}{q}\left(\frac{1}{n}\frac{dn(x)}{dx}\right) = q\left(\frac{k_B T}{q}\mu_n\right)\frac{dn(x)}{dx}$$

$$\boxed{J_{nx} = \sigma_n \frac{d(F_n/q)}{dx} \rightarrow qD_n \frac{dn(x)}{dx} \qquad D_n \equiv \frac{k_B T}{q}\mu_n}$$

M.S. Lundstrom Diffusion current

As shown above, we begin with the definition that relates the electron density to the quasi-Fermi level, solve for the quasi-Fermi level, and then take its gradient. We're assuming there is no electric field, which means E_c is spatially uniform, so we won't get any contribution to the derivative there. We differentiate the logarithmic term on the top line and we get an expression for the gradient of the quasi-Fermi level. Put that in our current expression, rearrange some terms, and we see that the current is q times some terms in parentheses, times dn/dx. That looks like our diffusion current. Indeed it is, but we have discovered that the diffusion coefficient and the mobility are related. The diffusion coefficient is k_BT/q times the mobility. This is a very important relation that you'll want to remember. The relationship is called the **Einstein relation**. The Einstein relation was discovered by Albert Einstein in 1905. The way to remember it is the rhyme "D over mu equals kT over q", so if we know μ_n (μ_p), which is easy to measure, then we know D_n (D_p).

To summarize, we have seen that concentration gradients can cause current to flow, and electric fields can cause currents to flow, but these are just special cases of the fact that what really causes current flow is a gradient in the quasi-Fermi level.

19.5 Summary

To summarize, we have the general near-equilibrium current equation for bulk semiconductors which expresses the electron and hole current densities as being proportional the gradient of their respective quasi-Fermi levels. When the electric field is zero, this general equation becomes:

$$J_{nx} = \sigma_n \frac{d(F_n/q)}{dx} \to J_{nx} = qD_n \frac{dn}{dx}$$
$$J_{px} = \sigma_p \frac{d(F_p/q)}{dx} \to J_{px} = qD_p \frac{dp}{dx}\,.$$

We have also discovered that there is an intimate relation between the diffusion coefficient and thew mobility, the Einstein relation:

$$\frac{D_n}{\mu_n} = \frac{D_p}{\mu_p} = \frac{k_B T}{q}\,.$$

Often in semiconductor problems, we will either be dealing with a case where only the electric field is important or only the concentration gradients is important, so we will often focus on drift currents or on diffusion currents. When both are important, we could add the two, or we could use our more general, more fundamental current expression, which says that current is proportional to gradient in the quasi-Fermi level.

In the last the two lectures, we have talked about what happens when there's an electric field, and what happens when there's a concentration gradient. In general, we have both. That's the subject of the next lecture.

19.6 Reference

For much more discussion about the diffusion current in semiconductors, see:

[1] R.F. Pierret, *Advanced Semiconductor Fundamentals*, (2nd Ed.), Pearson Education, Inc., 2003 (see Ch. 6, Sec. 6.2).

Lecture 20

Drift-Diffusion Equation

20.1 Introduction

The **drift-diffusion equation** is widely-used in semiconductor work. As the name implies, it is simply the sum of the drift and diffusion currents. In this short lecture, we'll discuss the drift-diffusion equation and its connection to the more fundamental description of electron transport.

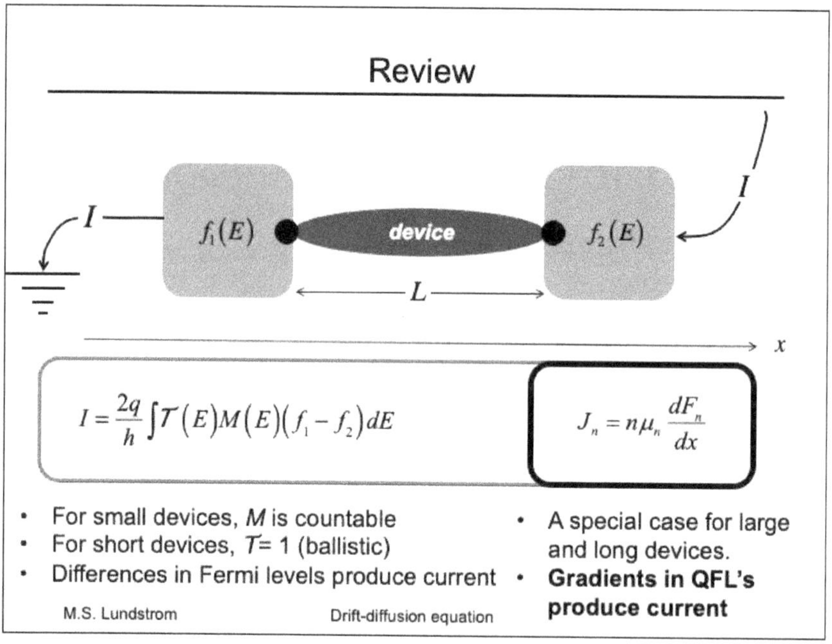

20.2 Review of the general current equation

In Lecture 17, we discussed a general expression for the current in a two terminal device using what is sometimes called the **Landauer approach** to transport, which involves some fundamental constants and quantities like **transmission, channels**, and **Fermi window**. For small devices, we can actually count the number of channels, so the **current is quantized.** For short devices, the transmission is one; we call those **ballistic devices.** The important point is that **differences in the Fermi levels of the two contacts is what causes current to flow.**

We also showed how we can use this same approach for very long devices. In this case, the contacts drop out of the picture, and we end up with the equation shown above on the right, a general expression for current in a large bulk semiconductor under a low applied voltage. For bulk semiconductors, the current is proportional to the gradient of the quasi-Fermi level, (the quasi-Fermi level is analogous to the Fermi level in equilibrium.) In the remainder of these lectures, we will largely be concerned with large, bulk semiconductors for which the **current is proportional to the gradient of the quasi-Fermi level.**

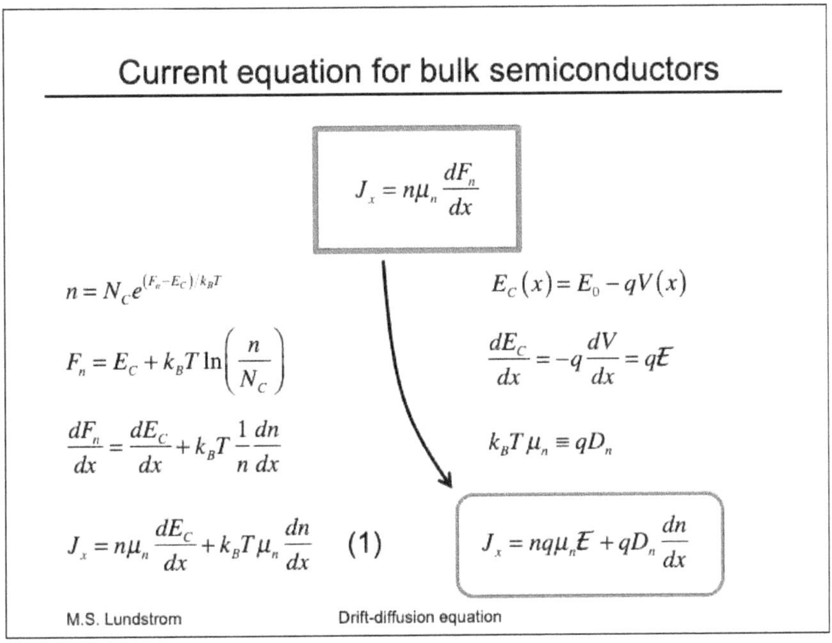

20.3 The drift-diffusion equation

The current equation for a near-equilibrium bulk semiconductor is the first equation shown above. We relate the quasi-Fermi level to the non-equilibrium carrier density, solve for the QFL, take its gradient, insert that into the current equation and end up with eqn. (1) above. This equation has two components, a drift component with the driving force being the gradient of E_c and a diffusion component.

The bottom of the conduction band goes down when the electrostatic potential goes up, so when we take the gradient of the conduction band minimum, we find that it is q times the electric field. We can also define $k_B T$ times mobility to be q times the diffusion coefficient. Finally, we end up with the equation shown above on the lower right.

The **drift-diffusion equation**, is often the starting point for semiconductor device analysis. The first term describes the drift of carriers in a small electric field, and the second term describes the diffusion of carriers down a small concentration gradient. This equation follows from the more general current equation, but we should understand that our general current equation assumes isothermal (uniform temperature) conditions.

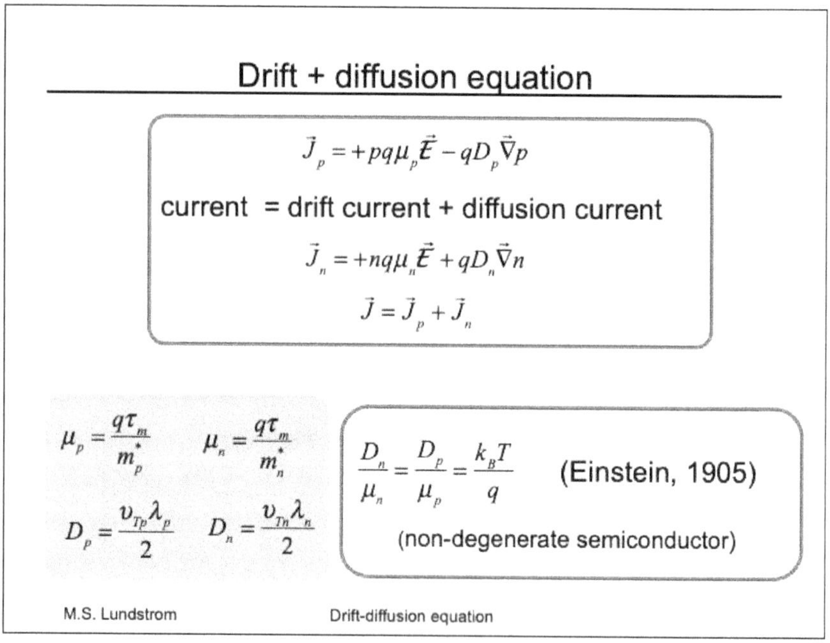

As shown above, there are two drift diffusion equations, one for holes, and one for electrons. In general, there are both electron and hole currents, so to find the total current, we simply add the two. Oftentimes, we'll find one dominates over the other.

We have also seen that we can relate the parameters in the drift diffusion equation to some microscopic parameters, things like momentum relaxation times, effective masses, thermal velocities, and mean free paths. We also understand now that the mobility and diffusion coefficient are intimately related by the Einstein relation. (If we are careful, we can show that the ratio of the microscopic definitions of D and μ reduce to the Einstein relation.)

It's relatively easy to measure the mobility of a semiconductor [1]. That's why you'll typically find data for the mobility of a semiconductor. If we need the diffusion coefficient, then we invoke the Einstein relation, and we deduce the diffusion coefficient from the measured mobility.

Mobility and diffusion coefficient

$$\mu_n = 1360 \frac{\text{cm}^2}{\text{V-s}}$$

$$\mu_p = 480 \frac{\text{cm}^2}{\text{V-s}}$$

pure (undoped) Si at 300 K

$$\frac{D_n}{\mu_n} = \frac{D_p}{\mu_p} = \frac{k_B T}{q} = 0.026 \text{ V}$$

$$D_n = \frac{k_B T}{q}\mu_n = 35.4 \frac{\text{cm}^2}{\text{s}} \qquad D_p = \frac{k_B T}{q}\mu_p = 12.5 \frac{\text{cm}^2}{\text{s}}$$

$\sqrt{D_n t} \approx 60\ \mu\text{m}$ (in 1 microsecond)

M.S. Lundstrom Drift-diffusion equation

Let's do an example using numbers for pure undoped silicon at room temperature. We invoke the Einstein relation and compute the diffusion coefficient for electrons. We find that it is about 35 cm^2/s. The diffusion coefficient for holes is less because the mobility is less. It's about 12.5 cm^2/s. How do we calibrate ourselves as to what those numbers mean?

It's useful to remember that when you solve a diffusion equation, the distance that particles move due to random thermal motion in a concentration gradient is roughly the square root of diffusion coefficient times time. How far can electrons diffuse in one microsecond? If we plug in numbers, we find that that's about 60 micrometers, which is a very significant distance. Charge carriers in semiconductor can diffuse quite far in a very short period of time.

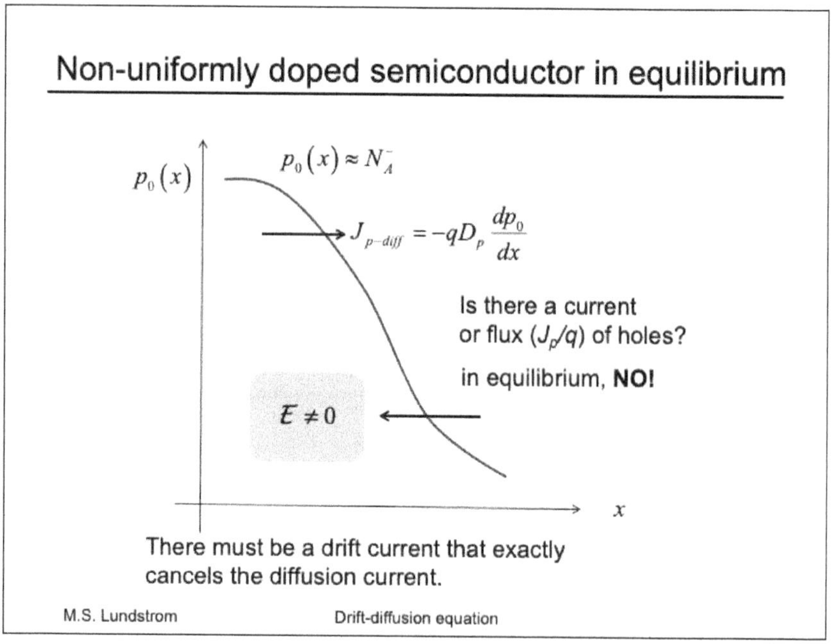

20.4 Built-in electric fields

Consider the nonuniformly doped P-type semiconductor in equilibrium. As shown above, this semiconductor is doped more heavily near $x = 0$ than in the bulk, so there is a gradient in the hole concentration. Holes are mobile, so they will diffuse down the concentration gradient. There'll be a diffusion flux, but there can be no current because we are in equilibrium.

How can there be no current flowing when we have a diffusion current? There must be a drift current exactly equal but opposite in sign to the diffusion current. The electric field must point in the $-x$ direction because it will exert a force on the holes that pushes them back and stops them from diffusing and keeps $p_0(x) \approx N_A(x)$. To solve for the electric field, we would simply set the drift-diffusion equation to zero. Equilibrium electric fields are called **built-in electric fields**. In PN junctions, the concentration gradient is very large, leading to large built-in electric fields.

Alternatively, we could have begun with the fact that the current is proportional to the gradient of the quasi-Fermi level. Since we are in equilibrium, the quasi-Fermi level is the Fermi level, but the Fermi level is constant in equilibrium, so there is no gradient and, therefore, no current.

20.5 Summary

The equations below summarize near-equilibrium current flow in bulk semiconductors under isothermal conditions.

$$
\boxed{\begin{aligned} J_{nx} &= \sigma_n \frac{d(F_n/q)}{dx} \rightarrow J_{nx} = nq\mu_n\mathcal{E} + qD_n\frac{dn}{dx} \\ J_{px} &= \sigma_p \frac{d(F_p/q)}{dx} \rightarrow J_{px} = pq\mu_p\mathcal{E} - qD_p\frac{dp}{dx} \,. \end{aligned}}
$$

"Near-equilibrium" means that the gradients of the quasi-Fermi levels, electrostatic potential, and carrier densities are small. The currents are proportional to the gradient of a quasi-Fermi level, or we can separate the currents into drift and diffusion components. We have also discovered that there is an intimate relation between the diffusion coefficient and the mobility:

$$
\boxed{\frac{D_n}{\mu_n} = \frac{D_p}{\mu_p} = \frac{k_BT}{q} \,.}
$$

The Einstein relation is a very important relation that allows us to determine the diffusion coefficient from the readily measured mobility.

The voltages applied to semiconductor devices usually vary in time. The time-dependent conduction currents are

$$
\begin{aligned} J_{nx}(t) &= \sigma_n(t)\frac{d(F_n(t)/q)}{dx} = n(t)q\mu_n\mathcal{E}(t) + qD_n\frac{dn(t)}{dx} \\ J_{px}(t) &= \sigma_p(t)\frac{d(F_p(t)/q)}{dx} = p(t)q\mu_p\mathcal{E}(t) + qD_p\frac{dp(t)}{dx} \,. \end{aligned}
$$

Although these are the equations that are typically used, they are not quite correct; we should add the **displacement current** from Maxwell's equations:

$$
J_{tot}(t) = J_n(t) + J_p(t) + \frac{\partial D}{\partial t} \,,
$$

where $D = \epsilon\mathcal{E}$ is the dielectric displacement vector.

In Lectures 17–20, we've discussed some fundamental concepts of current flow in semiconductors. There are some other important out of equilibrium processes. In the next lecture, we'll discuss one, carrier **recombination**.

20.6 Reference

For a deeper dive into carrier transport in semiconductors see:

[1] Mark Lundstrom *Fundamentals of Carrier Transport*, (2nd Ed.), Cambridge University Press, 2000.

Lecture 21

Carrier Recombination

21.1 Introduction

Now that we understand carrier transport, it is time to discuss two other important nonequilibrium processes, recombination and generation. This lecture is about carrier **recombination**; net recombination can occur when a semiconductor is out of equilibrium.

In equilibrium there is no external input of energy, for example, from a battery or power supply connected to the semiconductor or from light shining on a semiconductor. Simple expressions relate the equilibrium carrier concentrations to the Fermi level:

$$n_0 = N_c e^{(E_F - E_c)/k_B T}$$
$$p_0 = N_v e^{(E_v - E_F)/k_B T} .$$

We can also express the out of equilibrium (but still near equilibrium) carrier concentrations in terms of the relevant quasi-Fermi level:

$$n = N_c e^{(F_n - E_c)/k_B T}$$
$$p = N_v e^{(E_v - F_p)/k_B T} .$$

Another way to write the out of equilibrium carrier densities is as the equilibrium carrier density plus an excess carrier concentration:

$$n = n_0 + \Delta n$$
$$p = p_0 + \Delta p .$$

Note that the excess carrier concentration can be positive or negative. There can be more carriers than in equilibrium, but it's also possible to have fewer carriers than in equilibrium, so Δn and Δp can be either positive or negative. In equilibrium, the electron and hole densities are independent of time. The question for this lecture is: "How do the excess carrier concentrations vary with time?"

Consider uniformly doped P-type Si in equilibrium doped at 10^{17} cm^{-3}. The equilibrium hole concentration is $p_0 = 10^{17}$ cm^{-3}. The minority electron concentration is $n_0 = n_i^2/p_0 = 10^3$ cm^{-3}. Now let's introduce some excess minority carrier electrons at time $t = 0$ and ask what happens. We will assume that Δn is spatially uniform, so there will be no tendency for the electrons to diffuse. In this lecture, we'll also assume $\Delta n \approx \Delta p$ (i.e. we assume approximate space charge neutrality).

Intuitively, if we perturb a system away from equilibrium, we expect the system to respond by trying to return to equilibrium. Oftentimes, this return to equilibrium is an exponential process. So we might guess that if we perturb this system by adding minority carriers at time $t = 0$, that the number of minority carriers will decay exponentially with time as the system returns to equilibrium. If we perturb this system by removing minority carriers at time $t = 0$, then the minority carrier concentration will grow exponentially with time as the system returns to equilibrium. We will see that this is indeed what happens.

There are three important, different, and common ways that this recombination can occur. The first is **band-to-band radiative recombination**, which gives off light. The second is **band-to-band Auger recombination**, which does not give off light, and the third involves recombination through defects, which is called **defect-assisted or SRH recombination** after the initials of the three people who first described this process, Shockley, Read, and Hall.

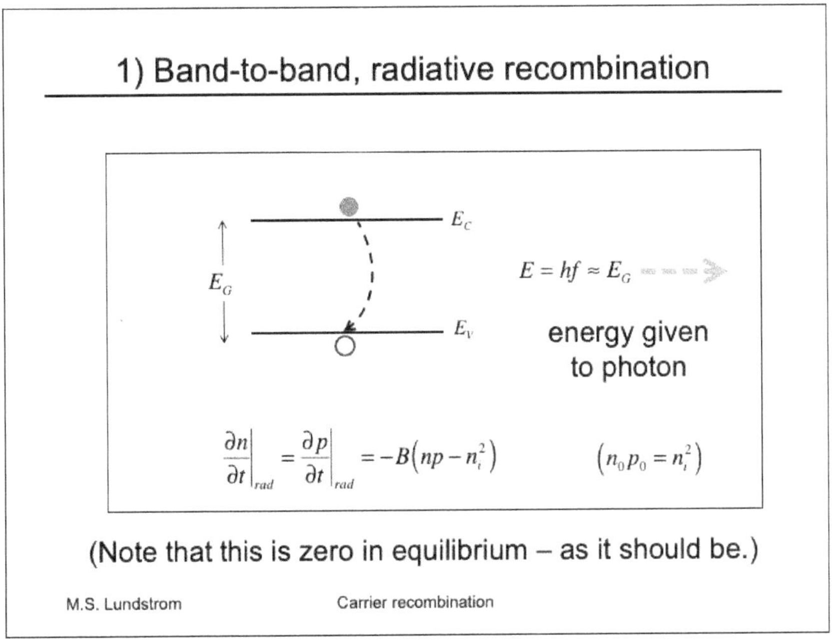

21.2 Band-to-band radiative recombination

Shown above is **band-to-band radiative recombination.** An electron in the conduction band drops down into the valence band and fills up a hole. An electron and hole have "recombined." The energy released in a radiative recombination event is given to a photon with an energy of about the band gap because electrons in the conduction band are near the bottom of the conduction band, and holes in the valence band are near the top. Radiative recombination changes the concentrations of electrons and holes. We describe this process with the equation on the lower left in the slide above. The change in concentration is negative when $np > n_i^2$, so this process reduces the concentration of excess carriers. In order for an electron and a hole to recombine, we need electrons and we need holes, so the rate at which they recombine is proportional to the product np. The constant of proportionality is B. The **B coefficient** is an important material parameter for a semiconductor. There will be no net recombination in equilibrium because $(np - n_i^2) = 0$ in equilibrium.

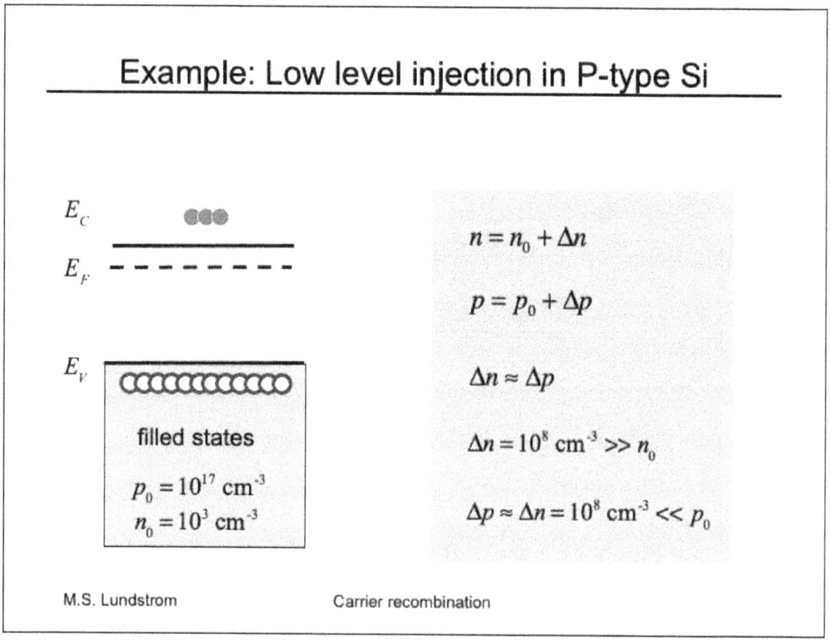

There's an important approximation that greatly simplifies the analysis of these problems. The approximation is called **low level injection**. Low level injection means that the concentration of minority carriers is small compared to the equilibrium majority carrier concentration, but large compared to the equilibrium minority carrier concentration.

Consider the example shown above, low level injection in a P-type semiconductor doped with acceptors that produce a concentration of $p_0 = N_A = 10^{17}$ cm^{-3} holes in equilibrium. Using $n_0 p_0 = n_i^2$, we see that the equilibrium minority electron concentration is 10^3 cm^{-3}, 14 orders of magnitude smaller than the hole concentration in this P-type semiconductor.

Now let's introduce excess electrons with a concentration of 10^8 cm^{-3}. This is five orders of magnitude larger than the minority electron concentration in equilibrium, so these excess carriers overwhelm the minority carriers that were there in equilibrium. But you'll also notice that 10^8 cm^{-3} is 9 orders of magnitude smaller than the equilibrium hole concentration, so it's insignificant compared to the majority carrier concentration. That's what we mean by low level injection. **Low level injection has negligible influence on the majority carriers, but it has a huge influence on the minority carriers.**

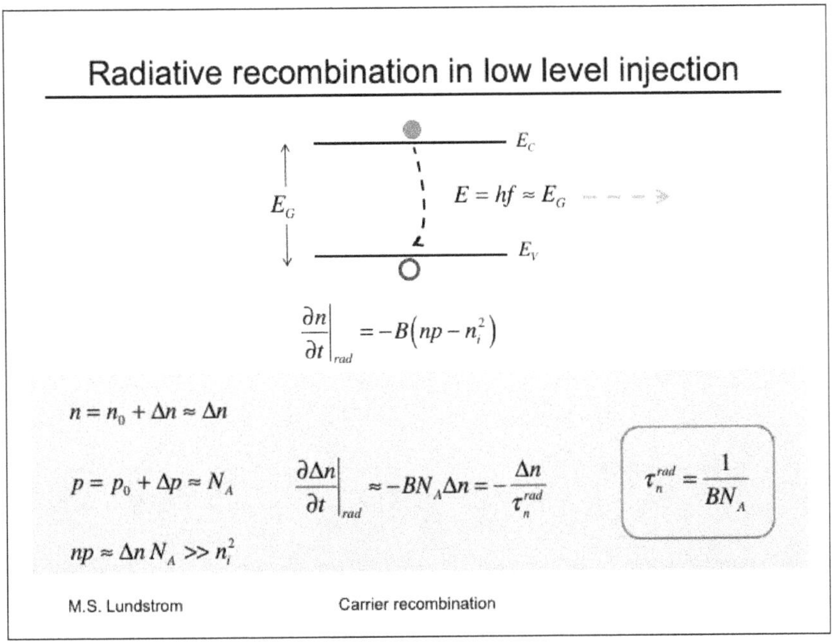

Let's examine band-to-band radiative recombination in a P-type semiconductor under low level injection conditions. The excess carriers overwhelm the minority carriers that were there in equilibrium, so the electron concentration is about equal to the excess electron concentration.

There was a large number of holes to begin with, so the excess hole concentration (assumed to be equal to the excess electron concentration) is insignificant. The hole concentration is about equal to the doping density, very nearly to what it was before the excess carriers were introduced.

Using these approximations, we evaluate the recombination rate and find that it is Δn over a characteristic recombination time. The characteristic time is the **minority carrier lifetime** due to radiative recombination. The **radiative lifetime** is one over the B-coefficient of the semiconductor times the doping density (or hole concentration).

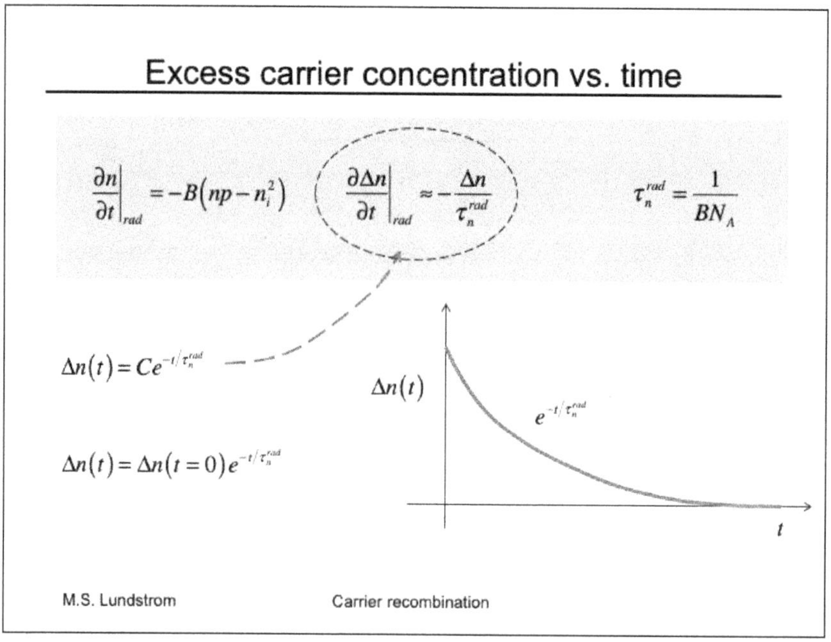

As summarized above in the gray box, to determine how an excess concentration of electrons varies with time, we began with our general expression for radiative recombination and simplified it for low level injection. As shown inside the dashed line oval above, this gives us a differential equation that describes how the excess carrier concentration evolves with time. You've probably seen equations like this before, and you can guess that an exponential solution will satisfy it. The solution is $C\exp[-t/\tau_n^{rad}]$, where C is a constant determined by the initial condition. Plug this assumed solution into the differential equation, and you'll see that it works.

The solution, $\Delta n(t)$, tells us that the excess carrier concentration decays exponentially with time. Band-to-band radiative recombination is one possible mechanism by which carriers can recombine.

Consider an example. For N-type GaAs, $B \approx 2 \times 10^{-10}$ cm$^3/s$ for $n_0 = N_D \approx 4 \times 10^{18}$ cm^{-3}. What is the minority hole radiative lifetime? A simple calculation gives $\tau_p^{rad} = 1/BN_D = 1.3$ nsec.

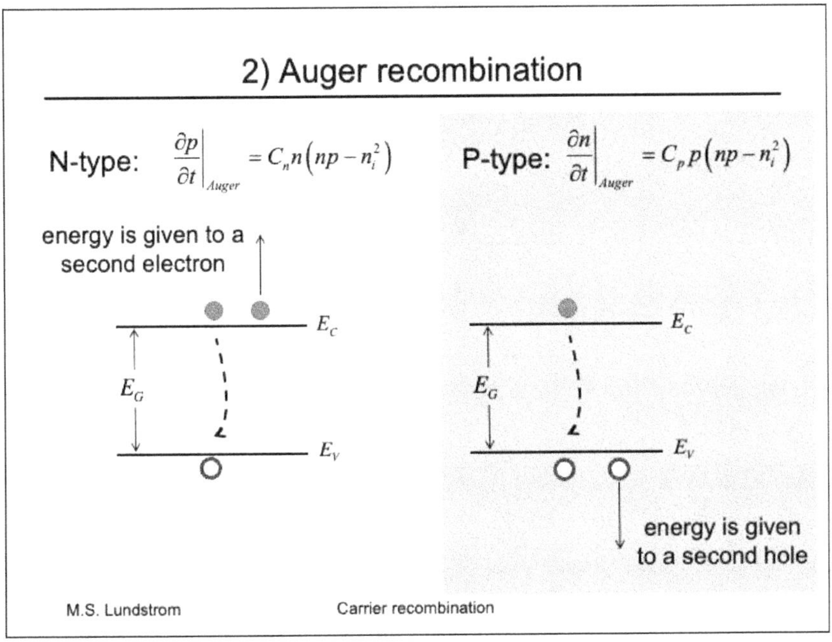

21.3 Auger recombination

Excess carriers can recombine in other ways. For example, when an electron and a hole recombine, the energy released doesn't necessarily have to go to to a photon. As shown above, in an N-type semiconductor, the energy lost in the recombination process might be given to a nearby electron, which kicks that electron to a higher energy in the conduction band. This process is called **Auger recombination**. (The term, "auger" is pronounced "Oh-zhay".) For an N-type semiconductor, the recombination rate is proportional to the product, np because we need an electron and a hole for an electron-hole recombination event, but it is proportional again to n because we need an electron nearby to give the energy to. So this process is proportional to n^2p, and a constant C_n describes the strength of the process.

For a P-type semiconductor, the Auger process requires a nearby hole that the energy can be given to. This process is described by the equation shown at the right above. Again, the $(np - n_i^2)$ terms ensure that there is no net recombination in equilibrium.

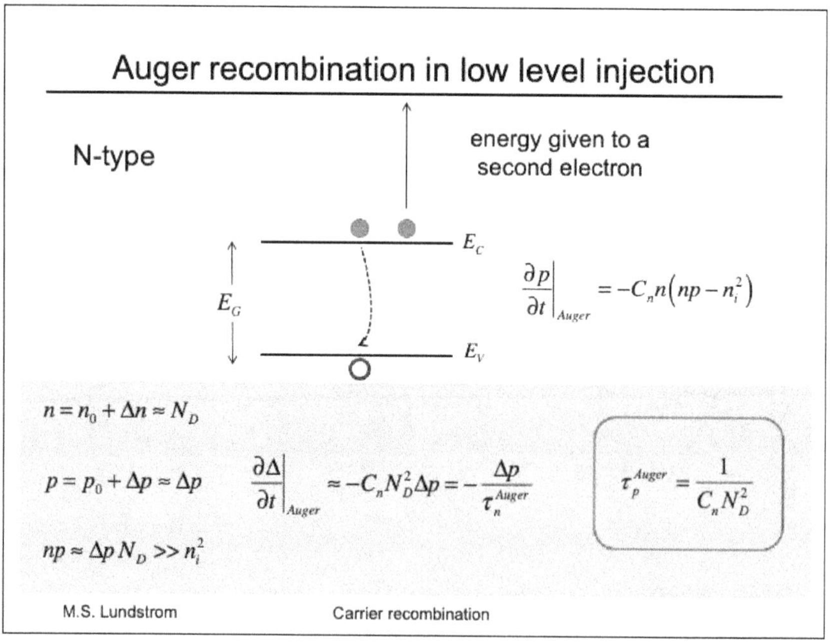

Let's take a closer look at Auger recombination under low level injection conditions in an N-type semiconductor. We begin with the general expression shown above to the right of the energy bands. We can safely assume that $n = N_D$ and that $p = \Delta p$ and that their product is much greater than n_i^2. The resulting differential equation is similar to the previous one, there is just a different characteristic time to describe the exponential decay of minority carriers. The characteristic time, the **Auger recombination lifetime**, is one over the Auger constant, C_n, times N_D^2. You can see that if the doping density is very large, the Auger recombination lifetime can become very small. If it's much smaller than the radiative lifetime, then the Auger recombination dominates.

Consider Auger recombination in N-type Si for which $n_0 = N_D = 10^{20}$ cm^{-3}. Assuming $C_n \approx 3 \times 10^{-31}$ cm^6/s, we find the minority hole lifetime due to Auger recombination as $\tau_p^{Auger} = 1/C_n N_D^2 = 0.3$ ns. The Auger lifetime can be very short when the carrier concentration is very high.

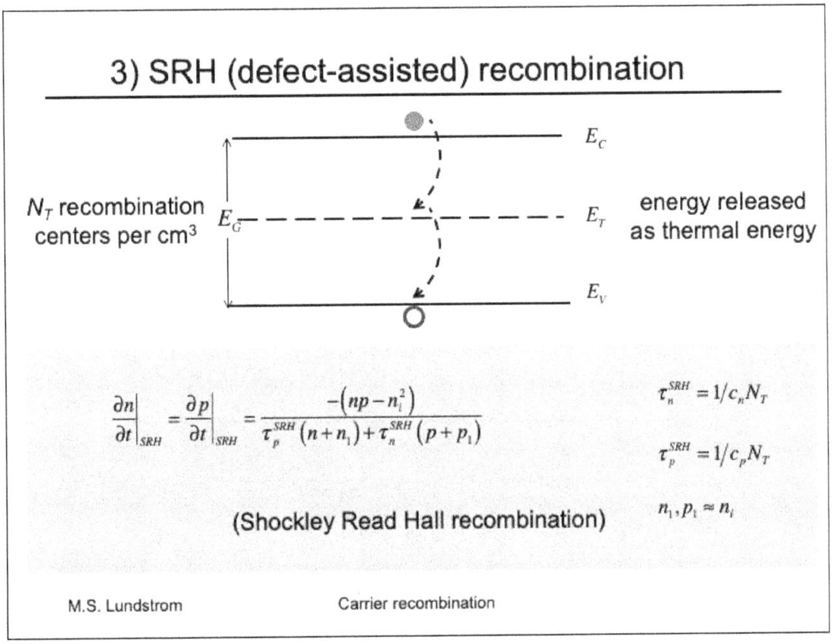

21.4 Defect-assisted (SRH) recombination

A recombination process that is common in semiconductors like silicon involves carriers recombining through defect states in the forbidden gap. Crystal defects and impurities such as heavy metals like copper and iron tend to introduce electronic states in the forbidden gap. These states are localized near the defects. Current can't flow through these defect states, but they can mediate recombination processes. An electron in the conduction band can be captured by a defect. It can then drop down to the valence band and annihilate a hole. This **defect-assisted recombination** is important when the other processes aren't important. The energy that's released in this process is thermal energy that heats up the semiconductor. The theory of defect-assisted recombination was worked out by Shockley, Read, and Hall many years ago. This kind of recombination is described by the equation shown above. There is an $(np–n_i^2)$ term to make sure that there is no net recombination under equilibrium conditions. The characteristic times τ_n^{SRH} and τ_p^{SRH}, are related to the concentrations of defects, N_T, and the parameters n_1 and p_1, have units of carrier density. Under typical conditions, they have a magnitude that is close to n_i, so they're very small.

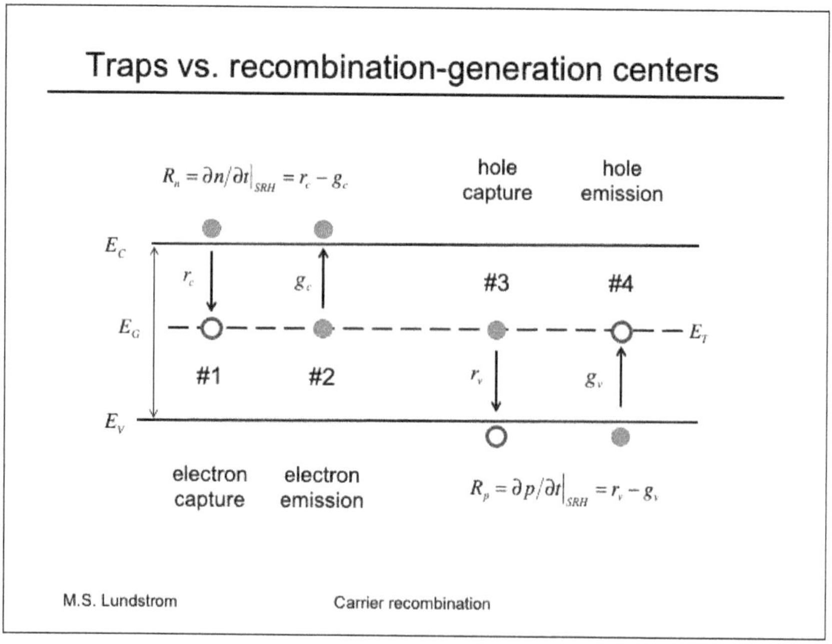

The slide above indicates how the SRH derivation begins. Process #1 is the capture of an electron in the conduction band by a defect. The rate at which this decreases the electron concentration is r_c. Process #2 is the emission of an electron in a defect to the conduction band. The rate at which the electron concentration increases is g_c.

In process #3, an electron in a defect drops down and fills a hole in the valence band. A hole in the valence band has been captured by the defect. The rate at which the concentration of holes in the valence band decreases is r_v. In process #4, an electron in the valence band hops up, filling an empty state on the defect and leaving a hole in the valence band. We call this a hole emission process, and the rate at which the hole concentration in the valence band increases is g_v.

If the defect communicates only with the conduction band, then only processes #1 and #2 are important; the defect is an **electron trap**. If the defect communicates only with the valence band, then only processes #3 and #4 are important; the defect is a **hole trap**. If the defect communicates with both bands, it can facilitate transitions between bands, and the defect is a **recombination-generation (R-G) center**. In steady-state, $R_n = R_p = R$ as given by the SRH expression. See [1] for the complete derivation.

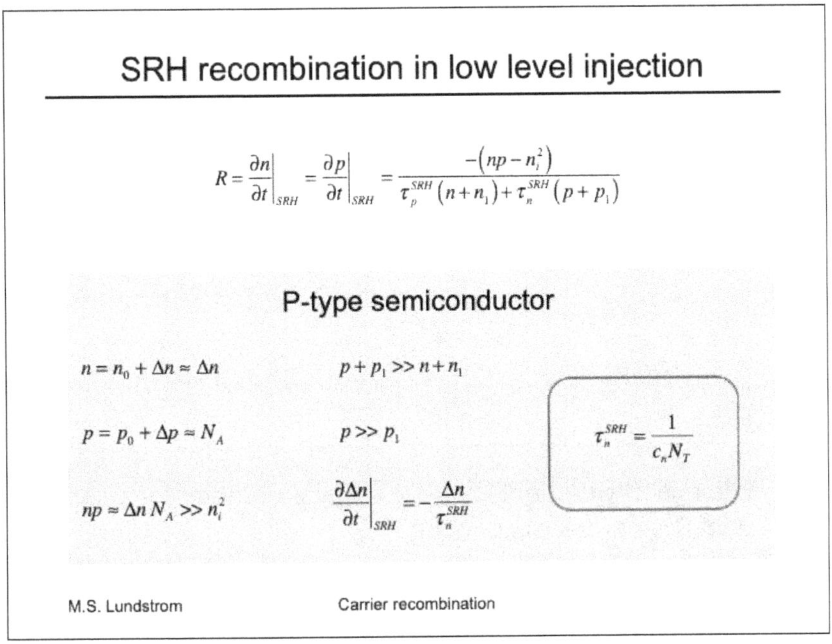

Let's apply the SRH expression to low level injection in a P-type semiconductor. The equilibrium minority electrons will be overwhelmed by the excess electrons, so $n = \Delta n$. The hole density will be $p = p_0 = N_A$, assuming low level injection. We also have $np = \Delta n N_A \gg n_i^2$. Since this is a P-type semiconductor, $p + p_1 \gg n + n_1$. In addition, since p_1 tends to be on the order of n_i, $p \gg p_1$. Carry that all through and you'll find that under low level injection conditions in a P-type semiconductor, the SRH recombination rate is described by the same differential equation we saw previously but now with a characteristic time that is related to the concentration of defects. The parameter, c_n is the electron capture coefficient of the defect state.

The minority carrier lifetimes for defect-assisted recombination depend on the concentration of defects and can vary over a broad range depending on the purity of the semiconductor. Minority carrier lifetimes on the order of 1 μs to 1 ms are observed in silicon.

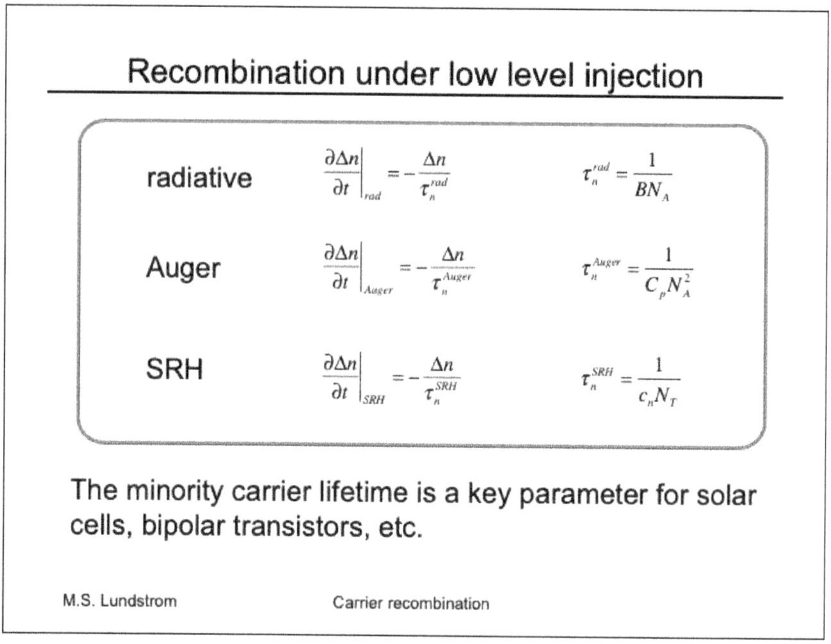

21.5 Multiple recombination processes

We have described three different ways that electrons and holes can recombine, band-to-band radiative recombination, band-to-band Auger recombination, and defect-assisted Shockley-Read-Hall recombination. We've shown that under low level injection the minority carrier recombination rates are all described by the same kind of differential equation, but with different characteristic decay times. The characteristic times for these three processes are three different **minority carrier lifetimes**, which are very important parameters in devices like solar cells or bipolar transistors that depend on minority carriers for their operation.

The slide above summarizes low injection, minority carrier electron recombination in a P-type semiconductor. Similar considerations apply to minority carrier hole recombination in an N-type semiconductor.

In general, all three of these processes will be operating at the same time. What then is the overall minority carrier lifetime?

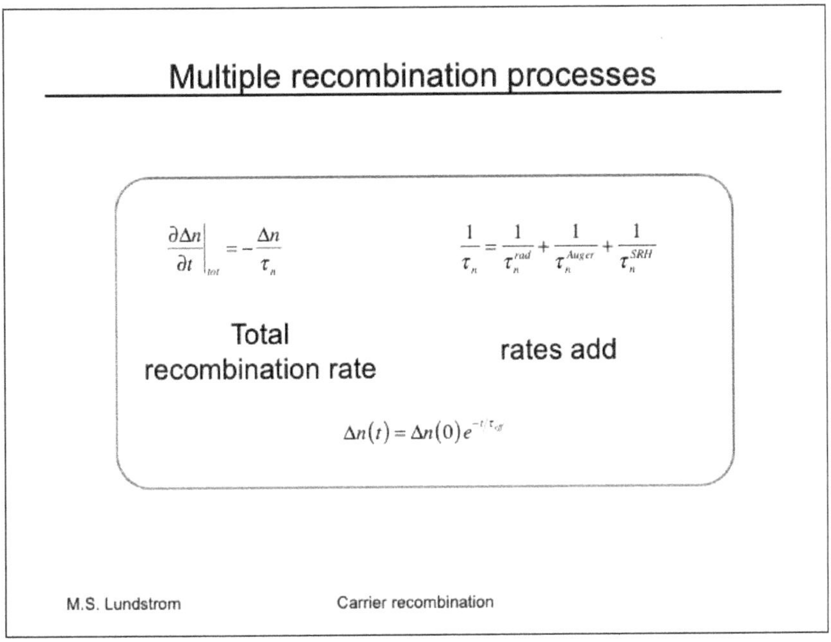

Very often we find that one recombination process dominates, and the others can be neglected, but there are conditions under which all of the processes are important. When they're all important we **add the rates** of the processes together, which means that we add the inverse lifetimes. You can see that this is the right thing to do because if we had a process that had a one nanosecond lifetime and another process that had a one microsecond lifetime, then if both processes occurred at the same time, we would expect a characteristic time less than the shorter of the two individual lifetimes. The way we get the shortest lifetime is by adding the inverses of the lifetimes. The minority carrier lifetime defined above describes the effects of all of the processes that are important for that particular semiconductor.

When are the various processes we have discussed important? The Auger process becomes important when we have very heavily doped materials. Shockley-Read-Hall recombination is important when the defect density is high, or when there aren't many defects present, but when none of the other effects are important. Radiative recombination is important in direct gap semiconductors, but not in indirect gap semiconductors. Why is that so?

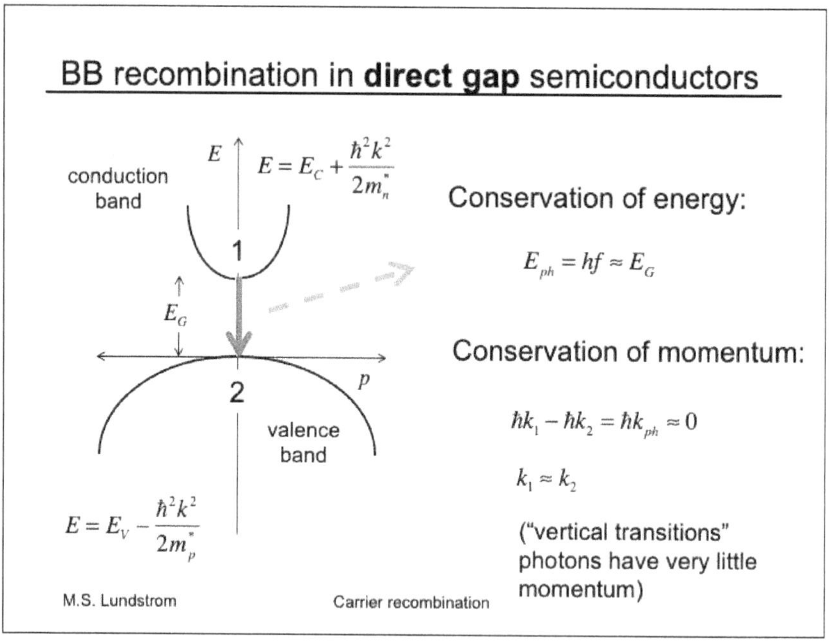

21.6 Band structure and recombination

Radiative recombination processes are strong in direct gap semiconductors. As shown above, an electron in the conduction band drops down and fills a hole in the valence band. This recombination process gives off light. The process has to conserve energy, so the energy released when the electron drops from the conduction band or the valence band is given to the photon. The band-to-band recombination process also has to conserve crystal momentum.

The momentum of the electron in the conduction band was $\hbar k_1$. After it drops to the valence band, it is $\hbar k_2$. The difference in momentum is the momentum of the photon. If you compute the momentum of the photon, you'll find that it carries very little momentum, so the difference in momentum between the initial state of the electron and the final state is nearly zero. This transition can take place "vertically" on the $E(k)$ diagram. An electron near $k = 0$ in the conduction band can drop down to a state near $k = 0$ at the top of the valence band. These are very strong recombination processes in direct gap semiconductors.

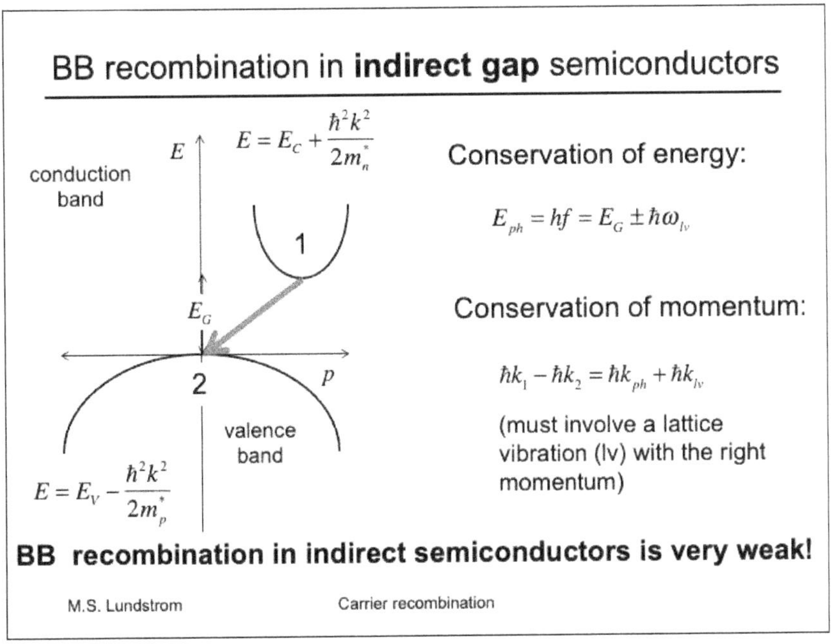

Radiative recombination processes are weak in indirect gap semiconductors. In an indirect gap semiconductor, electrons at the bottom of the conduction band recombine with holes at the top of the valence band, but the initial and final states have different crystal momenta. In order to conserve crystal momentum, something is needed that matches the difference in momenta between the initial and final state. That something is a lattice vibration (a phonon). The lattice vibration that provides the crystal momentum needed for momentum conservation will also have some energy. The transition must conserve both energy and crystal momentum. The need to find a phonon with the right momentum lowers the probability of band-to-band recombination in indirect gap semiconductors. In a direct gap transition, a lattice vibrations to satisfy momentum conservation is not needed, so band-to-band radiative processes are more probable in direct gap semiconductors.

We can generally ignore band-to-band radiative recombination in indirect gap semiconductors because the probability of a radiative recombination event is small (roughly 100 times smaller than for a direct band gap semiconductor), so defect-assisted processes dominate.

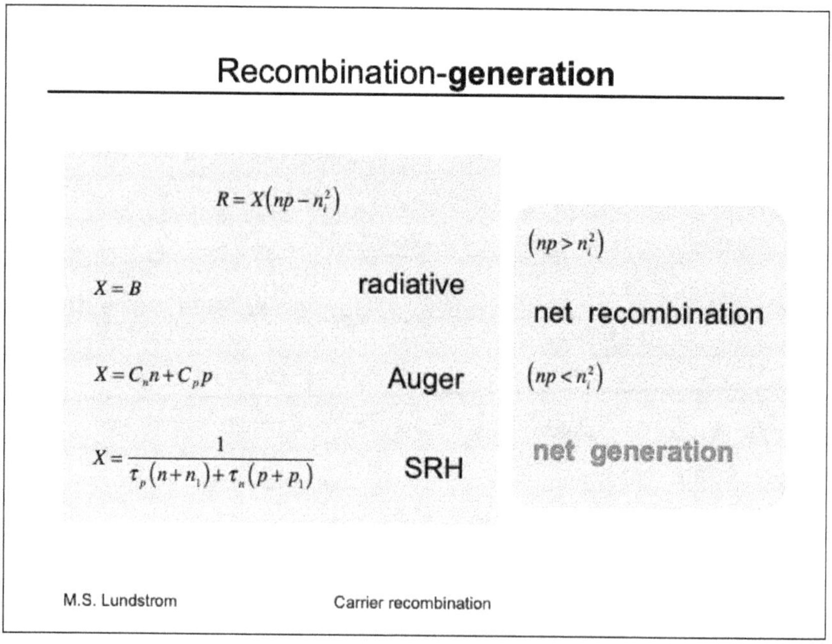

21.7 Recombination and generation

We have discussed three types of recombination. Band-to-band **radiative recombination**, which is important in direct gap semiconductors and which we can put to use to make semiconductor lasers and light emitting diodes. Band-to-band **Auger recombination** becomes important when carrier densities are very high. Auger recombination can degrade the performance of devices like semiconductor lasers where the carrier densities are very high. **Defect-assisted** or **Shockley-Read-Hall recombination** is the dominant process in indirect gap semiconductors because band-to-band recombination is weak. In a direct gap semiconductor, if there are large concentrations of defects, then SRH recombination can also become important.

Each of the recombination processes we have discussed is described by equations that have an $(np–n_i^2)$ term, which ensures that the net recombination is zero in equilibrium. We have been talking about situations in which $np > n_i^2$, so we have a net recombination. It is possible, however, that $np < n_i^2$. The recombination rate is negative, which is actually a carrier **generation** process.

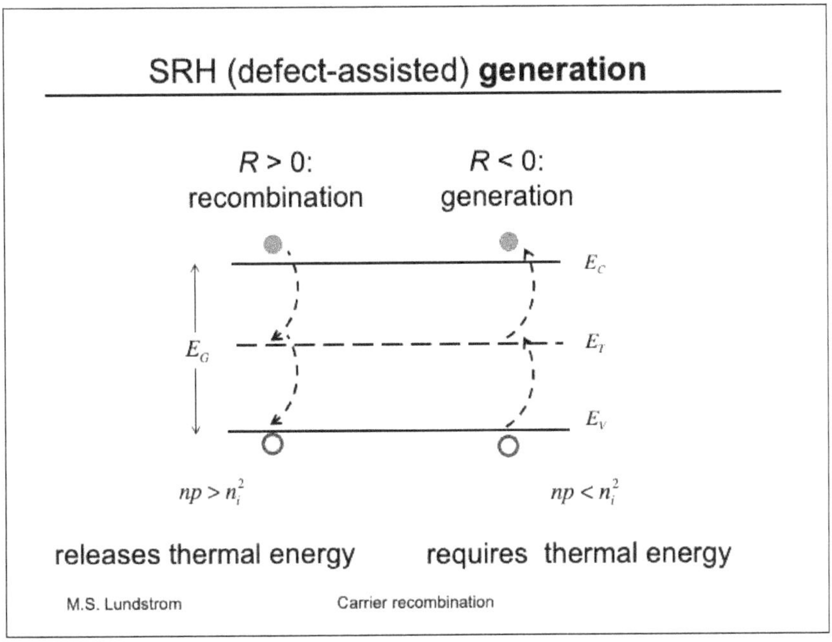

What does a defect-assisted generation process look like? As shown above on the left, for a recombination process, an electron in the conduction band drops down to a defect, and then drops down again to the valence band. In the process we lose one electron in the conduction band and one hole in the valence band, and thermal energy is released. This recombination process occurs when $np > n_i^2$. There is, however, an inverse process. For example, as shown above, an electron in the valence band can hop up to the defect level leaving behind a hole. Then it could hop from the defect up into the conduction band creating an electron in the conduction band. This occurs when $np < n_i^2$ and requires thermal energy. For this reason, this process is called **thermal generation**.

We see that the mechanisms that we have been discussing are actually both recombination and generation processes depending on whether R is greater than zero or less than zero. For this reason, the defects shown near mid-gap in the figure above are sometimes called **R-G centers**.

Consider a case where $n \ll n_1$, $p \ll p_1$, and $np \ll n_i^2$. The SRH formula gives $R = -n_i/(\tau_n^{SRH} + \tau_p^{SRH})$ which is a generation rate of $G = +n_i/(\tau_n^{SRH} + \tau_p^{SRH})$.

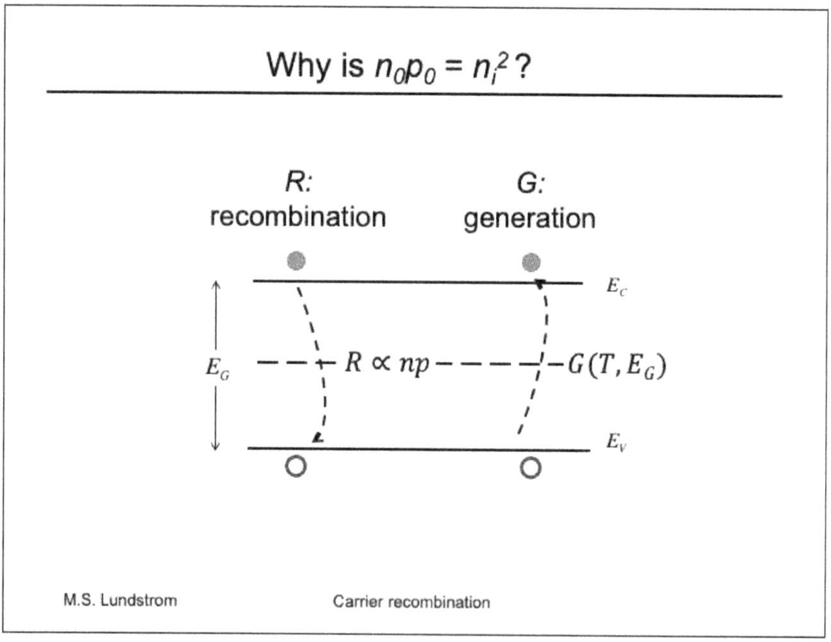

Now we can understand why $n_0p_0 = n_i^2$ is independent of doping. Of course, we could just multiply our expressions for n_0 and p_0 and see that the Fermi level drops out, but how do we explain this physically?

For an intrinsic semiconductor, $n_0 = p_0 = n_i$, but microscopically, carriers are continually being generated at a rate, G, and are recombining at a rate, R. The principle of detailed balance states that in equilibrium every process and its inverse must balance, so $G = R$. The balance of generation and recombination results in the steady-state intrinsic concentrations.

Consider thermal R-G. The generation rate depends on the temperature, which supplies the thermal energy, and the band gap. The larger the band gap, the harder it is to generate carriers. Recombination is proportional to np because it takes both an electron and hole to recombine. We conclude that

$$G(T, E_G) \propto np\,.$$

Now dope the intrinsic semiconductor N-type. The electron concentration increases greatly, but np cannot change because recombination must balance generation, and the doping has not changed the generation rate. The concentration of holes must drop to keep np constant. That is why n_i^2 is independent of doping.

21.8 Summary

To summarize, when we introduce excess carriers into a semiconductor, the semiconductor responds by trying to restore equilibrium, which means removing the excess carriers. The simplest and most common case is when the perturbation decays exponentially in time with a characteristic time we call the **minority carrier lifetime**, Assuming low level injection in a P-type semiconductor, the time-dependent excess minority electron concentration is

$$\Delta n(t) = \Delta n(t=0)e^{-t/\tau_n}\,.$$

The minority carrier lifetime, τ_n, is controlled by different types of physical processes. Depending on the particular semiconductor, the doping density, and the concentration of defects, the dominant process might be radiative processes characterized by the radiative lifetime, τ_n^{rad}, Auger processes characterized by the Auger lifetime, τ_n^{Auger}, or defect assisted processes characterized by the Shockley-Read-Hall lifetime, τ_n^{SRH}. It's also possible that all of those processes might be important at the same time in which case we find the overall minority carrier lifetime, τ_n, by adding the rates of recombination:

$$\frac{1}{\tau_n} = \frac{1}{\tau_n^{rad}} + \frac{1}{\tau_n^{Auger}} + \frac{1}{\tau_n^{SRH}}\,.$$

An analogous set of equations applies to minority carrier holes in N-type semiconductors.

Note also that the recombination rate can be less than zero. When that happens, we have a generation process. This occurs when the excess minority carrier concentration is negative, and the semiconductor generates minority carriers to return to equilibrium. Thermal generation processes of this kind are often the dominant component of a PN diode in reverse bias.

In addition to the recombination-generation processes we have discussed, there are also some explicit generation processes. For example, if we shine light on a semiconductor, we generate electron-hole pairs. If we apply a large electric field, we can also generate electron-hole pairs. The subject of the next lecture, the final lecture in this unit, is a discussion of **generation processes**.

21.9 References

Minority carrier recombination is important for devices such as bipolar transistors and solar cells, which rely on minority carriers for their operation. For a much more thorough discussion of recombination, including a derivation of the SRH formula for defect-assisted recombination, see:

[1] R.F. Pierret, *Advanced Semiconductor Fundamentals*, (2nd Ed.), Pearson Education, Inc., 2003 (see Ch. 5).

For another thorough and excellent discussion of carrier recombination and generation, see:

[2] J.A. del Alamo, *Integrated Microelectronic Devices*, Pearson Education, Inc., 2018 (see Ch. 3).

Lecture 22

Carrier Generation

22.1 Introduction

Part 4 is about charge carriers (electrons in the conduction band and holes in the valence band) under out-of-equilibrium conditions – the normal operating condition of a semiconductor device. We have discussed the electron and hole current densities, carrier recombination, and a little about carrier generation. In this final lecture of Part 4, we take deeper look at carrier generation.

As discussed in the previous lecture, if we inject excess minority carriers so that $np > n_i^2$, the semiconductor responds with recombination processes that restore equilibrium. If we remove carriers so that $np < n_i^2$, the recombination rate is negative, which means that carriers are being generated. Thermal generation of electron-hole pairs breaks covalent bonds and produces the intrinsic carriers. In thermal equilibrium, the balance of thermal generation and recombination is what makes $n_0 p_0 = n_i^2$.

In this lecture, we'll discuss the **optical absorption** of light incident on a semiconductor; optical absorption is the inverse of radiative recombination. We'll also discuss two mechanisms that generate electron-hole pairs in the presence of a high electric field. The first is **impact ionization**, which is the inverse of Auger recombination, and the second is **field ionization**.

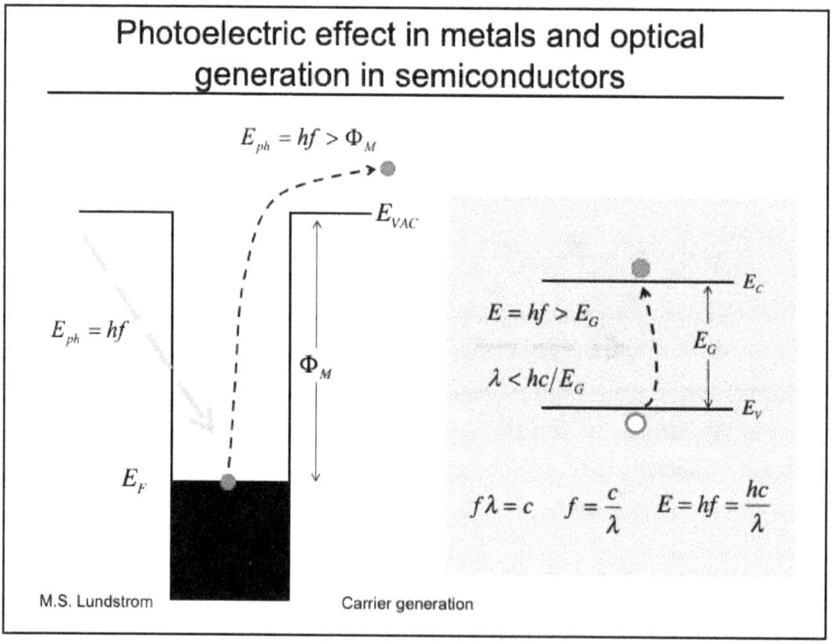

22.2 Optical absorption

When we shine light on a metal, electrons can be ejected out of the potential well that holds them in the metal into the vacuum and collected by a plate with a positive voltage on it. This **photoelectric effect** was explained in 1905 by Albert Einstein using the idea that light behave as particles. Photoemission occurs when the energy of the photons, $E = hf$, is greater than the **work function** of the metal. For a discussion of the photoelectric effect see `http://en.wikipedia.org/wiki/Photoelectric_effect`.

Shown above on the right is a simple energy band diagram of a semiconductor with a band gap E_G. If photons with energy, $E = hf > E_G$ are incident on a semiconductor, then there is sufficient energy to break the covalent bond, move an electron from the valence band into the conduction band, and create two mobile charge carriers for each photon, one electron and one hole.

Frequency times wavelength is the speed of light, $f\lambda = c$ from which we can express the photon energy in terms of the wavelength, as shown above on the right. To generate electron-hole pairs, the frequency of the light has to be high enough or, equivalently, the wavelength has to be short enough.

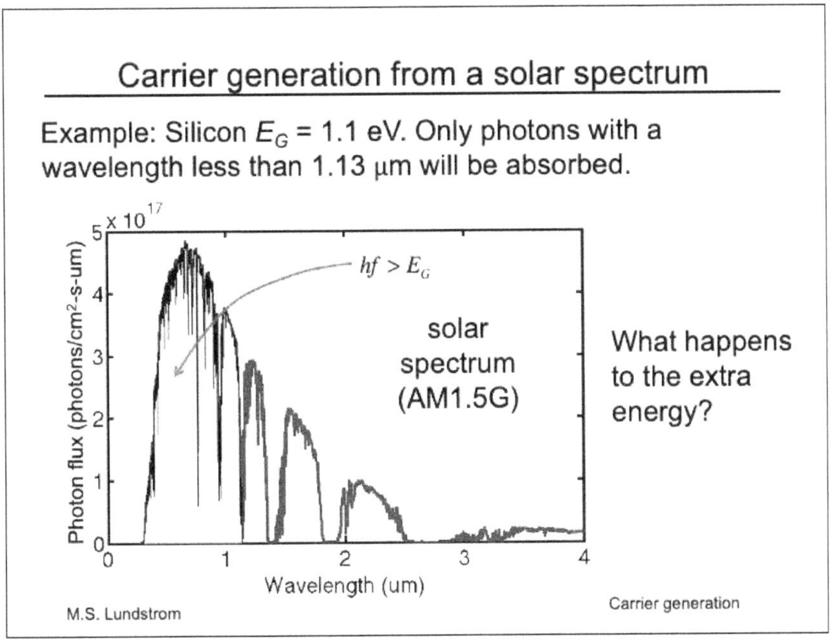

Shown above is a solar spectrum as measured on the surface of the Earth. (This particular solar spectrum is typical of the mid latitudes in the United States.) The plot is of the intensity (the number of photons per square centimeter per second per micrometer of the incident spectrum that is incident on an area) as a function of the wavelength of the light. The spectrum has some strong notches in it due to various absorption processes in the atmosphere that remove photons with some specific energies.

For light with a wavelength less than 1.13 micrometers, the photons have enough energy to create electron-hole pairs in Si. The fraction of the solar spectrum shown in yellow above can be used to create electron-hole pairs in Si. It is these photons that allow silicon solar cells to convert sunlight into electricity.

Many of the photons in the solar spectrum don't have sufficient energy to create electron-hole pairs; they simply transmit through the semiconductor. Many of the photons have more than enough energy to create electron-hole pairs. What happens to the extra energy?

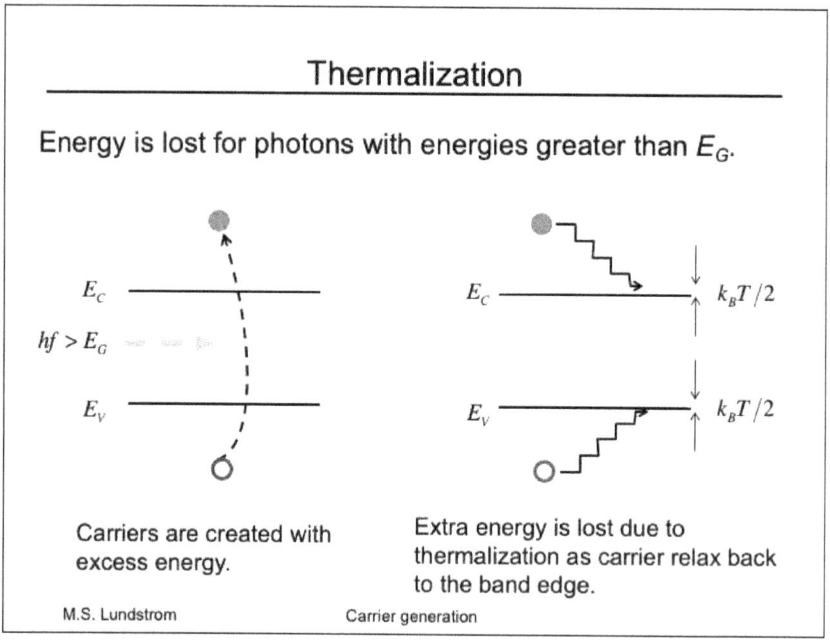

If the absorbed photon has more energy than is necessary, it can excite an electron from deep in the valence band and place the electron deep in the conduction band. This is shown above on the plot on the left. What happens then?

Particles, including charge carriers in semiconductors, tend to move down in energy. The extra initial energy is shed by various scattering processes. Very quickly (tens of femtoseconds or so), the electrons will end up at the bottom of the conduction band. As we've discussed before, in equilibrium, the peak in $n(E)$ is about $k_BT/2$ from the bottom of the conduction band.

What about holes? Remember that energy band plots are plots of the electron energy. Electron energy increase as we go up and decreases as we go down. For holes, it's just the opposite. When holes move up on an energy band diagram, it means that their energy is decreasing. So what happens is that the initial energetic hole quickly dissipates its excess energy, and the hole ends up near the top of the valence band.

The process shown above on the right is known as **thermalization**. It happens very quickly. After thermalization, the electrons and the holes are near the edges of the bands, just as they are in equilibrium.

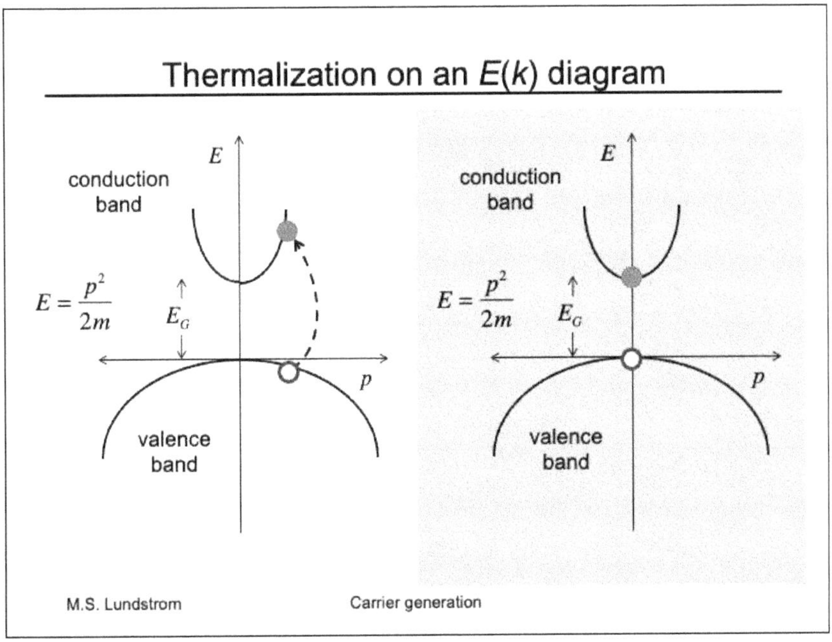

We can also show hot (energetic) carrier generation and thermalization process on an $E(k)$ plot. Shown above on the left are the excited hole and electron on the $E(k)$ plot. What happens next? Electrons will move down to lower their kinetic energy and holes will move up to shed their kinetic energy.

After a few 10's of femtoseconds, we have the situation shown above on the right. Electrons have moved down in potential energy (down on the $E(k)$ plot), and end up near the bottom of the conduction band. Holes have moved down in potential energy (up on the $E(k)$ plot), and end up near the top of the valence band. The excess energy released during the thermalization process would typically be released as heat. That's an undesirable side-effect of this process. This above band gap energy is wasted in a solar cell, just as the energy of the sub-band gap photons is wasted.

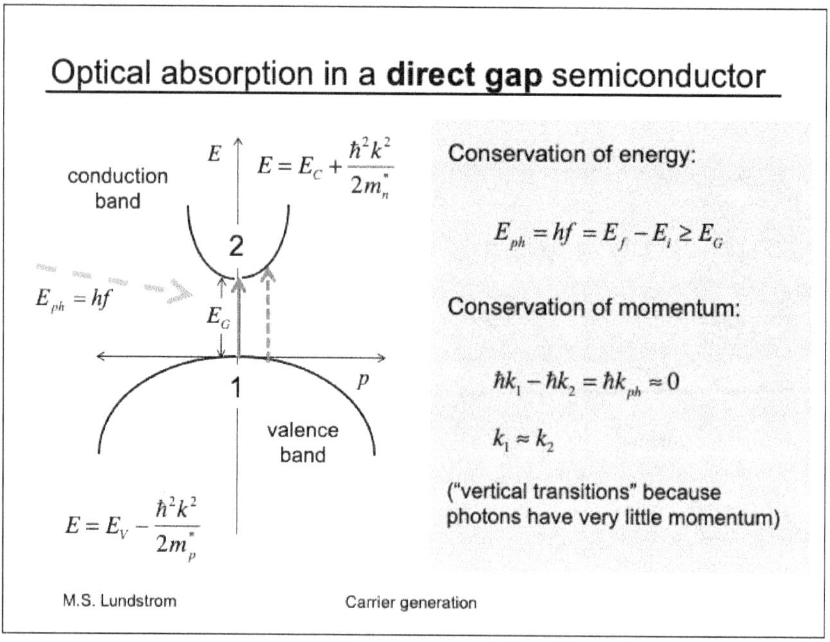

The slide above illustrates optical absorption in a direct-gap semiconductor. For direct gap semiconductors, the maximum energy in the valence band is at crystal momentum zero. The minimum energy in the conduction band is at also at crystal momentum zero. If we shine light on the semiconductor, we can move an electron from state 1 in the valence band to state 2 in the conduction band and create an electron-hole pair. This process must conserve energy and crystal momentum. The energy that it takes to create that electron-hole pair comes from the photon. The initial momentum of the electron in the valence band is $\hbar k_1$. The final momentum of the electron in the conduction band is $\hbar k_2$, which is very close to $\hbar k_1$ because the momentum of the photon is very small. We refer to these transitions as **vertical transitions**. The solid line is a photon with energy, $E = E_G$, and the dashed line is one with $E > E_G$.

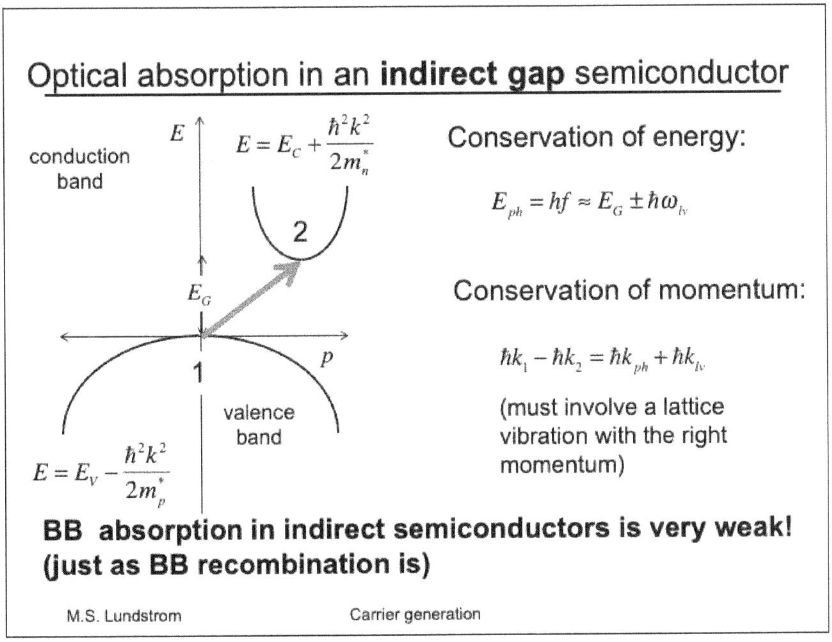

Things are different for indirect-gap semiconductors. As shown above, in an indirect-gap semiconductor, the maximum of the valence band is typically still at $k = 0$, but now the minimum of the conduction band is somewhere else, typically near the edge of the Brillouin zone. There is now a significant difference in the crystal momentum of state 1 in the valence band and state 2 in the conduction band. During the absorption process, an electron moves from the valence band to the conduction band. As shown above, the transition is not vertical for indirect gap semiconductors. To conserve momentum, a lattice vibration is needed, because the photon carries little momentum. A quantized lattice vibration, a **phonon**, will supply the additional momentum that is needed. That lattice vibration will also have some energy, so we have to account for that in energy conservation as well.

Optical absorption in an indirect-gap semiconductor is less probable than in a direct gap semiconductor because we need to find a phonon with a proper momentum to facilitate the transition. That reduces the probability that a photon with enough energy can be absorbed. We can see this very clearly by examining typical results for optical absorption in direct gap and indirect gap semiconductors.

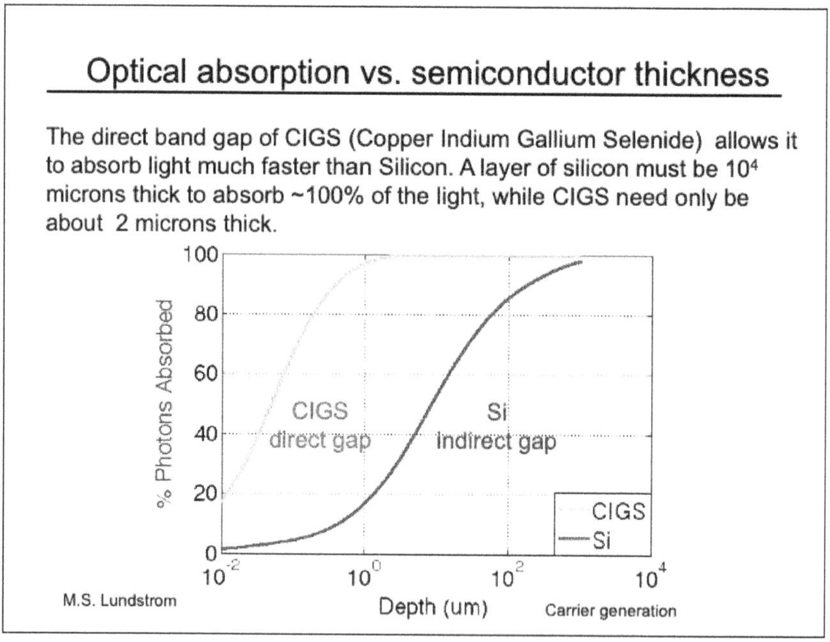

The plots above show optical absorption in two semiconductors; one is copper indium gallium selenide (CIGS), a direct-gap semiconductor used for solar cells. The other is silicon, also commonly used for solar cells. Silicon is an indirect-gap semiconductor. The incident illumination is a solar spectrum (AM1.5G). The plots show the percent of the photons with energy above the band gap absorbed in a given thickness of the semiconductor. For the direct-gap semiconductor, absorption occurs very rapidly. Only one micrometer of CIGS is needed to absorb most of the photons that can be absorbed (i.e. those with $E > E_G$). The process is much weaker in silicon. For Si, hundreds, actually thousands, of micrometers are needed to absorb every photon with an energy above the band gap.

Direct gap semiconductors are efficient optical absorbers, indirect gap semiconductors are not. Optical absorption and radiative recombination are inverse processes. Radiative recombination is strong in direct gap semiconductors and weak in indirect gap semiconductors.

Absorption coefficient

Incident flux: Φ_0

Flux at position, x: $\Phi(x) = \Phi_0 e^{-\alpha(\lambda)x}$

optical absorption coefficient:

$$\alpha(\lambda) > 0 \quad \text{for} \quad E > E_G \; (\lambda < hc/E_G)$$

Generation rate at position, x

:

$$G(x) = -\frac{d\Phi(x)}{dx} = \Phi_0 \alpha(\lambda) e^{-\alpha(\lambda)x}$$

$\alpha(\lambda)$ direct gap → higher alpha

M.S. Lundstrom Carrier generation

The optical **absorption coefficient**, α, is an important material parameter for semiconductors used to make photodetectors or solar cells. An incident flux of light with photons having an energy greater than E_G is absorbed as it penetrates into the semiconductor. The intensity of the light decreases exponentially with distance into the semiconductor as photons are absorbed. The characteristic length is $1/\alpha$, where α is the absorption coefficient in inverse meters. One over α is approximately the depth at which all of the light is absorbed. The absorption coefficient is greater than zero only for photons with an energy larger than the band gap.

The **optical generation rate** of electron-hole pairs is what we're interested in. The intensity of the optical flux decreases as photons are absorbed and electron-hole pairs are created. The optical generation rate is minus the derivative of the optical flux. The expression shown above gives the generation rate of electron-hole pairs in terms of the incident optical flux and the absorption coefficient of the semiconductor. The absorption coefficient is an important parameter; it's very large in direct gap semiconductors and much smaller in indirect gap semiconductors

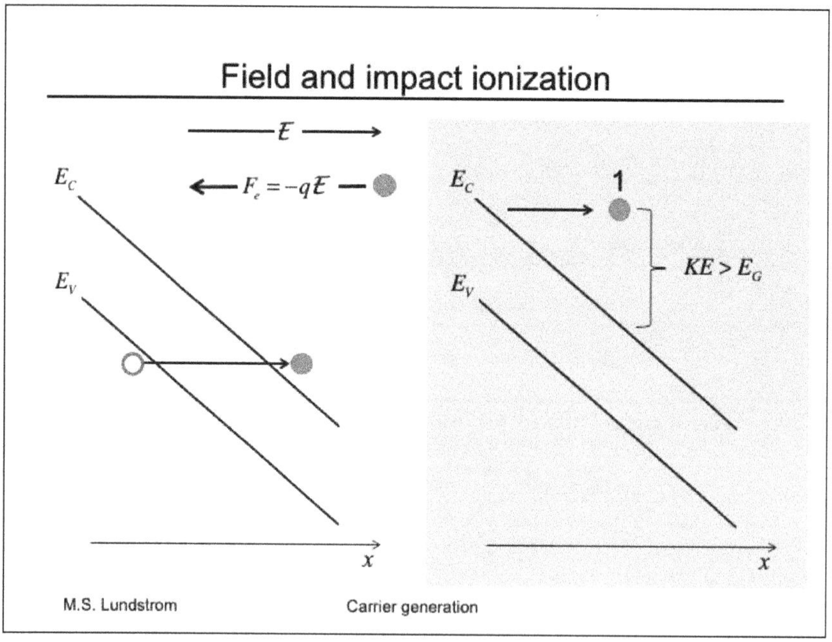

22.3 Field and impact ionization

Electron-hole pairs can also be generated by a strong electric field. An electric field exerts a force on electrons of $F_e = -q\mathcal{E}$, which changes the energy of electrons. As we'll discuss in Part 5, the slope of the energy band diagram is proportional to the electric field. Shown above are energy band diagrams for a semiconductor with a strong electric field.

If the electric field is strong enough, it can break covalent bonds and produce electron-hole pairs. As shown above on the left, this **field ionization** process occurs when an electron in the valence band quantum mechanically tunnels through the barrier of height, E_G and produces an electron in the conduction band. This process is known as **Zener tunneling.** Another process that generates electron-hold pairs under high electric fields is **impact ionization.** As shown above on the right, the electric field exerts a force on electrons in the conduction band, and electrons gain energy as they accelerate under the influence of this force. As shown on the next slide, the electron labeled 1 above can initiate an **impact ionization** event because $E_1 - E_c > E_G$.

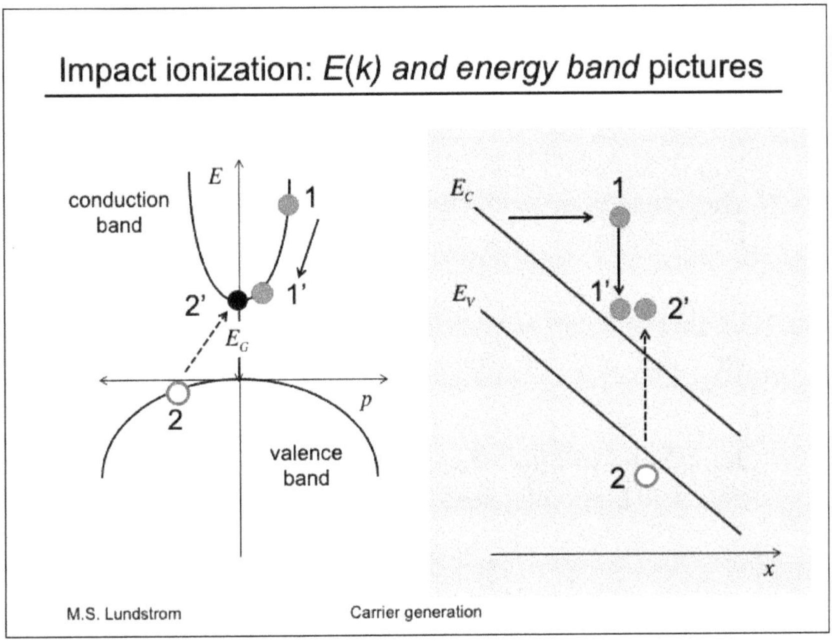

Impact ionization is shown above at the left on an $E(k)$ plot. In equilibrium electrons are near the bottom of the conduction band, but a very strong electric field can give electrons lots of kinetic energy and move them well above the bottom of the conduction band, as shown above for electron 1. If $(E_1{-}E_c) > E_G$, an electron hole pair can be created. Electron 1 undergoes a collision with the lattice, which breaks a bond and moves an electron from the valence band to the conduction band. The result of that process now is we have created an electron hole pair, 2 and 2'. The energy from the initial electron has been dissipated in that process, so electron 1 ends up near the bottom of the conduction band at 1'. Impact ionization is shown above at the right on an energy band diagram. A very strong electric field gives electron 1 lots of kinetic energy. Again, if $(E_1{-}E_c) > E_G$, an electron hole pair can be created. Electron 1 undergoes a collision with the lattice, which breaks a bond and moves an electron from the valence band (2) to the conduction band (2'). The energy from the initial electron now has been dissipated in that process, so electron 1 ends up near the bottom of the conduction band at 1'.

Finally, we should note that the holes can also be accelerated, gain energy, impact ionize, and create additional electron-hole pairs.

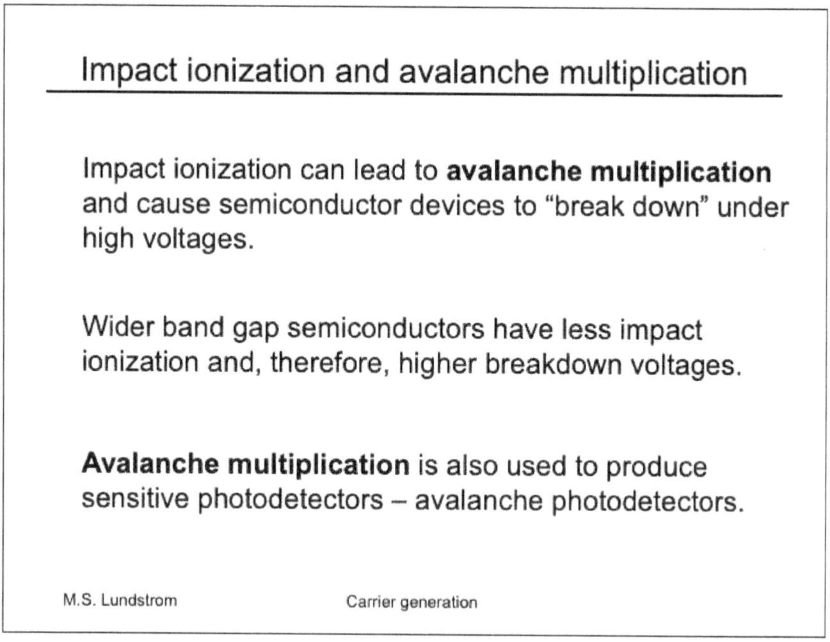

An energetic electron can initiate an impact ionization (II) event and produce a less energetic electron and an electron hole pair. Energetic holes can also initiate II events. As the process continues, more and more electron-hole pairs are generated leading to **avalanche multiplication**. This is one of the mechanisms that can cause a semiconductor device to break down under high voltages (the other is Zener tunneling). Impact ionization is analogous to what happens in the atmosphere in a lightning strike where the atmosphere breaks down in the same way. Wider band gap semiconductors are more tolerant of these processes because it takes more energy to create electron-hole pairs. As a result, wide band gap semiconductor devices have higher breakdown voltages due to impact ionization. If we're looking to use semiconductor devices at high voltages, we can make use of semiconductors that have large band gaps.

Avalanche multiplication can be a detrimental process when it leads to breakdown, but it can also be useful. For example, there are devices that are called **avalanche photomultipliers**. These devices can be so sensitive that a single electron creating a single electron hole pair can, through this avalanche multiplication process, create large enough currents that one can detect a single photon.

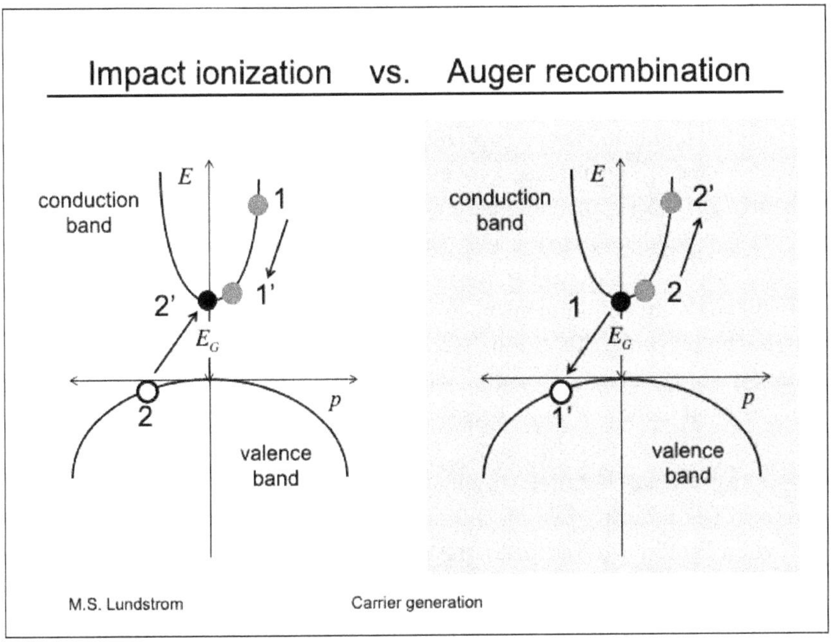

Shown at the left above is impact ionization initiated by an energetic electron, 1, which loses an energy greater that E_G and moves to a state near the bottom of the conduction band, 2. The energy difference, $E_1 - E_{1'} \geq E_G$ moves an electron 2 from a state in the valence band to 2', a state in the conduction band.

Shown above at the right is the inverse process, which is actually Auger recombination. In this case, electron 1 recombines with a hole in the valence band, 1'. The energy released in this process, $E_1 - E_{1'} \geq E_G$, is given to electron 2, which moves it up in the conduction band to 2'. The figures on the slide above are for an N-type semiconductor. See if you can draw the corresponding figures for a P-type semiconductor.

22.4 Summary

In this lecture, we discussed optical generation, the inverse process of radiative recombination. As mentioned in the previous lecture, radiative recombination is strong in direct gap semiconductors and weak in indirect gap semiconductors. The inverse process, optical absorption, takes place very efficiently in direct-gap semiconductors, and it takes place less efficiently in indirect gap semiconductors. Optical absorption in semiconductors is used to create photodectors and solar cells.

Field and impact ionization are processes that generate electron-hole pairs in the presence of a strong electric field. Impact ionization is actually the inverse of the Auger recombination process. Impact and field ionization can cause semiconductors to break down under high applied voltages. Impact ionization can also be used to created very sensitive photon detectors.

In the previous lecture, we mentioned carriers can recombine through defects in the middle of the forbidden gap, but carriers can also be generated when they're excited from the valence band to the defect and then to the conduction band. This thermal generation is the inverse of the Shockley-Read-Hall recombination process.

This wraps up Part 4 of these lectures. You should now:

- Understand the general current equation and what is meant by transmission, channels, Fermi window, and ballistic vs. diffusive transport.
- Understand how the current in a nano-device is related to the Fermi levels in the two contacts and how the current in the bulk is related to the gradient of the quasi-Fermi level.
- Be able to compute the drift velocity from the electric field in a uniform semiconductor and understand conductivity and resistivity.
- Be able to relate the mobility to effective mass and momentum relaxation time and compute the diffusion coefficient given the mobility.
- Understand the drift-diffusion equation, its relation to the general current equation, and what is meant by high-field transport.
- Be familiar with important recombination processes, low-level injection, and how to determine the minority carrier lifetime when several processes occur simultaneously.
- Understand thermal generation, impact ionization, and optical absorption in direct and indirect gap semiconductors.

We are now ready for Part 5, which pulls together everything we've discussed in these lectures.

22.5 Reference

For a deeper look at optical absorption including the different absorption coefficients, $\alpha(E)$, for direct and indirect gap semiconductors, see Ch. 7 in the reference below. For a deeper look at impact ionization including how energy and crystal momentum conservation leads to a threshold energy for impact ionization that is a little larger than the band gap, see Ch. 9, Sec. 3.

[1] Charles M. Wolfe, Nick Holonyak, Jr., and Gregory E. Stillman, *Physical Properties of Semiconductors*, Prentice Hall, 1989, 2003 (see Ch. 7).

PART 5

The Semiconductor Equations

Lecture 23

The Semiconductor Equations

23.1 Introduction

Part 5 will wrap up our basic introduction to semiconductor physics by pulling together key points discussed in previous parts. After completing this part, you should be able to understand the operation of almost any type of semiconductor device.

Part 5 begins with the mathematical formulation of the **semiconductor equations**, a set of three coupled, nonlinear partial differential equations that describes electrons and holes in semiconductor devices under equilibrium or non-equilibrium conditions. The second lecture is about **energy band diagrams**, plots of $E_c(x)$ and $E_v(x)$, which are a powerful tool because they provide qualitative solutions to the semiconductor equations. In the third lecture, we return to the topic of **quasi-Fermi levels** and relate them to energy band diagrams. Part 5 concludes with a discussion of the **minority carrier diffusion equation**, a simplification of the full set of semiconductor equation that is the starting point for the analysis of minority carrier devices such as solar cells and bipolar transistors. Upon completion of Part 5, you will:

- Be familiar with the three "semiconductor equations" that are the starting point for the analysis of most semiconductor devices.
- Understand how to qualitatively solve the semiconductor equations by drawing and reading energy band diagrams.
- Be able to explain what the electron and hole quasi-Fermi levels are and how they are used.
- Know when it is appropriate to simplify the semiconductor equations to the minority carrier diffusion equation (MCDE), how to do so, and how to solve the MCDE for important, simple cases.

Solving semiconductor problems

For a **uniform semiconductor in equilibrium**, we can assume space-charge neutrality:

$$\rho_0 = q\left(p_0 - n_0 + N_D^+ - N_A^-\right) = 0$$

(We also assume that the electric field is zero). Since we are assuming equilibrium:

$$n_0 p_0 = n_i^2$$

In general, neither of the above assumptions is valid. How do we solve for the electron and hole concentrations and for the electric field too?

M.S. Lundstrom Mathematical formulation

23.2 The unknowns in a semiconductor problem

In Part 3, we discussed how to solve semiconductor problems that involved uniform semiconductor in equilibrium. In general, even the equilibrium problem requires a numerical solution, but for uniform semiconductors, it's reasonable to assume that the mobile electrons and holes neutralize the ionized dopant charge, so that we can assume that the net space-charge density is zero. As shown above, this gives us one equation for two unknowns, p_0 and n_0.

We also know that in equilibrium $n_0 p_0 = n_i^2$, which gives us a second equation. Putting the two equations together, we can solve for the electron and hole concentrations for a uniform semiconductor in equilibrium. (If the dopants are not fully ionized, things get a little more complicated.) In general, however, the semiconductor may not be uniform or we may be out of equilibrium. How do we solve for the electron and hole concentrations more generally? That is the topic for this lecture. We'll also see that we're going to need to solve for the electric field (or electrostatic potential) as well as for p and n.

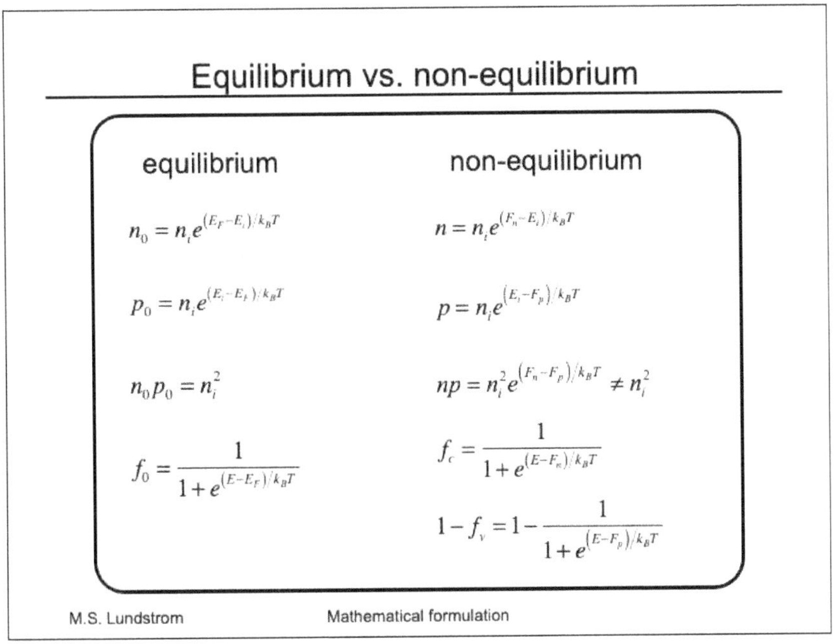

As shown above, there is a simple relation between the equilibrium electron and hole concentrations and the Fermi level and the intrinsic level. In equilibrium, we also have $n_0 p_0 = n_i^2$. The equations shown above on the left follow from the fact that in equilibrium the probability that a state is occupied is given by the Fermi function, f_0.

The equations look similar out of equilibrium, but the electron density is related to the quasi-Fermi level (QFL) for electrons, which is analogous to the Fermi level in equilibrium. Similarly, the hole concentration is related to the QFL for holes. The np product may not (or may) equal n_i^2. By multiplying the first two expressions on the right, we see that np depends on the separation of the QFL's.

The probability that a state in the conduction band is occupied, f_c, is no longer given by the equilibrium Fermi function, f_0. We will assume that it is still given by a Fermi function but with the Fermi level replaced by the QFL for electrons. The probability that a state in the valence band is empty is $1 - f_v$, but f_v is not the Fermi function, instead, it is a Fermi function with the Fermi level replaced by the QFL for holes. We will discuss QFLs in more detail in Lecture 25, but the question now is: "How do we solve for the carrier densities or equivalently for the QFLs out of equilibrium?"

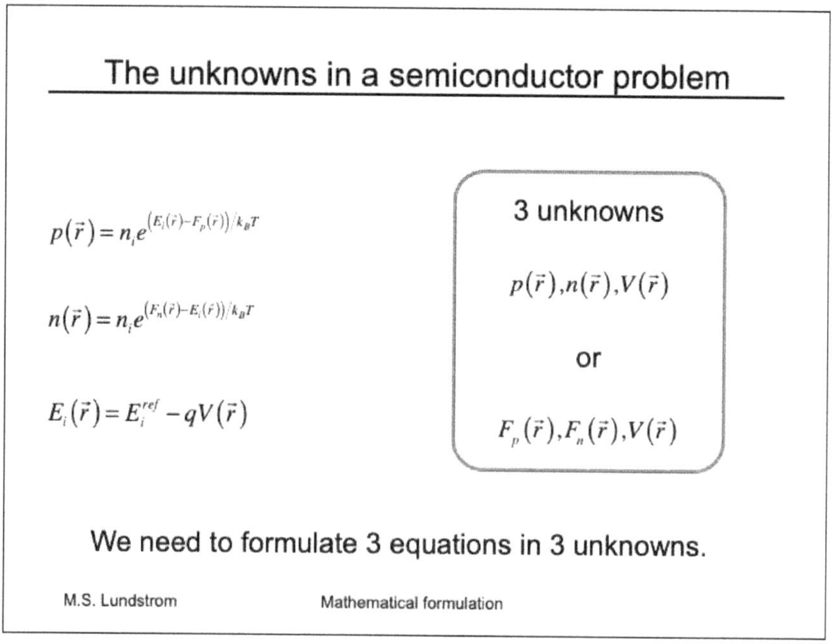

Two of the unknowns in semiconductor problems are the carrier concentrations, which are related to the quasi-Fermi levels and to the intrinsic level (or to E_c or E_v). When we have an energy, then as shown above, we have to define a reference, but there's also an electrostatic potential due to the electric fields that are set up by the charged carriers and ionized dopants. The position-dependent electrostatic potential means that the intrinsic level (as well as E_c and E_v) varies with position.

We conclude that there are three unknowns in our semiconductor problem, two carrier densities and the electrostatic potential (or the gradient of the electrostatic potential, which is the electric field). Alternatively, we could take our three unknowns as the two quasi-Fermi levels and the electrostatic potential.

In either case, we need three equations for the three unknowns. Those three equations are the mathematical description of a semiconductor. They are what we call **the semiconductor equations**.

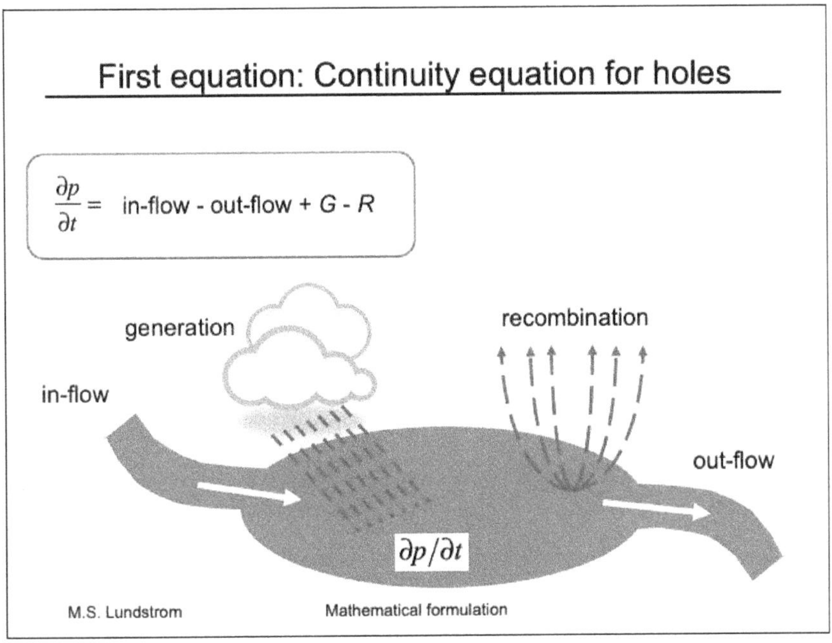

23.3 The continuity equation

The first of the semiconductor equations is an equation for the hole density. The result is a **continuity equation**. The slide above shows how to think about a continuity equation. Assume there is a lake with a water level that may be changing with time. We seek an equation for the time-dependent water level. Let $p(t)$ denote the water level, because the water level is analogous to the hole concentration in our semiconductor.

If a stream flows into the lake, the inflow of water increases the level of the lake, but if the stream flows out the other side, the outflow lowers the level. If the inflow is greater than the outflow, the level will rise.

There might also be generation processes analogous to the generation processes we have discussed for semiconductors. For example, rain would increase the water level in the lake. There might also be recombination processes. For example, evaporation would lower the level of the lake.

We are just writing a bookkeeping equation to account for all of the ways that the water level can change. The rate of increase is equal to the net flow in, plus the generation rate minus the recombination rate. This a continuity equation.

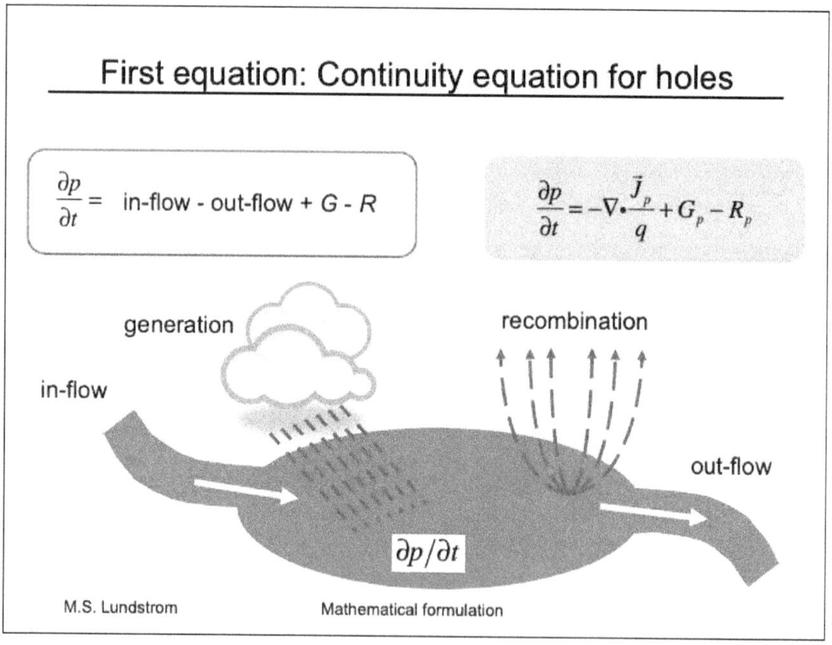

To continue our discussion, remember from vector calculus that **divergence of a vector** represents a net flow **out** of the region. We need a net flow **in** for our continuity equation. For the hole continuity equation, the net flow of holes into a region is minus the divergence of the hole flux, as shown above on the right. This is the first of our three equations, a book-keeping equation to account for how holes come and go from the region of interest. The current density, $\vec{J}_p$, is just the drift-diffusion equation. We use G_p to denote "external" generation processes, such as generation due to an incident optical illumination, and R_p to denote "internal" processes such as SRH recombination-generation. There is, of course, an analogous continuity equation for electrons, which is the second of our three equations. The only thing to be careful about is the sign of the charge on an electron. The electron current density is $\vec{J}_n$, and the flux of electrons is $\vec{J}_n/(-q)$, so the net flow of electrons out of the region of interest is $-\nabla \cdot \vec{J}_n/(-q) = +\nabla \cdot \vec{J}_n/(q)$. We conclude that the electron continuity equation is

$$\frac{\partial n}{\partial t} = -\nabla \cdot \frac{\vec{J}_n}{(-q)} + G_n - R_n = +\nabla \cdot \frac{\vec{J}_n}{q} + G_n - R_n \,.$$

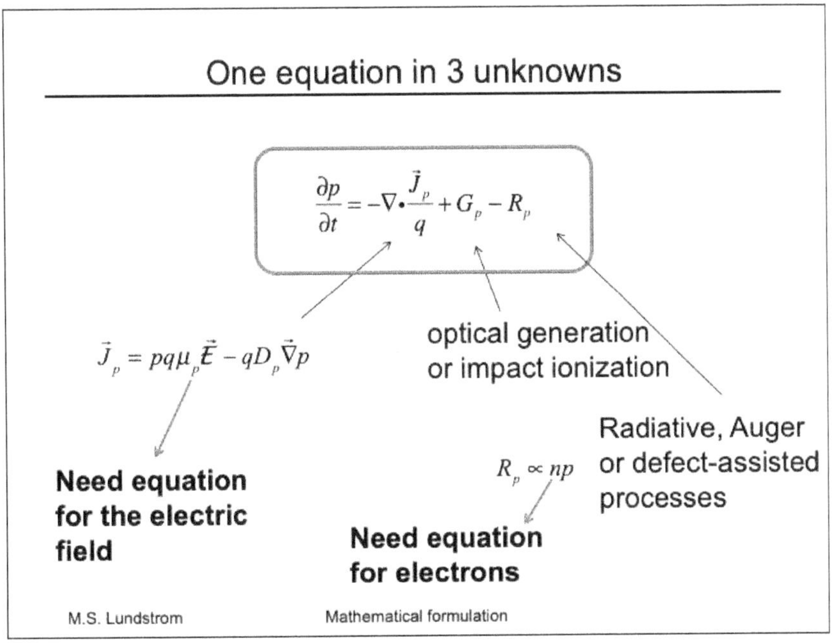

Let's take a closer look at the hole continuity equation shown above. The hole current density is on the RHS; we already have a hole current equation, it's the drift-diffusion equation, which involves the electric field, the hole density and its gradient. Because the hole continuity equation involves the electric field, we'll need to formulate an equation for the electric field, or for the electrostatic potential (minus the gradient of the electrostatic potential is the electric field).

As mentioned on the previous slide, the generation term on the RHS might describe the generation of holes due to light or impact or field ionization, for example. Finally, R_p, the recombination term, which, when it is negative, is also a generation term. The recombination mechanisms we discussed were proportional to $(np - n_i^2)$, so the hole continuity equation also involves the electron concentration.

The hole continuity equation involves the hole density, the electron density, and the electric field. We have one equation in three unknowns. The drift-diffusion equation contains the product of two unknowns, p and $\mathcal{E}$, so we have one nonlinear equation in three unknowns. Two more equations are needed – the first is the electron continuity equation that we already discussed, and the other equation that we need is one for the electric field.

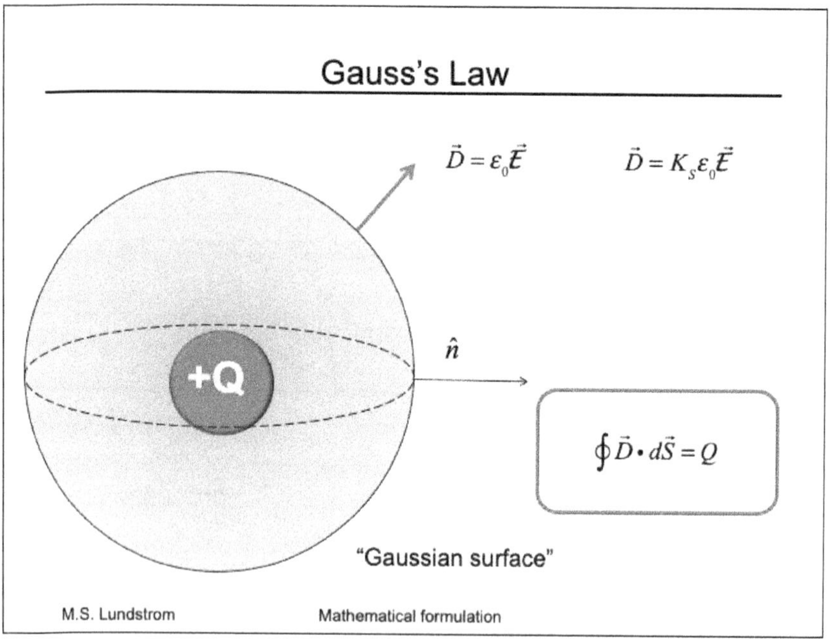

23.4 Gauss's law

Let's review Gauss's Law, which is typically discussed in a first-year physics course. Shown above is a charge inside a spherical volume enclosed by a **Gaussian surface.** The displacement vector, $\vec{D}$, emanating from that charge intersects the surface and is normal to it. In vacuum, the displacement vector is the electric field multiplied by the permittivity of free space, ϵ_0. In a semiconductor, the material will respond and polarize, and the permittivity of free space needs to be multiplied by the relative dielectric constant of the semiconductor, K_s, which is on the order of ten or so in common semiconductors. Gauss's law says that if we integrate the normal component of the displacement field over the surface that encloses the charge, the result will equal the charge enclosed. This is the integral form of Gauss' Law; there is also a differential form.

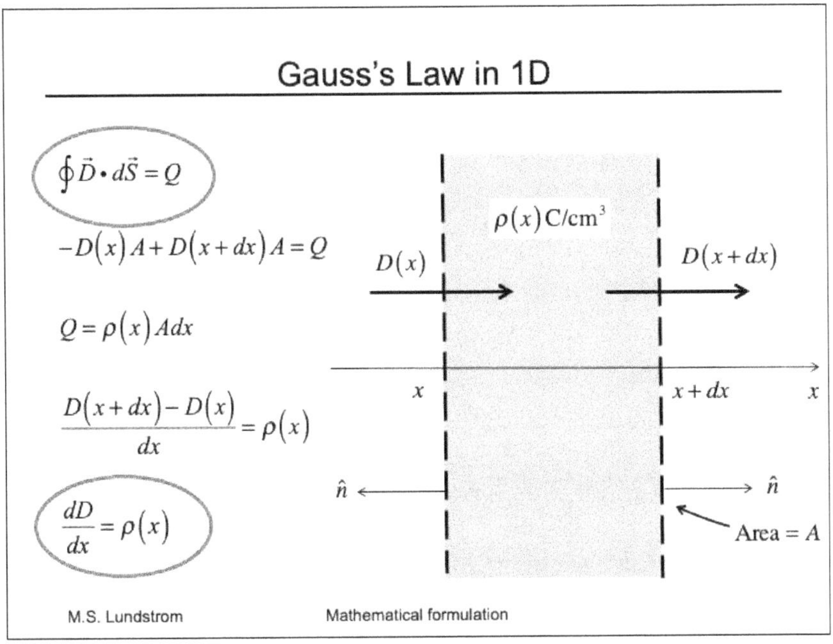

Shown above is Gauss's Law in one dimension. Assume a large plane coming out of the page with cross-sectional area, A. The first plane is located at x and there is another one at $x + dx$. These two planes enclose a charge, $\rho\, dx$, where ρ is the space-charge density in C/cm^3. At $x + dx$, the outward normal vector of the surface points in the same direction as $\vec{D}$. At x, the outward normal to the surface points in the $-x$ direction – the opposite direction that $\vec{D}$ is pointing. When we do the integral, we get a positive contribution from the surface at $x+dx$ and a negative contribution from the surface at x. Finally, by re-arranging the result, we obtain, $dD/dx = \rho(x)$.

We started with the **integral form of Gauss's law** at the top on the left and ended up with the **differential form of Gauss's law** on the bottom on the left. Since $\vec{D}$ is proportional to the electric field, we now have our third equation, an equation for the electric field.

It may look as though we've added another unknown because $dD/dx = \rho$, where ρ is the space charge density, but the space charge density is $\rho = q(p - n + N_D^+ - N_A^-)$, and we already have equations for p and n — the hole and electron hole continuity equations. The ionized donor and acceptor concentrations are known because we presumably know how we doped the semiconductor. So now let's put this all together.

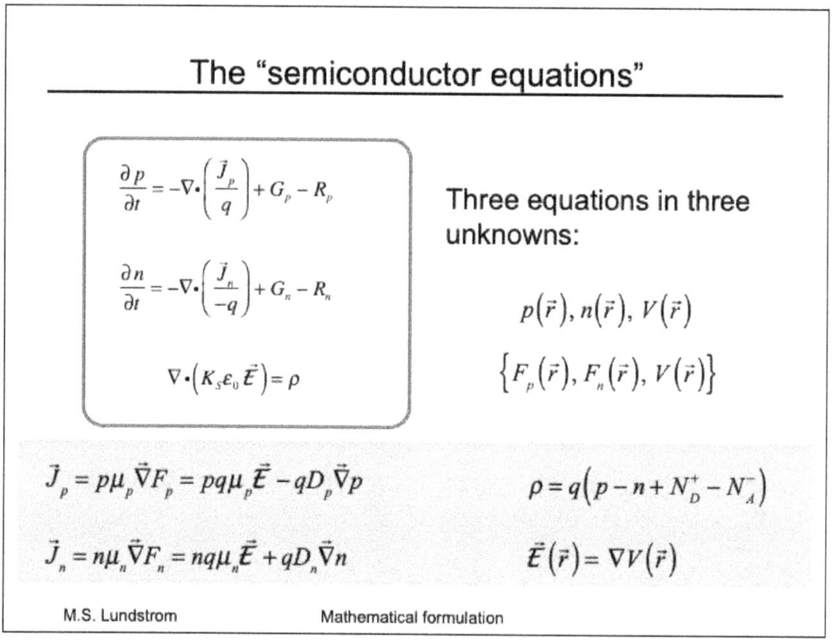

23.5 The semiconductor equations

The equations shown above are sometimes called the **semiconductor equations** because they're often taken as the mathematical description of a semiconductor device. There is a continuity equation for holes, a continuity equation for electrons, and Gauss's law for the electric field. We often use the fact that the electric field is minus the gradient of the electrostatic potential to write the last equation in terms of the electrostatic potential, in which case, it is called the **Poisson equation**. We can write:

$$\nabla \cdot \left(K_s \epsilon_0 \vec{\mathcal{E}}\right) = \nabla \cdot \left(K_s \epsilon_0 (-\vec{\nabla} V)\right) = -K_s \epsilon_0 \nabla^2 V = \rho\,,$$

where we have assumed that K_s is spatially uniform.

The equations in the gray box are constitutive relations that we use in the three semiconductor equations so that they are reduced to three equations in three unknowns. (Note that we would also need similar equations to express recombination and impact and field ionization in terms of n, p, and $\mathcal{E}$.)

We now have three equations and three unknowns; if we solve these equations, we can simulate the operation of a semiconductor device.

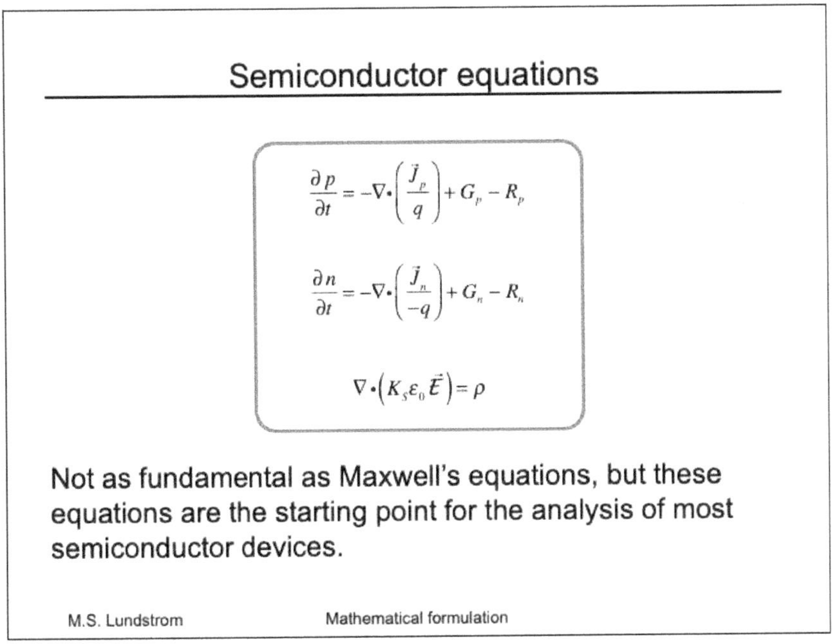

The semiconductor equations are not as as fundamental as Maxwell's equations. For example, Gauss's law (Poisson's equation) ignores the time dependences that give rise to electromagnetic waves, and the drift-diffusion equation is only valid near equilibrium and for time scales longer than the momentum relaxation time, but although we've made approximations, these equations work remarkably well for most semiconductor devices; they're usually our starting point when we encounter a new device. The semiconductor equations are three coupled, nonlinear partial differential equations. It's not trivial to solve them. Today, we have powerful computer simulation techniques that provide numerical solutions to these equations. If you work in semiconductors, you want to be familiar with using these **computer-aided design** or **CAD tools**, but you will also want a deep understanding of how these equations can be simplified under specific conditions that will allow you to *understand* the results produced by such computer simulations. It takes a good understanding of the underlying physics in order to simplify the equations appropriately; it also takes some intuition that comes with experience.

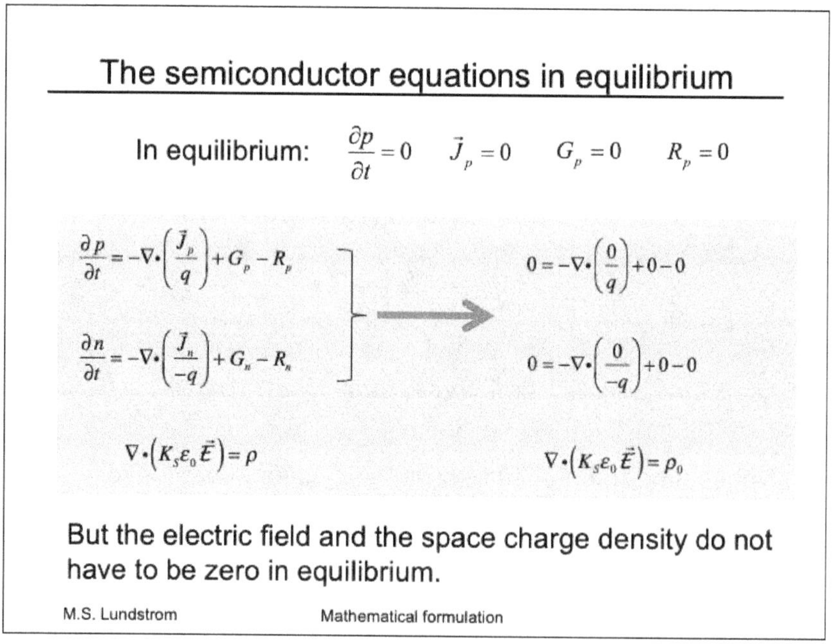

23.6 The semiconductor equations in equilibrium

In equilibrium the semiconductor equations simplify considerably. As shown above, nothing changes with time in equilibrium, so $\partial p/\partial t = \partial n/\partial t = 0$. There is no current in equilibrium, so $\vec{J}_p = 0$. There is no external generation of carriers such as by optical illumination, so $G_p = 0$. There are microscopic generation and recombination processes occurring, but according to **the principle of detailed balance**, each transition from the valence to conduction band is balanced by a transition from the conduction band to the valence band, so $R_p = 0$. The same considerations apply to electrons, so in equilibrium, the two continuity equations just reduce to $0 = 0$.

As shown on the next slide, however, there can be electric fields in equilibrium.

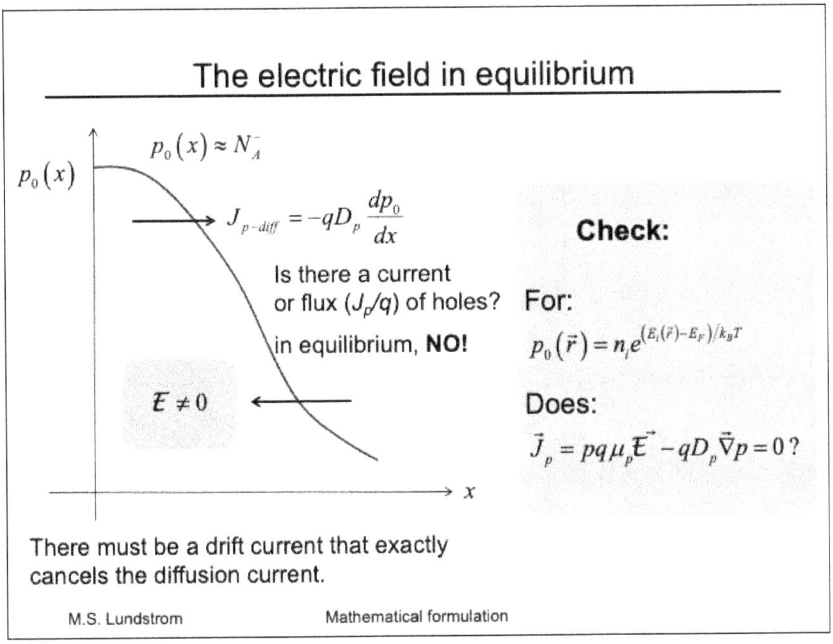

Consider a problem we discussed in Part 4, a nonuniformly doped semiconductor in equilibrium. As shown above, it's doped more heavily with acceptors near $x = 0$ than it is in the bulk, so there is a gradient in the hole concentration. Holes are mobile, they're in random thermal motion, so they will diffuse down the concentration gradient to the right. There'll be a diffusion flux, but is there a current, a net flux of holes? As shown in the gray box on the right, we can show that if we plug the known solution for $p_0(\vec{r})$ into the current equation, we find, $\vec{J}_p = 0$.

How can there be no current flowing when we have a diffusion current? There must be a drift current exactly equal and opposite in sign to the diffusion current that give us zero total current. There must be an electric field to produce this drift current.

When the doping is nonuniform as in this example, the mobile charge carriers are not able to completely cancel out the charge of the ionized dopants, so $p_0(\vec{r}) \approx N_A(\vec{r})$, and there is a small space charge that produces an electric field, and an electric field implies a spatially varying electrostatic potential.

How do we solve for $V(\vec{r})$ in equilibrium? From $V(\vec{r})$, we can take the gradient and find the equilibrium electric field.

Poisson-Boltzmann equation

$$\nabla \cdot \left(K_S \varepsilon_0 \vec{\mathcal{E}}\right) = \rho_0 \qquad \rho_0 = q\left(p_0 - n_0 + N_D^+ - N_A^-\right) \qquad \vec{\mathcal{E}}(\vec{r}) = -\vec{\nabla} V(\vec{r})$$

$$\nabla^2 V = -\left(\frac{q}{K_S \varepsilon_0}\right)\left(p_0 - n_0 + N_D^+ - N_A^-\right)$$

$$p_0(\vec{r}) = n_i e^{(E_i(\vec{r}) - E_F)/k_B T} \qquad n_0(\vec{r}) = n_i e^{(E_F - E_i(\vec{r}))/k_B T}$$

$$E_i(\vec{r}) = E_i^{ref} - qV(\vec{r}) \qquad E_i^{ref} = E_F$$

$$p_0(\vec{r}) = n_i e^{-qV(\vec{r})/k_B T} \qquad n_0(\vec{r}) = n_i e^{qV(\vec{r})/k_B T}$$

$$\nabla^2 V = -\left(\frac{q}{K_S \varepsilon_0}\right)\left(n_i e^{-qV(\vec{r})/k_B T} - n_i e^{+qV(\vec{r})/k_B T} + N_D - N_A\right)$$

M.S. Lundstrom Mathematical formulation

In equilibrium, only one of the three semiconductor equations needs to be solved — Gauss' Law as shown on the first line above. The electric field is minus the gradient of the electrostatic potential, so the Poisson equation can be written as in the second line. As shown on the third line, the equilibrium carrier densities are known in terms of the position-dependent intrinsic level, and the fourth line shows how the intrinsic level is position dependent because the electrostatic potential is position dependent.

Energies are measured with respect to an arbitrary reference. Choosing the reference for the intrinsic level to be the intrinsic Fermi level, $E_i^{ref} = E_F$, we can very simply relate $p_0(\vec{r})$ and $n_0(\vec{r})$ to the electrostatic potential, $V(\vec{r})$.

Finally, we insert the expressions for the carrier densities in the Poisson equation (the second equation from the top), assume complete ionization of dopants, and obtain the **Poisson-Boltzmann equation**, the one equation that we need to solve for semiconductors in equilibrium. The equation has "Boltzmann" in its name because we have assumed non-degenerate (i.e. Boltzmann) carrier statistics. The Poisson-Boltzmann equation is a non-linear partial differential equation for the one unknown in equilibrium, $V(\vec{r})$. For more on the Poisson-Boltzmann equation, see [2].

23.7 Summary

In this lecture, we mathematically formulated the semiconductor problem. The three semiconductor equations are:

$$\boxed{\begin{aligned} \frac{\partial p}{\partial t} &= -\nabla \cdot \frac{\vec{J}_p}{q} + G_p - R_p \\ \frac{\partial n}{\partial t} &= -\nabla \cdot \frac{\vec{J}_n}{(-q)} + G_n - R_n \\ \nabla \cdot \left(K_s \epsilon_0 \vec{\mathcal{E}}\right) &= \rho \,. \end{aligned}}$$

The first equation is a continuity equation for holes, the second a continuity equation for electrons, and the third is Gauss's law for the electric field (or the Poisson equation for the electrostatic potential). These are three equations in three unknowns — either $p(\vec{r}), n(\vec{r}), V(\vec{r})$ or $F_p(\vec{r}), F_n(\vec{r}), V(\vec{r})$, where $\vec{\mathcal{E}}(\vec{r}) = -\nabla V(\vec{r})$.

The right hand sides of the continuity equations must be expressed in terms of the unknowns, which means that the generation and recombination rates must be expressed in terms of the three unknowns. The hole and electron current densities are also functions of the three unknowns:

$$\begin{aligned} \vec{J}_p &= -p\mu_p \vec{\nabla} F_p = pq\mu_p \vec{\mathcal{E}} - qD_p \vec{\nabla} p \\ \vec{J}_n &= -n\mu_p \vec{\nabla} F_n = nq\mu_n \vec{\mathcal{E}} - qD_n \vec{\nabla} n \,. \end{aligned}$$

If we solve the semiconductor equations, we can simulate the operation of almost any semiconductor device. In order to solve these equations without a computer, we need to simplify them for the problem of interest.

Finally, we should mention that while these equations are commonly used to described the transient (i.e. switching) behavior of semiconductor devices, we have left something out. The electron and hole current densities refer to particle currents — there is also a **displacement current**,

$$\vec{J}_D = \frac{\partial \vec{D}}{\partial t} \,,$$

which becomes important when the rate of change is high. In general, the total current is the sum of the particle and displacement currents,

$$\vec{J}_{tot} = \vec{J}_p + \vec{J}_n + \frac{\partial \vec{D}}{\partial t} \,.$$

Before we discuss simplifying and solving these equations, we'll introduce in the next lecture a very powerful approach that provides us with insightful, qualitative solutions to these equations without doing any mathematics whatsoever. **Energy band diagrams** are the subject of the next lecture.

23.8 References

For a discussion of nonuniformly doped semiconductors in equilibrium, how to calculate the equilibrium electrostatic potential and carrier density, see:

[1] J.A. del Alamo, *Integrated Microelectronic Devices*, Pearson Education, Inc., 2018 (see Ch. 4, Sec. 5).

For a discussion of solving the Poisson-Boltzmann equation for semiconductors in equilibrium, see:

[2] Mark Lundstrom and Xingshu Sun, *Notes on the Solution of the Poisson-Boltzmann Equation for MOS Capacitors and MOSFETs*, 2nd Ed., 2012. `https://nanohub.org/resources/5339/download/Exact__MOS_Solution.pdf`

Lecture 24

Energy Band Diagrams

24.1 Introduction

In the previous lecture, we presented the **semiconductor equations** and saw that they are a fairly complicated set of mathematical equations. If we solve the semiconductor equations, we can simulate the operation of a semiconductor device. In this lecture, we'll show how to bypass mathematical complexities and get qualitative insight into the solution of these equations by using **energy band diagrams**. An energy band diagram is a plot of the bottom of the conduction band and the top of the valence band vs. position, $E_c(x)$ and $E_v(x)$ vs. x. Herbert Kroemer, a famous semiconductor physicist who won the Nobel Prize for his work on semiconductors, often stressed the importance of energy band diagrams. He made the following point in his 2000 Nobel lecture [1].

> "Whenever I teach my semiconductor device physics course, one of the central messages I try to get across early is the importance of energy band diagrams. I often put this in the form of Kroemer's lemma of proven ignorance:
>
> If, in discussing a semiconductor problem, you cannot draw an Energy Band Diagram, this shows that you don't know what you are talking about."
>
> He goes on to add a corollary:
>
> "If you can draw one, but don't, then your audience won't know what you are talking about."
>
> Let's see how to draw and read an energy band diagram.

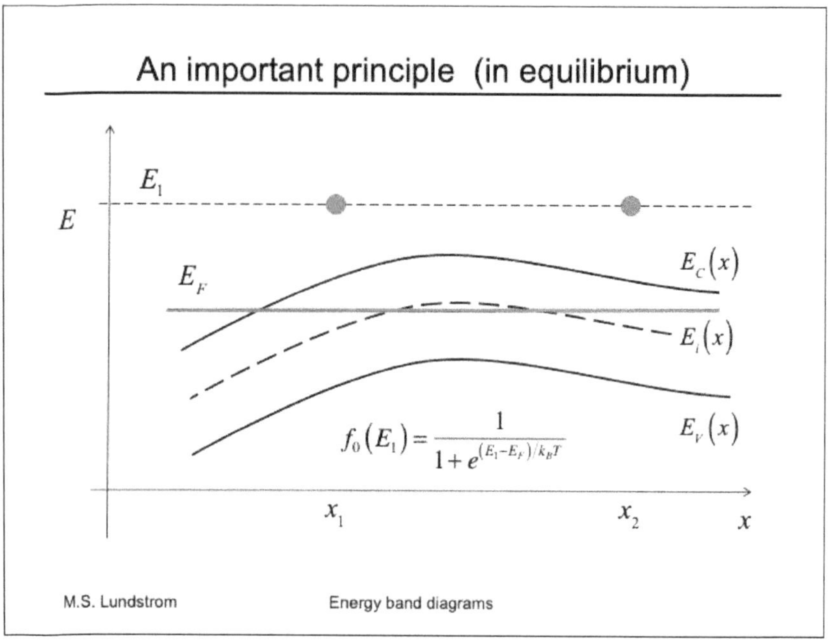

24.2 The Fermi level is constant in equilibrium

Equilibrium energy band diagrams are easy to draw because we can make use of a very important principle illustrated in the slide above. For this energy band diagram, the conduction and valence bands vary with position, and so does the intrinsic level, but E_F is constant.

Consider a state at x_1 with total energy E_1 and a state at x_2 with the same total energy. The probabilities that these two states are occupied must be the same, otherwise, electrons would tend to move from one state to the other and produce a current in the process, but since this is equilibrium, the current must be zero.

The probability that a state at x_1 is occupied is given by $f_0(E_1)$, and it depends only on the total energy of the state, E_1, and on the Fermi level, E_F. The probability that a state at x_2 is occupied must be the same. Its total energy is the same as for the state at x_1, so we conclude that E_F must at the same at x_1 and at x_2. **The Fermi level must be independent of position.**

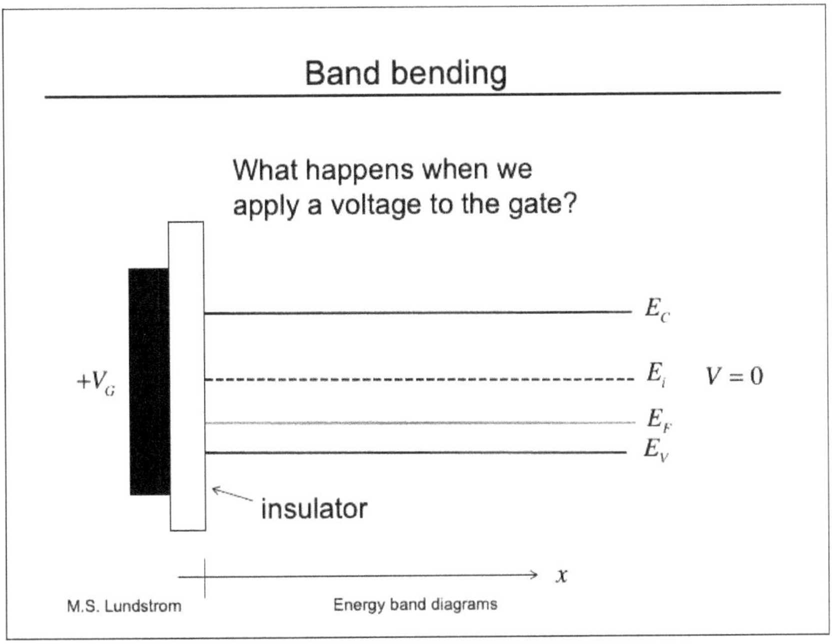

24.3 Band bending

Energy bands can bend in equilibrium, but the fact that the Fermi level is constant will help us draw equilibrium energy band diagrams. The concept of **band bending** is very important. Shown above is a uniform, P-type semiconductor in equilibrium. The Fermi level is below the intrinsic level, close to the valence band. The carrier concentrations are independent of position because the bands are not bending.

If we put an insulating layer on the surface of a semiconductor and a metal plate on top, then we have an MIS structure, metal-insulator-semiconductor. If we ground the right side of the semiconductor and then apply a positive voltage to the metal plate which we'll call the gate, then the bands will bend. This structure is actually the heart of **a MOSFET, a metal-oxide-semiconductor field-effect transistor.** In this example, the insulator is an oxide, typically silicon dioxide, SiO_2.

The positive voltage on the gate will induce a positive potential near the surface of the semiconductor, but deep in the semiconductor the potential will go to zero. To determine what happens to the energy band diagram, we need to know what an electrostatic potential does to the electron energy.

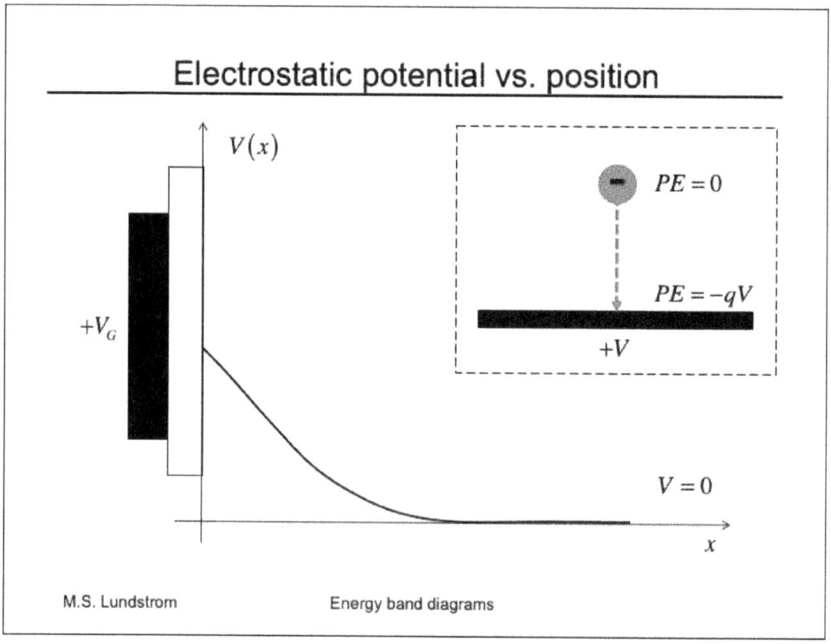

You may remember from first year physics that if we have a positive voltage on a metal plate, it will pull an electron towards the plate and lower its potential energy by an amount, $-qV(x)$. We can see this because potential energy is minus the integral of force over distance:

$$PE = -\int F_e dx\,,$$

where $F_e = -q\mathcal{E}$ is the force on electrons. By definition, the electrostatic potential is minus the integral of the electric field over distance, so we conclude that

$$PE = \text{constant} - qV(x)\,.$$

This is the principle we will use to understanding band bending in the MOS structure. A positive electrostatic potential lowers the electron energy.

A positive voltage, V_G, on the metal plate shown above will increase the electrostatic potential on left side of the semiconductor near the insulator-semiconductor interface. The electrostatic potential will be less and less positive as x increases into the semiconductor, and then will approach zero deep in the semiconductor. The plot sketched above shows how we expect $V(x)$ to vary when we put a positive voltage on the metal gate.

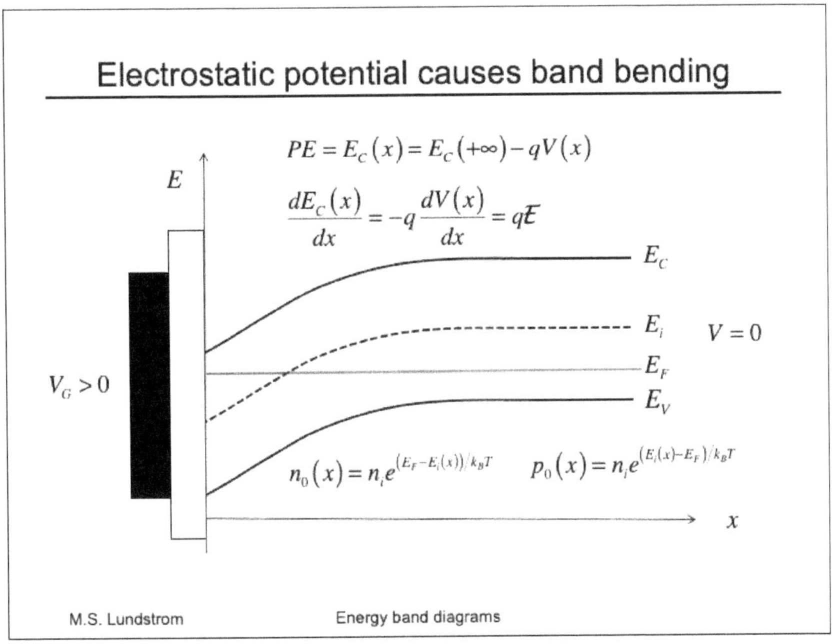

The figure above shows the effect of the positive gate voltage on the energy bands in the semiconductor. If the potential is more positive near the surface, the electron energy will be lower near the surface. So, if the electrostatic potential increases with position as we go towards $x = 0$, then the energy bands bend down near the surface. When we see energy bands bending as shown above, we know that there is a position-dependent electrostatic potential in the semiconductor. When we take the slope of $E_c(x)$, we only differentiate the electrostatic potential. As shown above, the slope of the conduction band is q times the electric field. The slope of the valence band also gives the electrostatic potential, and the slope of the intrinsic level would too. So, just by taking the slope, we can determine what the electric field is in the semiconductor. We can also deduce the carrier densities from the energy band diagram because the carrier densities are related to the position of the Fermi level with respect to the intrinsic level.

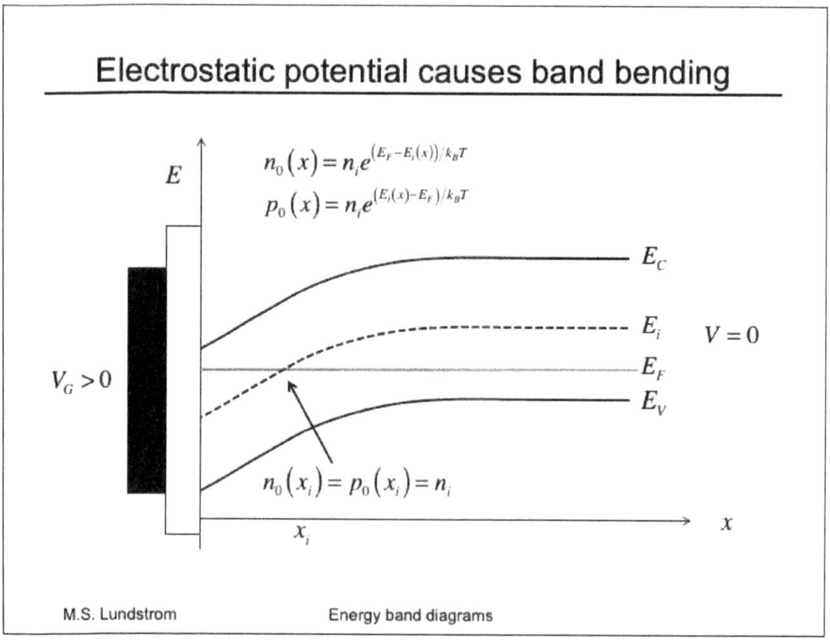

To continue the discussion, let's take a close look at the energy band diagram shown above.

Deep in the semiconductor, the Fermi level is close to the valence band, so there is a large concentration of holes. This semiconductor is P-type. As we move towards the surface, the bands bend down, and the Fermi level gets closer and closer to the conduction band. At the point where the Fermi level crosses the intrinsic level, the semiconductor is intrinsic and $n_0 = p_0 = n_i$.

As we continue to move towards the surface, the bands bend down more and more, and near $x = 0$, the Fermi level is significantly above the intrinsic level, closer to the conduction band than to the valence band so, this semiconductor actually has a high electron density near the surface, and a high density of holes in the bulk. We would say that the surface is **inverted**. We have inverted a P-type semiconductor and made it an N-type semiconductor near the surface. The effect is sometimes referred to as "electrostatic doping" because it is another way to change the equilibrium carrier concentrations, in this case, without introducing dopants.

Given an energy band diagram, we can read quite a lot right off of the figure.

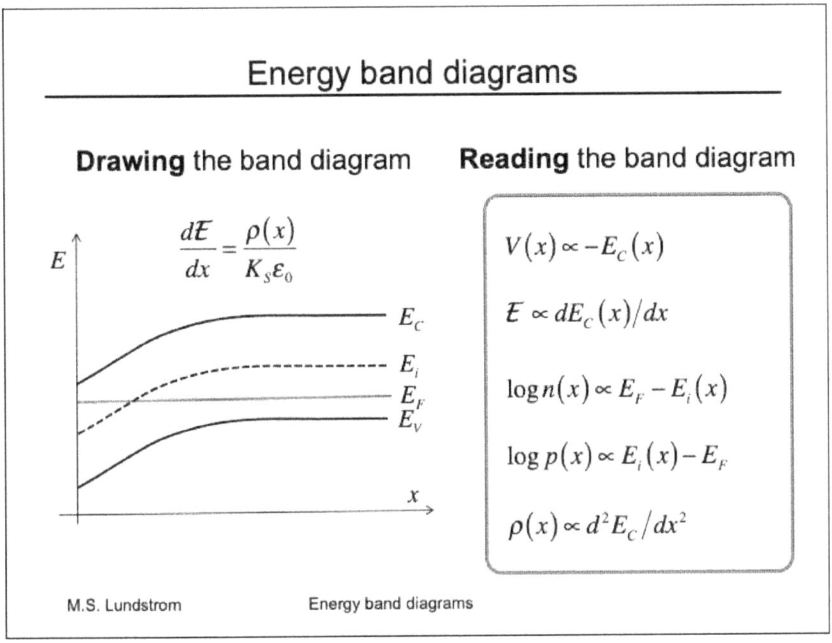

24.4 Energy band diagrams

The first thing to do when we encounter a semiconductor problem is to draw the energy band diagram. Given an equilibrium energy band diagram, we can then "read" it to extract the quantities and interest.

(1) When the electrostatic potential increases, the electron energy decreases, so to find the electrostatic potential vs. position, just flip E_c, E_v, or E_i upside down.
(2) The slope of E_c, E_v, or E_i is proportional to the electric field.
(3) The electron density is exponentially related to $E_F - E_c$, so the log of the electron concentration is proportional to $E_F - E_c$. Similarly, the log of the hole concentration is proportional to $E_v - E_F$.
(4) To find the space charge density, ρ, we can use the Poisson equation, which tells us that ρ is proportional to the slope of the electric field. Since $\mathcal{E}$ is proportional the slope of the energy band, $\rho \propto d^2E_c/dx^2$, but it is difficult to accurately take the second derivative from a sketch of an energy band diagram, so we might prefer to get the space charge density from the carrier densities and the doping density.

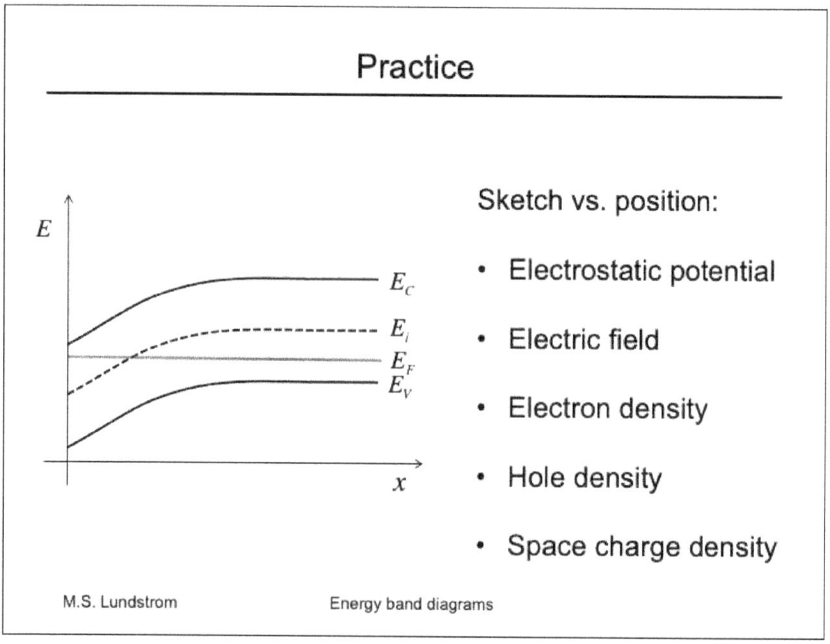

For practice, take the energy band diagram shown above and "read" it. Make a plot of the electrostatic potential, electric field, electron density, hole density, and space charge density vs. position. When you've done that, we will run through another example, so you could see how this all works.

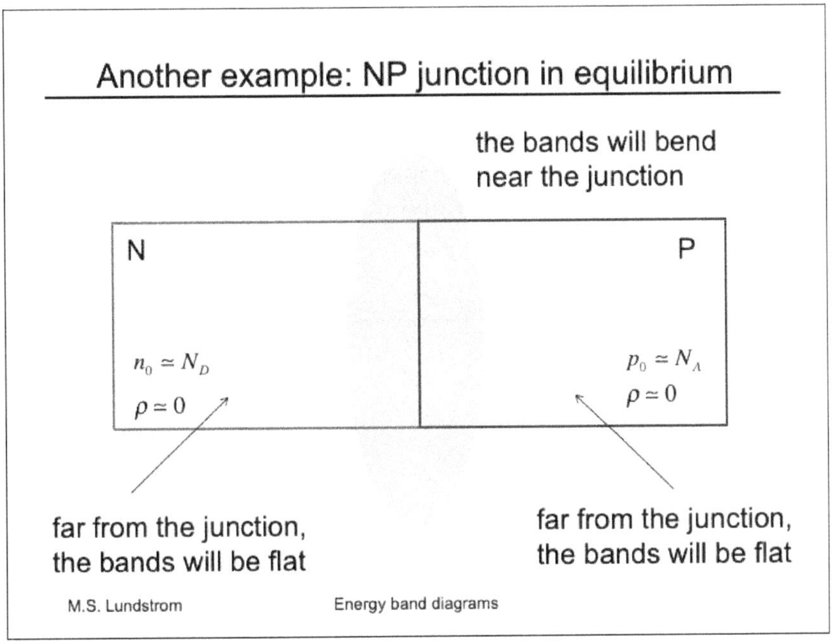

24.5 Energy band diagram of an NP junction

Shown above is a semiconductor device, a junction between a uniformly doped N-type semiconductor and a uniformly doped P-type semiconductor. This is a semiconductor diode that does useful things like rectify current, or if we shine light on it, it becomes a solar cell [3]. What does the equilibrium energy band diagram of this semiconductor device look like?

Far away from the junction, we just have a uniformly doped N-type semiconductor on the left and a uniformly doped P-type semiconductor on the right. Near the junction, the bands will bend.

To draw the equilibrium energy band diagram, remember that the Fermi energy is constant in equilibrium, so the first step is to draw a horizontal line and label it E_F.

The semiconductor is N-type far away from that junction on the left, so we'll show the conduction band near E_F, and the valence band will be one band gap energy below that. On the right, the semiconductor is P-type, so we'll draw a valence band near E_F, and the conduction band one band gap energy above that. Finally, we smoothly connect the energy bands across the transition region. The result is shown on the next slide.

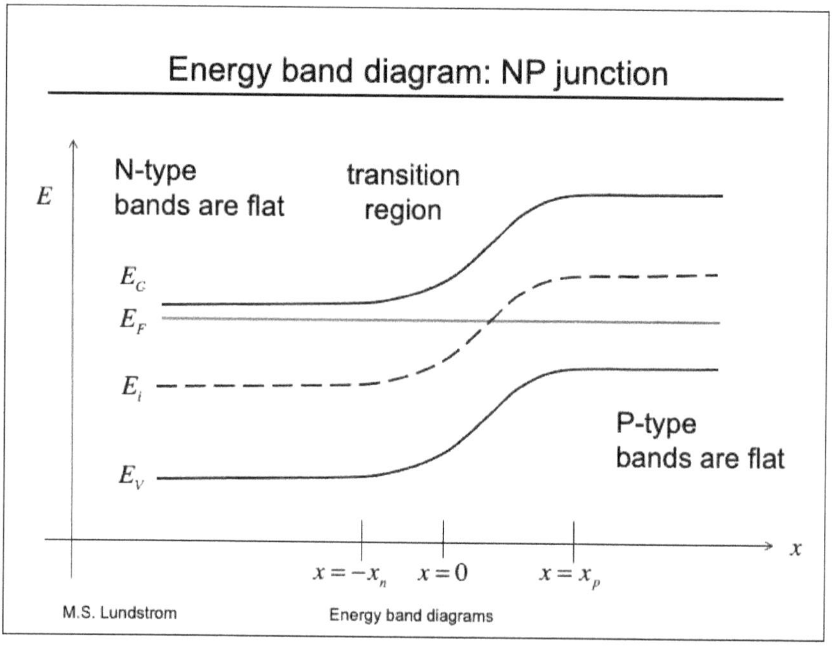

Shown above is the equilibrium energy band diagram for an NP junction that results from the process just described. Now we can read this energy band diagram and effectively solve the semiconductor equations. Things are simple for the N-type region on the left where the bands are flat and for the P-type region on the right, where the bands are also flat. Things get interesting in the transition region where the bands bend.

To see what happens in the transition region, let's "read" the energy band diagram and sketch:

(1) Electrostatic potential vs. position
(2) Electric field vs. position
(3) Electron and hole densities vs. position
(4) Space-charge density vs. position

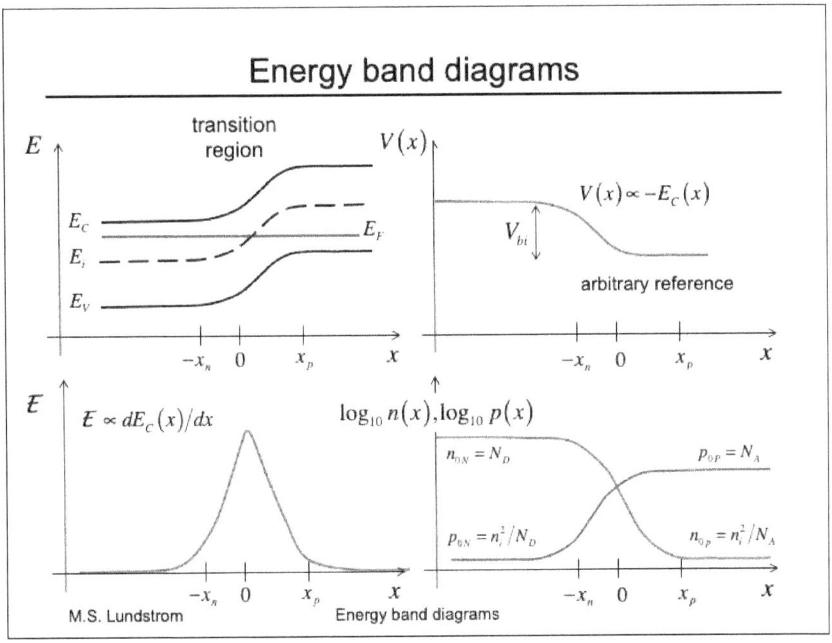

When the potential increases, the bands bend down; to get the electrostatic potential vs. position, we just flip the conduction band (or valence band or intrinsic level) upside down as shown on the upper right. Voltages are always with respect to a reference, only a difference in voltage matters. Although we have not applied an external voltage, we see that there is a voltage difference across the junction, the so-called **built-in voltage**, V_{bi}.

The electric field is proportional to the slope of the bands, so there is no electric field far from the junction, but there is a positive electric field in the transition region because the slope is positive there. As shown above at the lower left, the peak electric field occurs at the junction and the electric field decreases as we move away from the junction.

Carrier densities are related to the position of the Fermi level with respect to the intrinsic level. On the left, n_0 is large. As we move towards the junction, E_i moves towards and then above E_F, so $n_0(x)$ decreases. At the end of the P-type region, $E_F \ll E_i$, so n_0 is very small. The electron density goes smoothly from $n_0 = N_D$ on the left to $n_0 = n_i^2/N_A$ on the right. The hole density does the opposite. Remember that this is a log plot. In the transition region, the carrier densities are much less than the doping density.

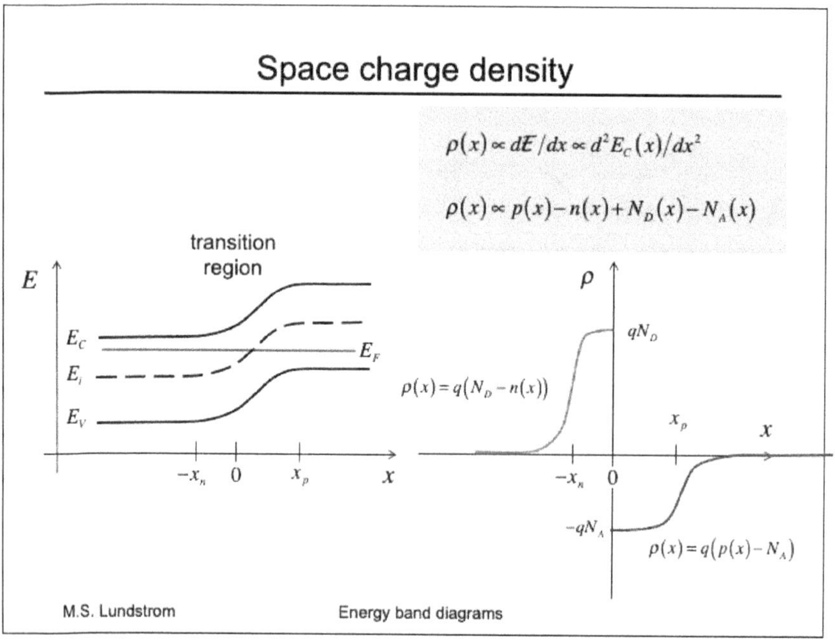

What about the space charge density vs. position? As shown above in the gray box, we could try to take the slope of the electric field vs. position, which requires taking the second derivative of $E_c(x)$, but taking a second derivative graphically difficult, so the result might not be very accurate.

Another way to get the space charge density is to use $\rho(x) = q\left[p_0(x) - n_0(x) + N_D^+(x) - N_A^-(x)\right]$. If we know the doping densities on the left and right sides, then we can use the carrier density profiles we've already sketched to obtain $\rho(x)$.

Shown above is the space charge density vs. position that results. Far to the left of the junction, we have a uniformly doped, neutral N-type semiconductor, so $\rho = 0$. Far to the right, we have a uniformly doped, neutral P-type semiconductor, so $\rho = 0$, but there is a significant space charge density in the transition region. Since the carrier densities are small in the transition region, the space charge is almost entirely due to ionized dopants. On the left of the junction, $\rho(x) \approx qN_D^+$, and on the right of the junction, $\rho(x) \approx qN_A^-$.

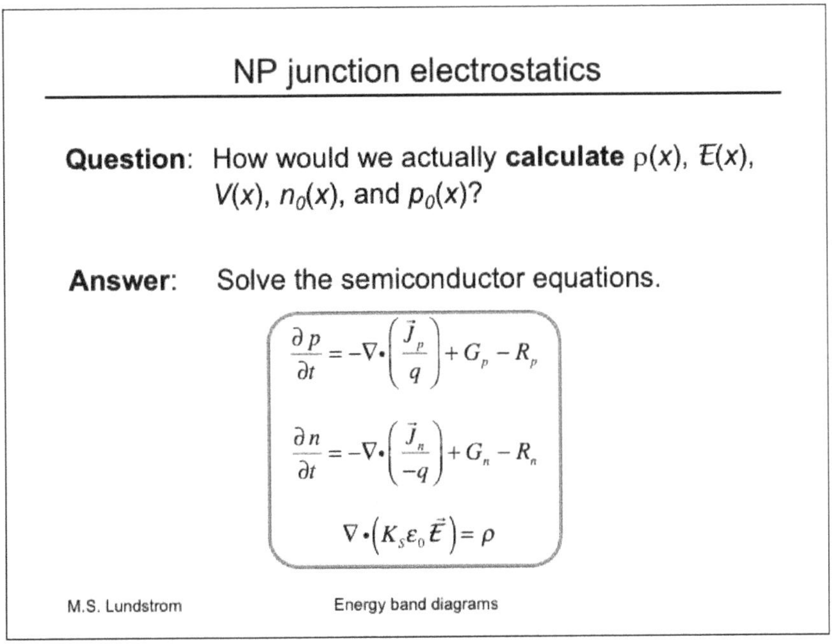

To get numerical values for the magnitude of the peak electric field, the width of the transition region, and the complete $V(x), \mathcal{E}(x), n_0(x), p_0(x)$, and $\rho(x)$ profiles, we need to solve the semiconductor equations. Since we assume equilibrium, we could solve the Poisson-Boltzmann equation, but for this particular case, there is a widely-used simplification of the semiconductor equations called the **depletion approximation**, that allows us to easily do approximate calculations [3]. For more complicated problems, we need to solve the full semiconductor equations, but before solving them, it's always useful to sketch an energy band diagram to understand qualitatively the nature of the solutions that are going to result.

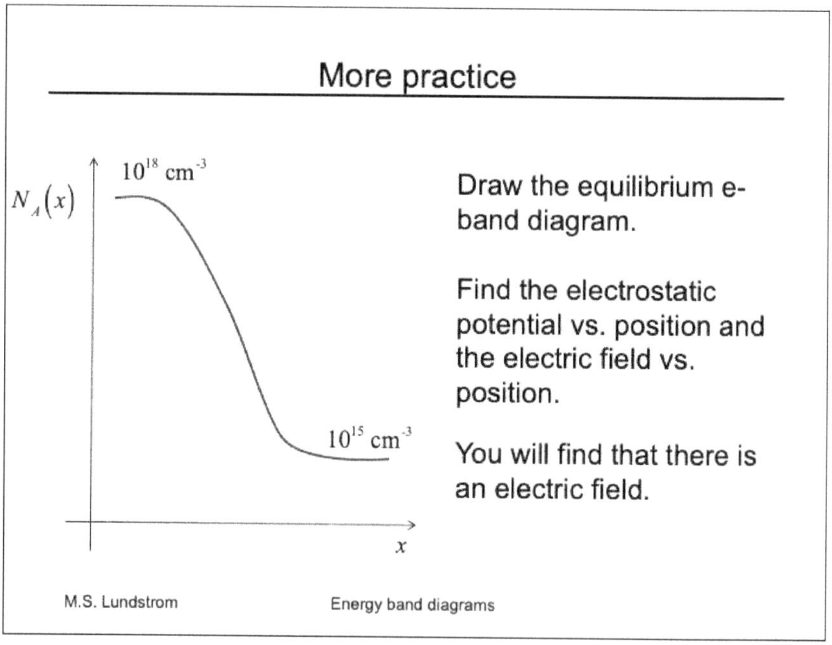

24.6 A graded doping example

Shown above is a problem that we've seen before and one that is also good practice for drawing and interpreting energy band diagrams.

The semiconductor shown above is doped very heavily P-type near the surface and less heavily doped as we move into the semiconductor. The position dependent doping density results in a position-dependent hole concentration. Draw the equilibrium energy band diagram for this case. You'll have to draw it to show uniform P-doping on the right and much more heavy P-type doping as we move towards $x = 0$. If you draw the energy band diagram and then read it, you will see that there is an electric field pointing in the $-x$ direction. The energy band diagram tells us that there is an electric field, but where does this electric field come from?

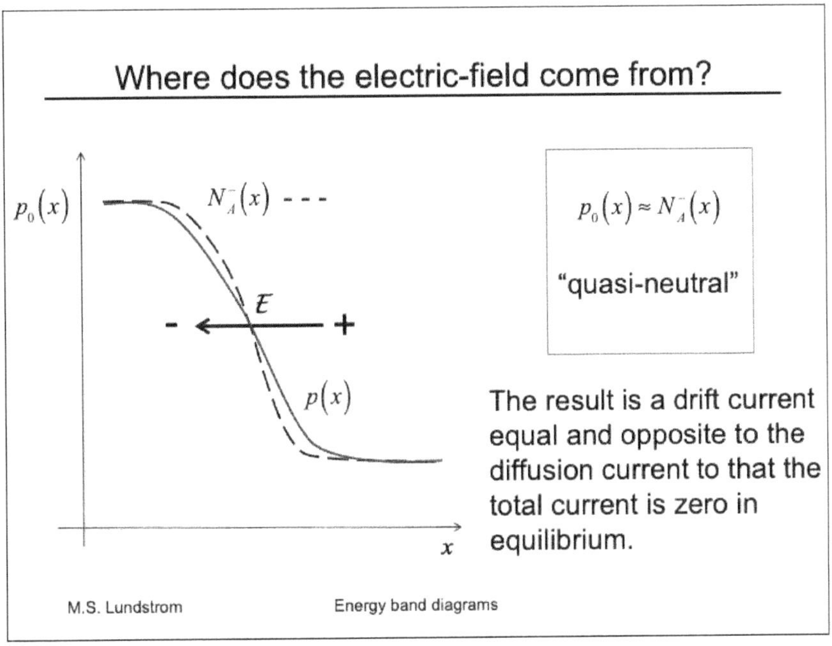

We can understand where the electric field comes from as follows. The nonuniform doping produces a nonuniform concentration of mobile holes, which are free to diffuse to the right — down the concentration gradient. Near $x = 0$, we'll have fewer holes than dopants because the holes have diffused away, but the dopants can't move. Because holes have diffused into the semiconductor, deeper into the semiconductor, there will be more holes than dopants. The result is a charge imbalance — a net negative charge near the surface and a net positive charge deeper into the semiconductor. The charge imbalance will set up an electric field that points from positive charge to negative charge, i.e. to the left. The electric field exerts a force on the holes that will push them back to the left and stop them from diffusing. When things have settled down, the hole concentration will be approximately equal to the doping density, but not exactly. We say that the semiconductor is **quasi-neutral**. It is also almost neutral, but there is a small charge imbalance that produces an electric field that keeps everything in place so that the mobile holes don't continue to diffuse. In equilibrium, we have zero net current, this is the electric field that produces a drift current to the left that balances the diffusion current to the right, so that we get zero current in equilibrium.

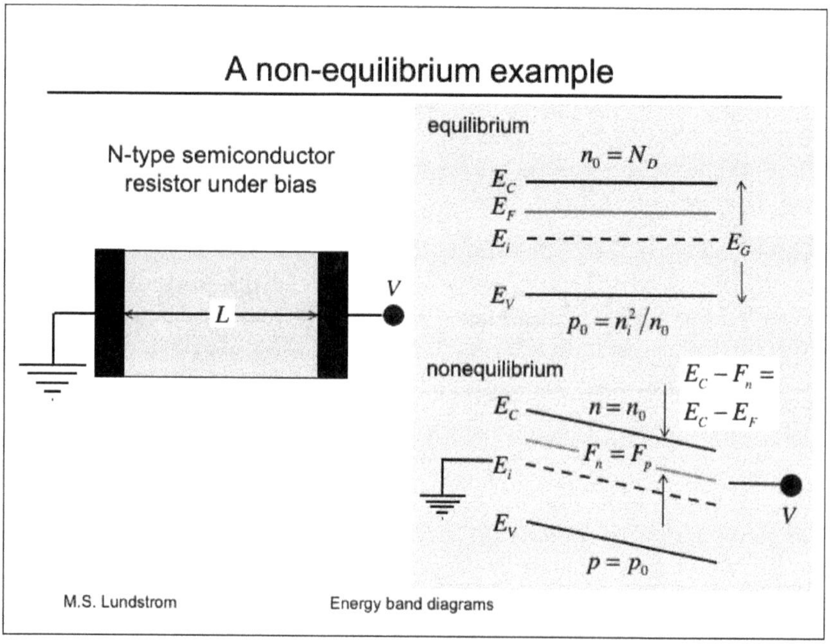

24.7 Non-equilibrium energy band diagrams

Out of equilibrium, drawing energy band diagrams is more complicated because there is no constant E_F reference. Shown above is an example, an N-type semiconductor with a positive voltage on the right contact. Shown above on the top right is the equilibrium energy band diagram. Because the semiconductor is assumed to be uniform, we expect that an applied voltage will drop linearly across the semiconductor, as shown on the lower right. The way to think about this is to understand that a good contact connects to the quasi-Fermi level. The positive voltage pulls the electron QFL down on the right. Electrons flow in from the battery and across the semiconductor, which they try to keep neutral, so $n_0 \approx N_D$, as in equilibrium. The conduction band moves down so that $E_c - F_n$ is the same as $E_c - E_F$ was in equilibrium. The slope in E_c tells us that there is an electric field pointing in the $-x$ direction, and the slope in F_n tells us that current is flowing. The device is a resistor.

Note that $F_n = F_p$ in this example because we assumed ideal contacts that only inject and extract majority carrier electrons. For this example, the semiconductor is out of equilibrium but np is still equal to $n_0 p_0 = n_i^2$.

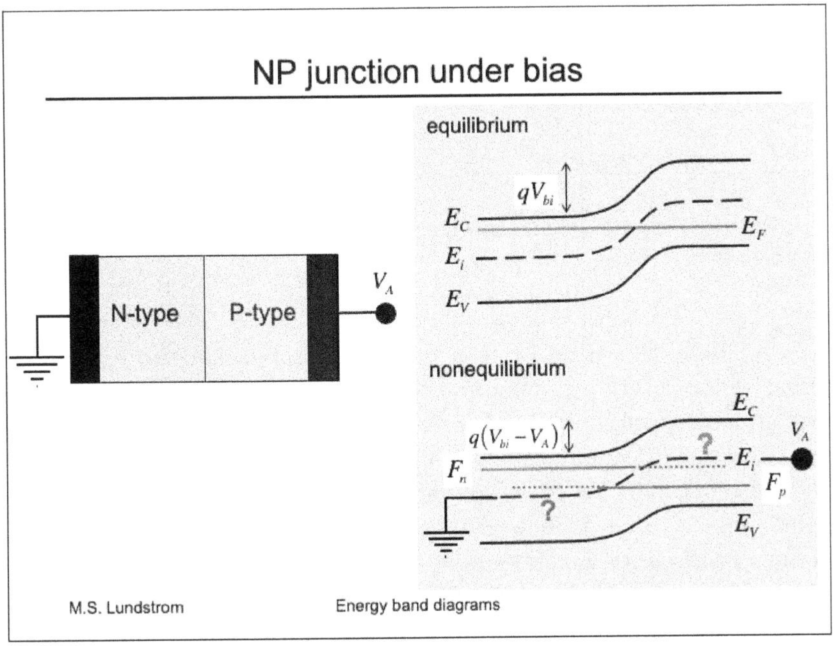

Consider a second example, an NP junction with $V_A > 0$ on the right contact. As shown above on the top right, in equilibrium E_c on the N-side is qV_{bi} lower in energy than on the P-side because of the built-in potential, V_{bi}. The applied voltage is labeled V_A to distinguish it from V_{bi}. To draw the band diagram under bias, begin away from the junction where on the left, the contact keeps F_n where E_F was on the N-side, so $n \approx n_0$ on the N-side. The right contact pulls F_p down by qV_A. Away from the junction, the hole concentration is very nearly what it was in equilibrium so $F_p - E_v = E_F - E_v$. We know F_n on the N-side and F_p on the P-side, but finding $F_n(x)$ and $F_p(x)$ across the device requires a solution to the semiconductor equations.

Away from the junction, we know E_c on the N-side because it adjusts itself with respect to F_n to maintain $n \approx n_0$, and we know E_v on the P-side because it adjusts itself with respect to F_p to maintain $p \approx p_0$. To find $E_c(x)$ and $E_v(x)$ across the device, we need a full solution to the semiconductor equations, but our intuition tells us that the voltage should drop across the most resistive part of the device, the transition region, where the carrier density is the lowest. As shown on the lower right — the potential difference between the N- and P-sides has been reduced to $q(V_{bi} - V_A)$.

24.8 Summary

The previous lecture introduced the semiconductor equations:

$$\boxed{\begin{aligned}&\frac{\partial p}{\partial t} = -\nabla \cdot \frac{\vec{J}_p}{q} + G_p - R_p \\ &\frac{\partial n}{\partial t} = -\nabla \cdot \frac{\vec{J}_n}{(-q)} + G_n - R_n \\ &\nabla \cdot \left(K_s \epsilon_0 \vec{\mathcal{E}}\right) = \rho\,.\end{aligned}}$$

To solve these equations, we can use sophisticated numerical techniques and solve them with a computer. Powerful software tools are available for this purpose, or we could simplify the equations so that they can be solved analytically. This simplification must be done on a case by case basis, and most of semiconductor device physics is about just this — appropriately simplifying the semiconductor equations for a problem of interest. In this lecture, we discussed a third way to solve the semiconductor equations — a graphical way. This powerful and widely-used **energy band diagram** approach provides us with insightful, qualitative solutions to these equations without doing any mathematics.

Drawing energy band diagram in equilibrium is straightforward because we can begin with the knowledge that E_F is constant. Drawing energy band diagrams out of equilibrium takes some intuition. We need to think about how an applied voltage would drop across the device and how the quasi-Fermi levels would vary with position. This intuition is developed with practice by understanding solutions for various semiconductor problems. It is a skill that is well worth developing.

24.9 References

In his lecture for the 2000 Nobel Prize in physics, Herbert Kroemer talks about the importance of energy band diagrams:

[1] Herbert Kroemer, *Quasi-Electric Fields and Band Offsets: Teaching Electrons New Tricks*, 2010. `https://www.nobelprize.org/prizes/physics/2000/kroemer/lecture/`.

For a gentle introduction to the MOSFET, the most common transistor currently in use, see:

[2] Mark Lundstrom, *Transistors!*, World Scientific Publishing Co., 2024.

For an introduction semiconductor PN junctions and devices based on them, such as bipolar transistors and solar cells, see:

[3] Robert F. Pierrt, *Semiconductor Device Fundamentals*, Addison-Wesley Publishing Co., 1996 (begin with Chapters 5, 6, 9, 10, and 11).

Lecture 25

Quasi-Fermi Levels

25.1 Introduction

In the previous lecture, we discussed energy band diagrams, a powerful tool for qualitatively understanding the operation of semiconductor devices. The first step is to draw the energy band diagram, and the first step in drawing an equilibrium energy band diagram is to draw the Fermi level, which is constant in equilibrium. We also want to draw out of equilibrium energy band diagrams to understand the operation of devices under bias for example, and that is more difficult.

Non-equilibrium semiconductor problems bring in **quasi-Fermi levels**. We have briefly discussed quasi-Fermi levels in earlier lectures. In this lecture, we say a few more words about them so that we're comfortable with the use of quasi-Fermi levels when we put them in energy band diagrams. We'll see an example of that in the next lecture.

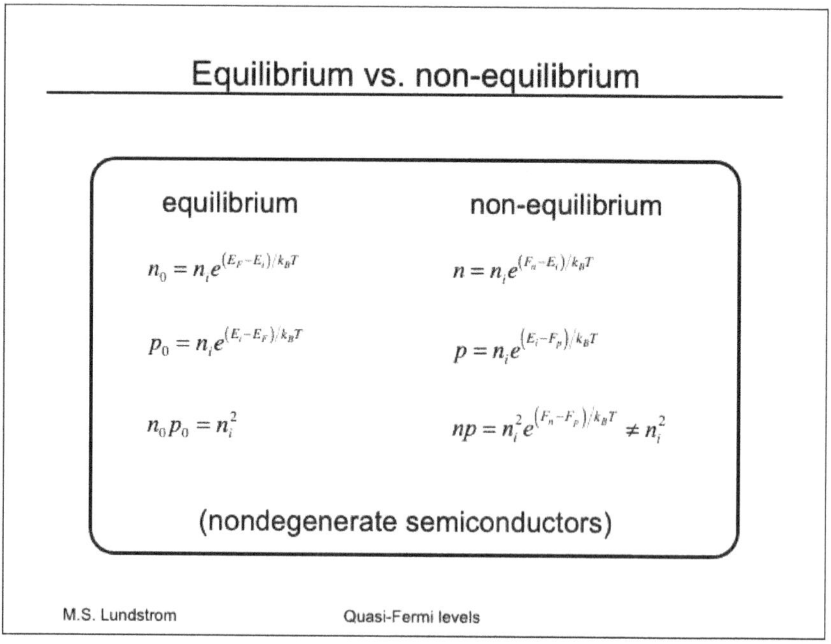

25.2 Equilibrium vs. non-equilibrium

The equilibrium carrier densities are related to the location of the Fermi level. Out of equilibrium, we introduce **quasi-Fermi levels**. The probability that the states in the conduction band are occupied is given by a Fermi function with a quasi-Fermi level for electrons, F_n:

$$f_c = \frac{1}{1 + \exp\left[(E - F_n)/k_B T\right]} .$$

The probability that the states in the valence band are empty is determined by a quasi-Fermi level for holes, F_p:

$$1 - f_v = 1 - \frac{1}{1 + \exp\left[(E - F_p)/k_B T\right]} .$$

Out of equilibrium, the np product depends on the splitting of the QFLs and may be less than, equal to, or greater than n_i^2. Using the expressions on the slide above, we find,

$$np = n_i^2 e^{(F_n - F_p)/k_B T} \neq n_i^2 .$$

In equilibrium, $F_n = F_p = E_F$ and the np product is independent of the location of the Fermi level, but this is only the case for a non-degenerate semiconductor in equilibrium.

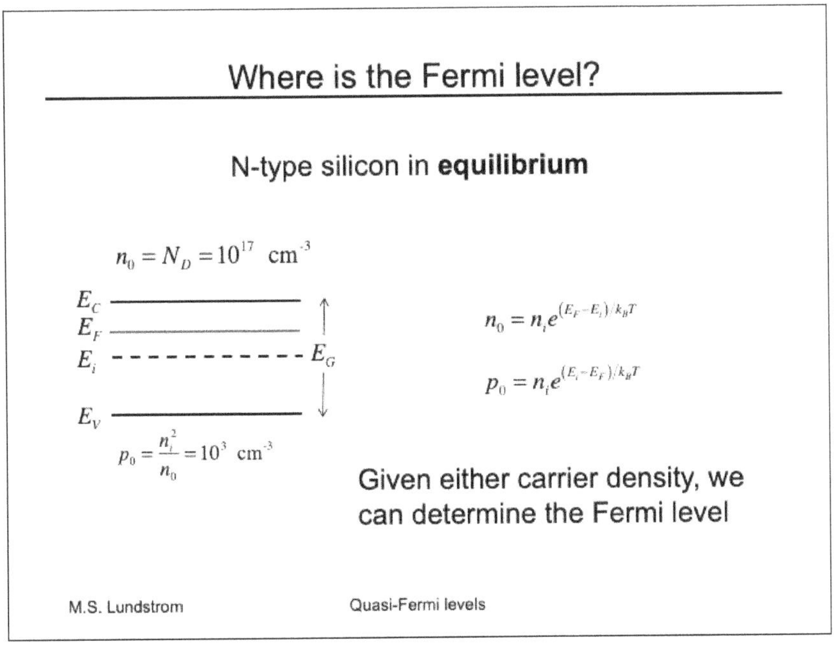

Let's review semiconductors in equilibrium with a simple example. Consider a problem like the one shown above, a uniform semiconductor in equilibrium and doped at $N_D = 10^{17}\,\text{cm}^{-3}$. We can easily solve for the equilibrium electron density; it's $n_0 = N_D = 10^{17}\,\text{cm}^{-3}$ assuming that the dopants are fully ionized at room temperature. The minority hole concentration is much smaller, $p_0 = n_i^2/n_0 = 10^3\,\text{cm}^{-3}$. Where is the Fermi level?"

There is a simple relation between the Fermi level and the carrier density, so knowing the carrier densities we can determine the Fermi level; we can find the location of the Fermi level from either the known electron or hole concentration. In this case, we find that the equilibrium Fermi level lies above the intrinsic level because it's an N-type semiconductor. It is closer to the bottom of the conduction band than to the top of the valence band because there are more electrons in the conduction band than holes in the valence band.

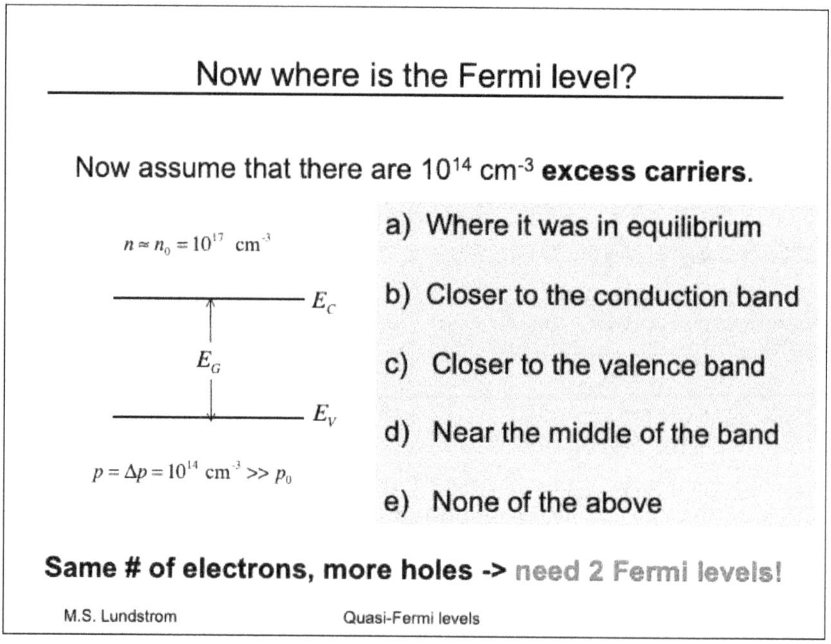

Consider next an out of equilibrium problem. Assume that we have introduced some excess carriers into the semiconductor; to be specific we'll introduce 10^{14} per cc of excess carriers over and above what we had in equilibrium. The density of excess carriers is only 1/1000th of the equilibrium electron concentration, so the excess carriers will have very little effect on the majority carrier electrons. But the density of excess carriers is 11 orders of magnitude greater than the equilibrium density of minority carrier holes, so the excess carriers will have a huge effect on the minority carrier hole concentration. This is **low-level injection**.

Where should we draw the Fermi level now that we have excess carriers? You could argue that it should be where it was in equilibrium because we haven't changed the electron density significantly. You could also argue that it should be closer to the valence band because we have many more holes now. It can't be in both places at once. We cannot describe these out of equilibrium carrier densities with a single Fermi level. We need to introduce separate "Fermi levels," which are known as the quasi-Fermi levels for electrons and for holes. These two quantities play the role of the one Fermi level in equilibrium.

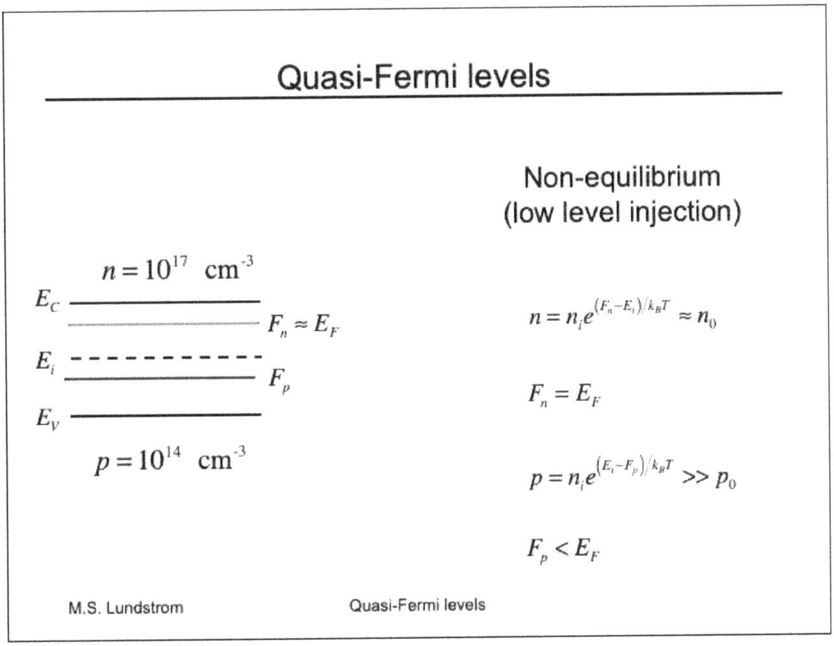

25.3 Quasi-Fermi levels

Let's first compute the location of the quasi-Fermi level for electrons. **Low level injection** is a fairly common case where the number of excess carriers doesn't significantly perturb the majority carriers. The majority electron concentration is almost exactly what it was in equilibrium. The electron density is related to the quasi-Fermi level for electrons, F_n. Since the electron density is almost exactly the same as it was in equilibrium, the quasi-Fermi level for electrons will be very nearly equal to the Fermi level in equilibrium, so we draw the quasi-Fermi level for electrons where the equilibrium Fermi level was.

The excess carrier concentration has an enormous effect on the minority carrier holes. We now have 11 orders of magnitude more holes. When we solve for the hole quasi-Fermi level using the third equation on the right, we find that it is much different from its value in equilibrium. In order produce 11 orders of magnitude more holes, the hole quasi-Fermi level has to be much closer to the valence band than the equilibrium Fermi level was. Finally, note that F_n is greater than F_p. We often find that the quasi-Fermi levels are split out of equilibrium.

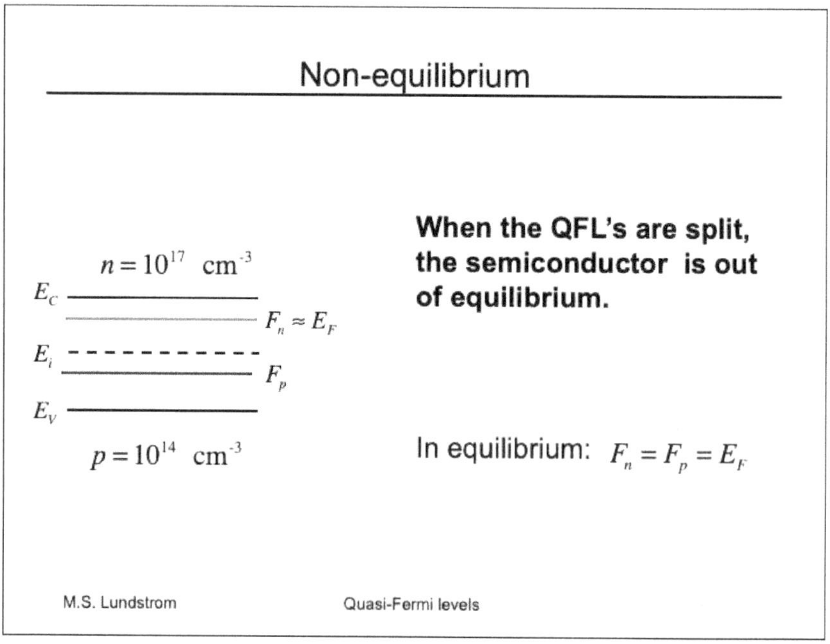

When the quasi-Fermi levels are split as shown above, it's an indication that we're dealing with a non-equilibrium situation. As discussed in Lecture 21, Sec. 2, if we introduce excess carriers at $t = 0$ and then wait long enough, the excess carriers will recombine and the hole quasi-Fermi level will relax to the electron quasi-Fermi level, and we'll have one Fermi level when we reach equilibrium. Alternatively, if we maintain the excess carrier concentration (for example, with steady-state illumination), then the QFL's will remain split.

Finally, we should also point out that it's possible to be out of equilibrium with no splitting in the quasi-Fermi levels, as was discussed in Lecture 24, Sec. 7 and as we'll discuss later in this lecture.

np product

Equilibrium

$$n_0 = n_i e^{(E_F - E_i)/k_B T}$$

$$p_0 = n_i e^{(E_i - E_F)/k_B T}$$

$$n_0 p_0 = n_i^2$$

Out of equilibrium

$$n = n_i e^{(F_n - E_i)/k_B T}$$

$$p = n_i e^{(E_i - F_p)/k_B T}$$

$$np = n_i^2 e^{(F_n - F_p)/k_B T}$$

$$F_n > F_p \rightarrow np > n_i^2$$

$$F_n < F_p \rightarrow np < n_i^2$$

M.S. Lundstrom Quasi-Fermi levels

As shown above in the gray box, when we multiply the equilibrium electron density times the equilibrium hole density, the Fermi level cancels out. The $n_0 p_0$ product, n_i^2, is independent of the Fermi level was, assuming, as we are, a non-degenerate semiconductor.

Out of equilibrium, when we multiply the electron density by the hole density, we find that $np = n_i^2 \times \exp[(F_n - F_p)/k_B T]$. If $F_n > F_p$, we have excess carriers and $np > n_i^2$. If $F_n < F_p$, then the excess carrier concentration is negative, which means that we have fewer carriers than in equilibrium and $np < n_i^2$. Either situation can occur in a semiconductor.

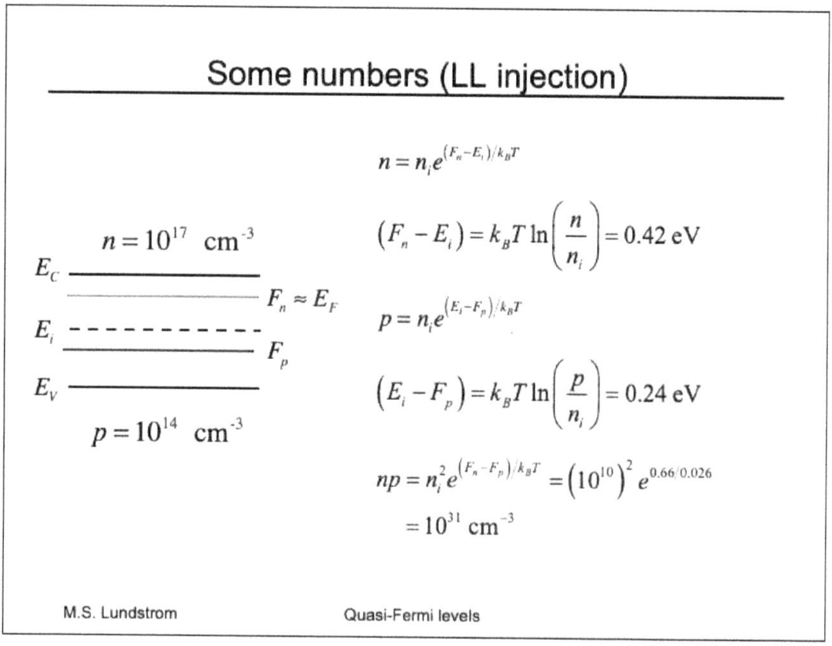

Let's put some numbers in, so we can see how this all plays out. Shown above is the example we have been discussing, a semiconductor doped at $N_D = 10^{17}\,\text{cm}^{-3}$. In equilibrium, we find $n_0 = 10^{17}$ and $p_0 = 10^3\,\text{cm}^{-3}$. We determine the location of the Fermi a level from the equilibrium carrier density, and find $(E_F - E_i) = 0.42$ eV; the Fermi level is located 0.42 eV above the intrinsic level. What happens out of equilibrium?

Now let's assume that we are out of equilibrium but in low level injection with $\Delta n = \Delta p = 10^{14}\,\text{cm}^{-3}$ as before. The effect on the electron density is insignificant, but the effect on the minority hole density is huge.

As shown above, the quasi-Fermi levels are related to the carrier densities. To find the electron quasi-Fermi level, we solve the first equation for $F_n - E_i$. Put numbers in, and we find that the quasi-Fermi level for electrons is very nearly where the Fermi level was in equilibrium. If we do the same calculation for holes, we solve the third equation for $E_i - F_p$ and find that F_p is about 0.24 eV below the intrinsic level, or 0.66 eV below F_n.

The bottom line is that $np = n_i^2 \times \exp[(F_n - F_p)/k_BT] = 10^{31}\,\text{cm}^{-3}$, which is 11 orders of magnitude greater than n_i^2 because n is unchanged but p has increased by 11 orders of magnitude.

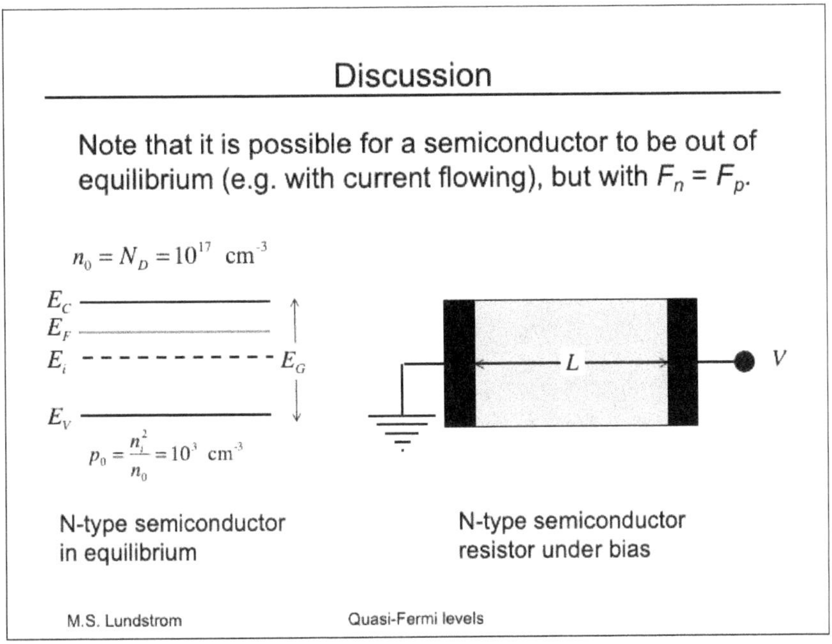

25.4 Quasi-Fermi level discussion

When we see quasi-Fermi level splitting, it's an indication that we are dealing with a non-equilibrium semiconductor, but it is also possible for the quasi-Fermi levels not to be split and to still be out of equilibrium. The example shown above was discussed in Lecture 24, Sec. 7. This is our same N-type sample, but now with metal contacts at each end and a voltage applied across the contacts. We're making a resistor with this N-type doped semiconductor. We will be out of equilibrium because current will flow through this resistor. What are the quasi-Fermi levels?

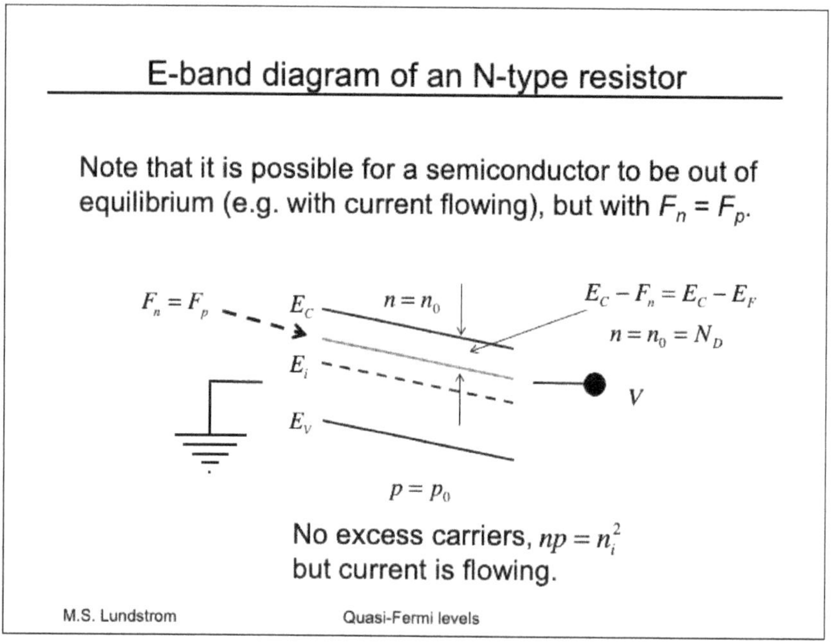

To draw the energy band diagram, note that when we put a positive voltage on the right contact that lowers all of the energies. If the electron density continues to be $10^{17}\,\text{cm}^{-3}$ as it was an equilibrium, which it will if the contacts are good, non-minority-carrier-injecting, ohmic contacts, then the separation between the bottom of the conduction band and the quasi-Fermi level for electrons now must be the same as it was in equilibrium everywhere. So the quasi-Fermi level must have the same slope that the bands have. The $10^{17}\,\text{cm}^{-3}$ electrons are just flowing from the first contact, through the semiconductor, out the second contact, through an external circuit, and coming back in the first contact again. Current is flowing and we are out of equilibrium, but if we're not injecting any excess electrons or holes, then $F_n = F_p$. The QFL's are not split, but they do have a slope, and the fact that they have a slope then indicates that current is flowing.

In this example, the electron quasi-Fermi level is where the Fermi level was in equilibrium (with respect to E_c), and the hole quasi-Fermi level is identical to the electron QFL, but the gradient of the QFLs indicates that we are out of equilibrium.

Assumptions

What assumption is involved in replacing the equilibrium Fermi level with two quasi-Fermi levels out of equilibrium?

In equilibrium, the probability that a state is occupied is given by:

$$f_0(E) = \frac{1}{1+e^{(E-E_F)/k_BT}}$$

We assume that the **same function** describes the occupation of states in the conduction and valence bands if we simply replace the Fermi level by the appropriate QFL:

$$f_C(E) = \frac{1}{1+e^{(E-F_n)/k_BT}}$$

$$1-f_V(E) = 1-\frac{1}{1+e^{(E-F_p)/k_BT}}$$

M.S. Lundstrom Quasi-Fermi levels

Finally, you might ask, "What assumptions are we making when we replace the Fermi level by a quasi-Fermi level?" In equilibrium, one can rigorously derive the Fermi function, an **occupation function** that describes the probability that states are occupied in equilibrium. Out of equilibrium we are assuming that the same type of function of energy describes the probability that a state out of equilibrium is occupied — we just need to replace the Fermi level with the apprpriate quasi-Fermi level. The form of the occupation function is assumed to be the same in both cases. In general, the occupation function can be a quite different function of energy out of equilibrium, especially when we are far from equilibrium. For these cases, we need to solve a more complicated equation called the **Boltzmann Transport Equation** to determine what the occupation function is. By assuming an occupation function that has the same form as the equilibrium Fermi function, we are assuming that we are not too far from equilibrium. This is sometimes called **quasi-equilibrium**.

25.5 Summary

To summarize, in equilibrium we have a Fermi level and a Fermi function, f_0, that determines how the states are occupied. Out of equilibrium, we still use a Fermi function, but with a quasi-Fermi level for electrons and a different quasi-Fermi level for holes. Our assumption is that we can still describe the occupation probability of states with a function of energy that has the same form as a Fermi function, but we must use the appropriate quasi-Fermi level. This is a near-equilibrium assumption that can fail if we are too far out of equilibrium.

We expect the quasi-equilibrium approximation to work as long as we are not too far away from equilibrium. Far from equilibrium conditions do occur when very large voltages or other inputs are applied. In those cases, the accuracy of this approximation should be considered carefully, but for most of the problems that we encounter, it is reasonable to assume that we can use electron and hole quasi-Fermi levels in a Fermi function to describe the occupation of states in the conduction and valence bands.

This lecture has been an opportunity to get comfortable with the concept of quasi-Fermi levels. In the next lecture we will mathematically solve the semiconductor equations for a few common out of equilibrium examples, and we will illustrate in one of those examples how we draw energy band diagrams with quasi-Fermi levels.

25.6 References

The term, quasi-Fermi level, was introduced by Shockley. Some prefer to use the term, electrochemical potential because the electrochemical potential is a thermodynamically well-defined quantity. For some discussion of this topic, see Sec. 4.1.2 in the classic book by Blakemore.

[1] J.S. Blakemore, *Semiconductor Statistics*, Dover Publications, Inc., 1987.

For a good discussion of quasi-Fermi levels and quasi-equilibrium, see Sec. 4.6 in the following book.

[2] J.A. del Alamo, *Integrated Microelectronic Devices*, Pearson, New York, 2018.

Lecture 26

Minority Carrier Diffusion Equation

26.1 Introduction

In this lecture we will solve the semiconducting equations under a set of specific simplifying assumptions. We will focus on minority carriers and show how we can mathematically solve the equations analytically by making suitable approximations.

The semiconductor equations are the starting point for the analysis of most semiconductor devices:

$$\frac{\partial p}{\partial t} = -\nabla \cdot \frac{\vec{J}_p}{q} + G_p - R_p$$

$$\frac{\partial n}{\partial t} = -\nabla \cdot \frac{\vec{J}_n}{(-q)} + G_n - R_n$$

$$\nabla \cdot \left(K_s \epsilon_0 \vec{\mathcal{E}}\right) = \rho\,.$$

These equations are difficult to solve in general. Energy band diagrams provide qualitative solutions. For quantitative solutions, we have two options.

(1) Direct numerical solutions by discretizing the equations and solving the resulting systems of equations numerically.
(2) Simplify the semiconductor equations appropriately so that they can be solved analytically.

In this lecture, we take the second approach. We will focus on **minority carriers** and will show how the semiconductor equations can be simplified to the **minority carrier diffusion equation** (MCDE), an important equation that describes PN junctions, bipolar transistors, and solar cells. Several examples of how to solve the minority carrier diffusion equation will be discussed.

Minority carrier diffusion equation (MCDE)

$$p = p_0 = N_A$$ (P-type semiconductor in low level injection)

$$\frac{\partial n}{\partial t} = -\nabla \bullet \left(\frac{\vec{J}_n}{-q} \right) + G_n - R_n$$ (electron continuity equation)

$$\frac{\partial n}{\partial t} = -\frac{d}{dx}\left(\frac{J_n}{-q} \right) + G_L - R_n$$ (1D, generation by light)

$$\frac{\partial \Delta n}{\partial t} = -\frac{d}{dx}\left(\frac{-qD_n \, d\Delta n / dx}{-q} \right) + G_L - \frac{\Delta n}{\tau_n}$$ (low-level injection, Δn >> n_0, no electric field)

$$\frac{\partial \Delta n}{\partial t} = D_n \frac{d^2 \Delta n}{dx^2} - \frac{\Delta n}{\tau_n} + G_L$$ (D_n spatially uniform)

M.S. Lundstrom Minority carrier diffusion equation

26.2 Minority carrier diffusion equation (MCDE)

Shown above is how we simplify the semiconductor equations to describe minority carriers. We assume a uniformly doped P-type semiconductor and focus on minority carrier electrons. We assume low-level injection, so the hole density is very nearly what it was in equilibrium; we don't need to solve an equation for p, which eliminates one of the three equations. We'll also assume that the semiconductor is neutral with no electric field, so we don't need to solve the Poisson equation, which eliminates a second equation. We do, however, need to solve the electron continuity equation for n.

We'll also assume that $n(x)$ varies only in one direction and write G as G_L to denote that the only generation process to be considered is the optical generation of minority carriers. These assumptions give us the electron continuity equation shown above in the third equation on the left.

For this problem with no electric field, the drift-diffusion equation only has a diffusion current component. Assuming that the diffusion coefficient is independent of position, we can pull it out of the derivative. The simplified electron continuity equation, shown above on the bottom line, is **the minority carrier diffusion equation**.

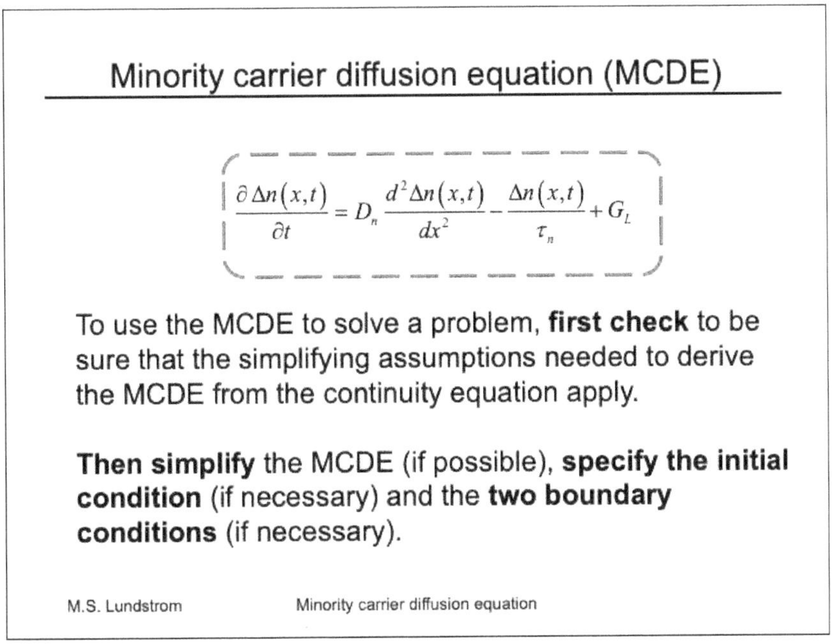

The equation shown above will be the starting point for the examples we will discuss. The minority carrier diffusion equation (MCDE) is a reasonably complicated partial differential equation itself. Since it has both time and space dependence, we'll need to specify an initial condition and two boundary conditions. When we encounter a minority carrier problem, we will go through the following four steps:

(1) Check that the simplifying assumptions we made to derive the MCDE are valid.
(2) Examine the problem and specify the problem-specific initial condition and boundary conditions.
(3) Simplify the MCDE for the specific problem being solved.
(4) Solve the MCDE.

Reminder: Low level injection

P-type semiconductor:

$$p(x,t) \approx p_0 = N_A$$

$$n(x,t) \approx \Delta n(x,t) >> n_0 = n_i^2 / p_0$$

$$\frac{\partial \Delta n(x,t)}{\partial t} = D_n \frac{d^2 \Delta n(x,t)}{dx^2} - \frac{\Delta n(x,t)}{\tau_n} + G_L$$

N-type semiconductor:

$$n(x,t) \approx n_0 = N_D$$

$$p(x,t) \approx \Delta p(x,t) >> p_0 = n_i^2 / n_0$$

$$\frac{\partial \Delta p(x,t)}{\partial t} = D_p \frac{d^2 \Delta p(x,t)}{dx^2} - \frac{\Delta p(x,t)}{\tau_p} + G_L$$

M.S. Lundstrom Minority carrier diffusion equation

Just a reminder that we are assuming low-level injection in a P-type semiconductor. Excess carriers have an insignificant effect on the majority carrier holes, but the excess carriers overwhelm the equilibrium density of minority carrier electrons. We will be solving for the position- and time-dependent minority carrier electron concentration in this P-type semiconductor.

For an N-type semiconductor in low-level injection, things are similar. The excess carriers don't perturb the majority carrier electron concentration, but they greatly perturb the minority hole concentration, and we would solve a minority hole MCDE to address this problem.

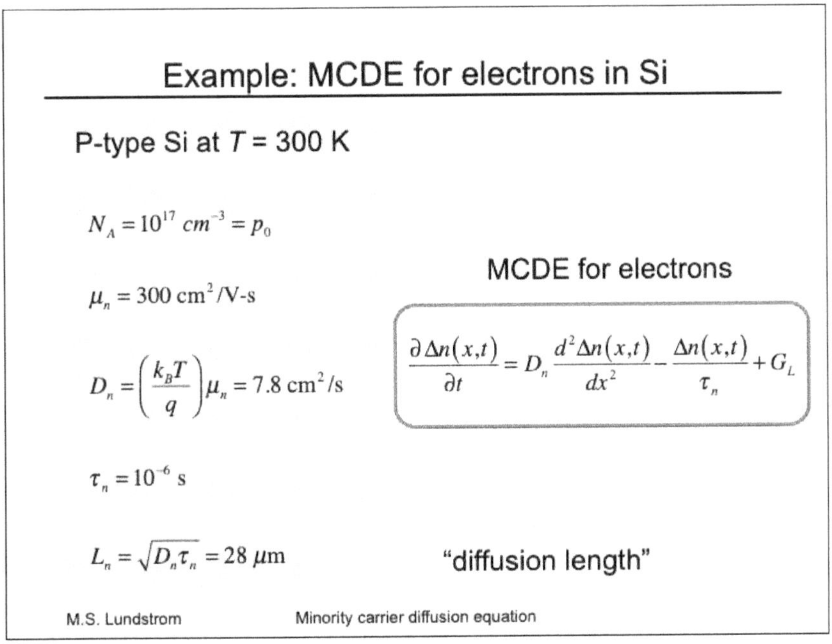

The examples in this lecture will be for minority carrier electrons in P-type silicon doped at $10^{17}\ \text{cm}^{-3}$. We'll assume a minority carrier electron mobility of 300 $\text{cm}^2/\text{V} - \text{s}$, which is a reasonable number for this doping density. We get the diffusion coefficient from the Einstein relation; it is 7.8 cm^2/s.

Because Si is an indirect gap semiconductor, the minority carrier lifetime is dominated by defect-assisted Shockley-Read-Hall processes. We'll assume that the minority carrier lifetime is one microsecond. This is not an extremely short lifetime, but it's also not particularly long either. (It is possible to get milliseconds lifetimes in high-quality crystalline silicon.)

We will find that a quantity called the **minority carrier diffusion length**, L_n, is very important in these problems. The diffusion length is approximately the distance a minority carrier can travel before recombining. The diffusion length, $L_n = \sqrt{D_n \tau_n}$, has units of meters in SI units. For the examples to be considered, $L_n = 28$ micrometers.

Let's now examine several examples of solving the MCDE using these material parameters.

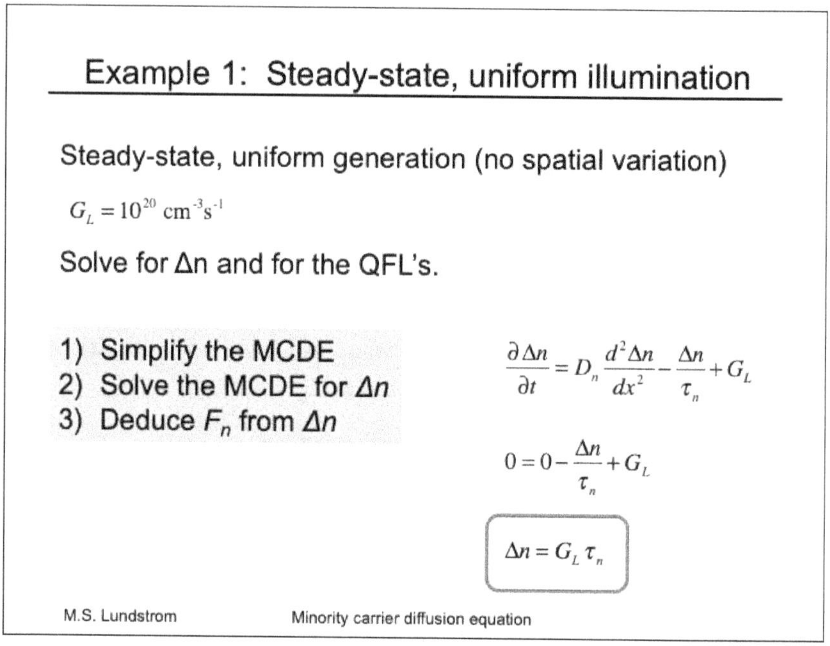

26.3 Ex. 1: Steady-state, uniform illumination

For Example 1, we assume steady state conditions ($\partial \Delta n/\partial t = 0$) while shining light on this P-type semiconductor. This is a uniform semiconductor with a uniform optical generation of carriers. The optical generation rate is $G_L = 10^{20}$ carriers/cm^3/s; we'd like to solve for the minority electron concentration and for the quasi-Fermi levels.

We first simplify the minority carrier diffusion equation for this specific problem, then solve it, and then deduce the quasi-Fermi level from the resulting excess minority carrier concentration.

We're in steady state, so there's no time dependence. This is a uniform problem, so there's no space dependence. As shown above on the right, the minority carrier diffusion equation simplifies greatly, and we can solve the equation easily. We find that the excess minority electron concentration is just the optical generation rate times minority carrier lifetime. It's worth remembering this result, because we will encounter it later.

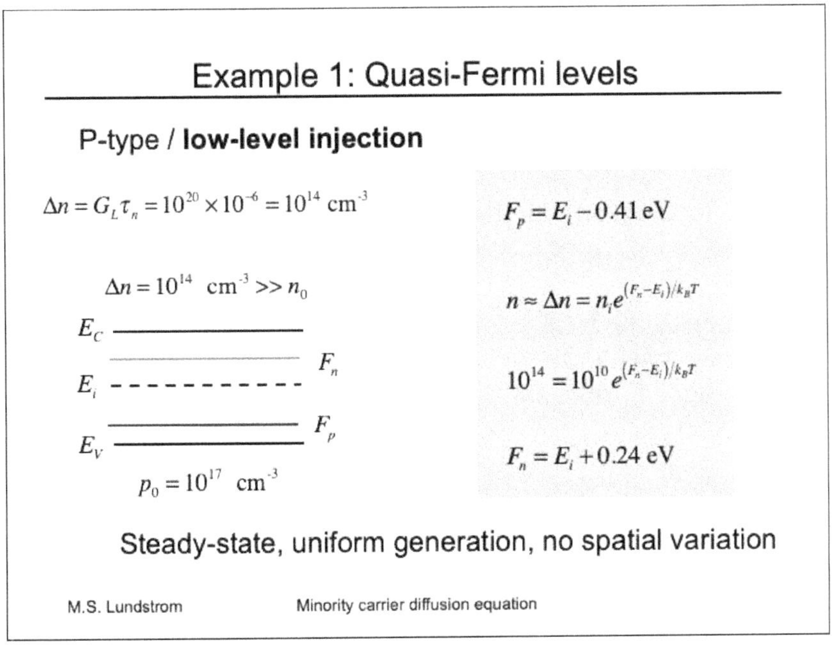

The excess minority electron concentration, $\Delta n = G_L \tau_n$, is spatially uniform and independent of time. Putting numbers in, we get 10^{14} excess carriers/cm^3. The uniform, steady-state optical generation has produced a uniform, steady-state concentration of minority carriers.

The next step is to find the quasi-Fermi level, but first, let's go back to the equilibrium case before we turned the light on. We had a P-type semiconductor doped at $N_A = 10^{17}\,\mathrm{cm}^{-3}$. The minority electron concentration in equilibrium was $10^3\,\mathrm{cm}^{-3}$. We can deduce the Fermi level from p_0 and find that E_F is 0.41 eV below the intrinsic level. This is the situation before we turn the light on.

Out of equilibrium nothing really changes for the holes, so F_p is where the Fermi level was in equilibrium. The hole quasi-Fermi level is 0.41 eV below the intrinsic level, just where the Fermi level was in equilibrium.

For electrons, there is a big change. We relate the electron density to the electron quasi-Fermi level and find that F_n is 0.24 eV above the intrinsic level. This reflects the fact that we have many more electrons now than we did in equilibrium. The quasi-Fermi levels are split; $F_n > F_p$, so $np \gg n_i^2$.

Example 2: Transient decay to equilibrium

Now turn off the light.
Transient, no generation, no spatial variation

Solve for $\Delta n(t)$ and for the QFL's.

1) Simplify the MCDE
2) Solve the MCDE for Δn
3) Deduce F_n from Δn

$$\frac{\partial \Delta n(x,t)}{\partial t} = D_n \frac{d^2 \Delta n(x,t)}{dx^2} - \frac{\Delta n(x,t)}{\tau_n} + G_L$$

$$\frac{\partial \Delta n}{\partial t} = 0 - \frac{\Delta n}{\tau_n} + 0$$

$$\frac{\partial \Delta n}{\partial t} = -\frac{\Delta n}{\tau_p}$$

$$\boxed{\Delta n(t) = \Delta n(0) e^{-t/\tau_n} = 10^{14} e^{-t/\tau_n}}$$

M.S. Lundstrom Minority carrier diffusion equation

26.4 Ex. 2: Transient decay to equilibrium

For Example 2 we'll assume that the semiconductor has been illuminated for a long time, and then we'll turn the light off at time, $t = 0$. This is a transient problem with no optical generation. We didn't have any spatial variation to begin with, and there's nothing to induce a spatial variation as time progresses, so we won't have any spatial variations here either. Again, we'll simplify the MCDE, solve it for $\Delta n(t)$, and then deduce the quasi-Fermi level as a function of time from $\Delta n(t)$.

The MCDE is the first equation above on the right. There is a time dependence, but no spatial dependence, and no optical generation, so the equation simplifies. We have a simple differential equation and recognize that the solutions go as $\exp(-t/\tau_n)$. The solution at $t = 0$ will be $\Delta n(t = 0) = 10^{14}\,\text{cm}^{-3}$, which we obtained when the light was on, and then it will decay as $\exp(-t/\tau_n)$.

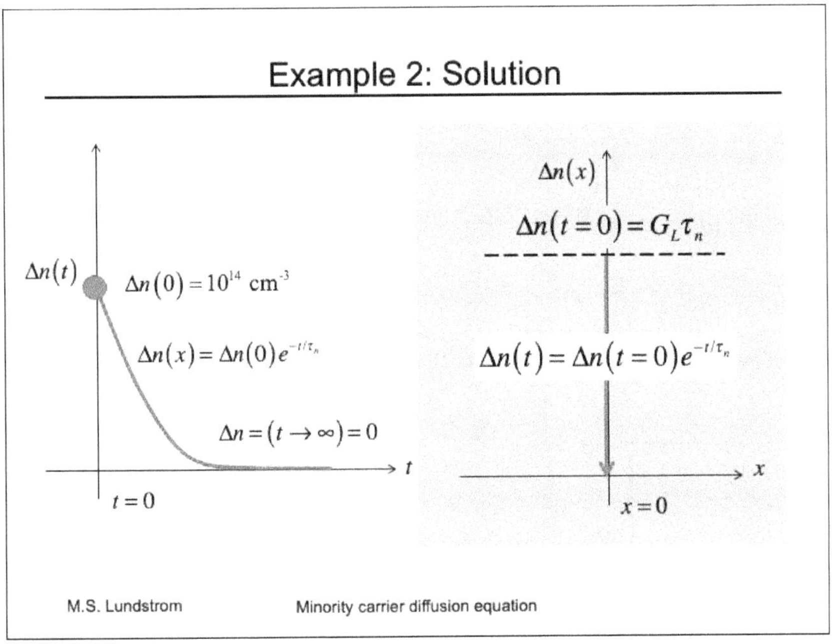

Shown above on the left, is a plot of the excess minority carrier concentration vs. time. It began at $10^{14}\,\text{cm}^{-3}$ when the light was on. When we turn the light off, Δn decays as $\exp(-t/\tau_n)$. Long after the light is switched off, equilibrium is restored.

Shown above on the right is a plot of the excess minority electron concentration vs. position. At time, $t = 0$, the minority carrier concentration was $10^{14}\,\text{cm}^{-3}$, and that number just decays as $\exp(-t/\tau_n)$ until the excess carriers are gone.

We have assumed a specific minority carrier lifetime of one microsecond, but for silicon, τ_n can vary from less than one microsecond to more than one millisecond depending on the concentration of defects.

The final question for Exercise 2 is how the quasi-Fermi levels vary with time.

Example 2: Solution

$$F_p = E_i - 0.41\,\text{eV}$$

$$n(t) \approx \Delta n(t) = n_i e^{(F_n(t)-E_i)/k_BT}$$

$$10^{14} e^{-t/\tau_n} = 10^{10} e^{(F_n(t)-E_i)/k_BT}$$

$$F_n(t) = E_i + k_BT \ln\left(10^4\right) - k_BT \frac{t}{\tau_n}$$

$$F_n(t) = F_n(t=0) - k_BT \frac{t}{\tau_n}$$

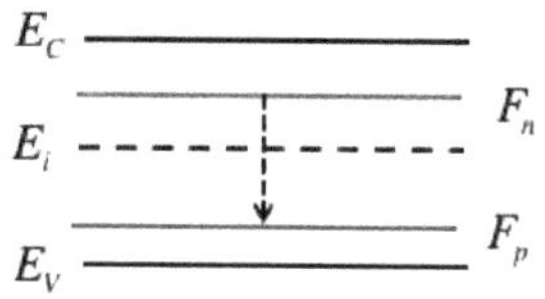

For long times, F_n should approach the equilibrium Fermi level. Explain what is wrong with our answer in the long time limit.

M.S. Lundstrom Minority carrier diffusion equation

Shown above on the right is the energy band diagram showing the quasi-Fermi levels when the light was turned on at time $t = 0$. We turn the light off and ask what happens. Nothing will happen to the hole quasi-Fermi level, because it's the same in and out of equilibrium, but when equilibrium is restored, the electron quasi-Fermi level has to end up where the hole quasi-Fermi level is and they both will be equal to the equilibrium Fermi level.

There's a relation between $n(t)$ and $F_n(t)$. We also know that $n(t)$ is, to a good approximation, equal to the excess minority electron concentration, $\Delta n(t)$ as long as $\Delta n(t) \gg n_0$. As shown above by the equation on the bottom left, we find that $F_n(t)$ decays linearly with time. Eventually, it reaches the quasi-Fermi level for holes and everything settles back into equilibrium.

The electron cannot drop linearly with time forever; it eventually must merge with F_p. What's missing in the equation for $F_n(t)$ above? See if you can figure out how to fix the result so that if behaves properly as time approaches infinity.

The general point is that after we solve the minority carrier diffusion equation we can always translate the minority carrier diffusion concentration vs. time to a quasi-Fermi level vs. time.

Example 3: SS diffusion in a long sample

Steady-state, sample is long (200 micrometers) compared to the diffusion length. No generation.

$$\Delta n(x=0)=10^{12}\ \text{cm}^{-3}$$
$$\Delta n(x=L)=0\ \text{cm}^{-3}$$

fixed

$$\frac{\partial \Delta n}{\partial t}=D_n\frac{d^2\Delta n}{dx^2}-\frac{\Delta n}{\tau_n}+G_L$$

Solve for Δn and for the QFL's.

$$0=D_n\frac{d^2\Delta n}{dx^2}-\frac{\Delta n}{\tau_n}+0$$

1) Simplify the MCDE
2) Solve the MCDE for *Δn*
3) Deduce F_n from *Δn*

$$\frac{d^2\Delta n}{dx^2}-\frac{\Delta n}{D_n\tau_n}=0$$

$$\frac{d^2\Delta n}{dx^2}-\frac{\Delta n}{L_n^2}=0 \qquad L_n\equiv\sqrt{D_n\tau_n}$$

diffusion length

M.S. Lundstrom Minority carrier diffusion equation

26.5 Ex. 3: Steady-state diffusion in a long semiconductor

Example 3 is a steady state problem. The semiconductor is very long; by which we mean long compared to the minority carrier diffusion length that we introduced earlier. We'll assume that the semiconductor is 200 micrometers long. There is no optical generation in this problem. We must specify two boundary conditions. For the first boundary condition, we fix the excess minority electron concentration at $x = 0$ to 10^{12} cm^{-3}. For the second boundary condition, we fix the excess minority electron concentration at $x = 200$ micrometers to zero. The problem is to solve for the electron concentration and the quasi-Fermi level vs. position.

We begin with the MCDE, the first equation shown above on the right. This is a steady state problem, so there's no time dependence. There will be a spatial dependence, so we have to retain the second derivative with respect to position. Recombination will occur, so we can't throw the recombination term out, but there is no generation, so we can set $G_L = 0$. If we rearrange the simplified equation, we can write it as shown on the bottom right, where we have introduced the **minority carrier diffusion length,** L_n.

Example 3: Continued

Steady-state, sample **long** compared to the diffusion length. No generation.

$$\Delta n(x=0) = 10^{12}\ \text{cm}^{-3}$$
$$\Delta n(x=L) = 0\ \text{cm}^{-3}$$

fixed

$$\frac{d^2\Delta n(x)}{dx^2} - \frac{\Delta n(x)}{L_n^2} = 0 \qquad L_n \equiv \sqrt{D_n \tau_n}$$

$$\Delta n(x) = Ae^{-x/L_n} + Be^{+x/L_n}$$

$$\Delta n(x) = Ae^{-x/L_n}$$

$$\Delta n(x) = \Delta n(0)e^{-x/L_n} = 10^{12}e^{-x/L_n}\ \text{cm}^{-3}$$

M.S. Lundstrom Minority carrier diffusion equation

For this example, we have two boundary conditions and a simple second order differential equation to solve. The first boundary condition assumes that an excess minority carrier concentration at $x = 0$ exists. This may be produced, for example, by shining light on the sample, which is absorbed very near the surface. The second boundary condition is that Δn at $x = L$ is zero. A properly designed metal contact can ensure that minority carriers that reach the contact quickly recombine.

For this example, the $L = 200\ \mu$m length of the semiconductor is much longer than the 28 μm diffusion length. We can think of L as approaching infinity, so we'll interpret the second boundary condition to be that $\Delta n(x)$ approaches zero as x approaches infinity.

The solutions to this differential equation are $\exp(+x/L_n)$ and $\exp(-x/L_n)$. Both of those solutions satisfy the second-order differential equation. Plug either one into the differential equation and you will see that it is a solution. For long L, only $\exp(-x/L_n)$ satisfies the second boundary condition that Δn approaches zero as L approaches infinity, so we can throw out $\exp(+x/L_n)$. We specify the value of A to satisfy the first boundary condition. The results is that $\Delta n(x) = \Delta n(x = 0)\exp(-x/L_n)$.

We have a simple exponential solution for this example.

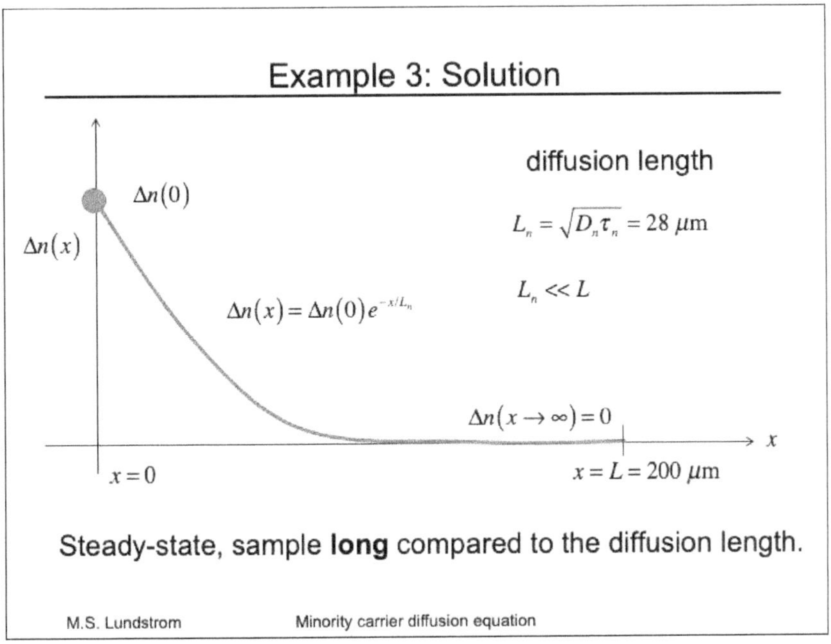

Shown above is a plot of the steady state excess electron concentration vs. position. When the length of the sample, L, is much greater than the diffusion length, L_n, then the solution simply decays from its value at $x = 0$ as $\exp(-x/L_n)$. The characteristic length for problems like this when carriers are diffusing spatially is the diffusion length, which can be interpreted as the average distance that a minority carrier diffuses before it recombines. When the semiconductor is very long, the spatial variations take place on the scale of the minority carrier diffusion length. **We find exponential solutions to the MCDE when the semiconducting region is long compared to a diffusion length.**

We could translate the excess minority carrier electron concentration into a quasi-Fermi level vs. position plot. I encourage you to think about that – what does $F_n(x)$ look like? What does $F_p(x)$ look like? We'll see this in another example later.

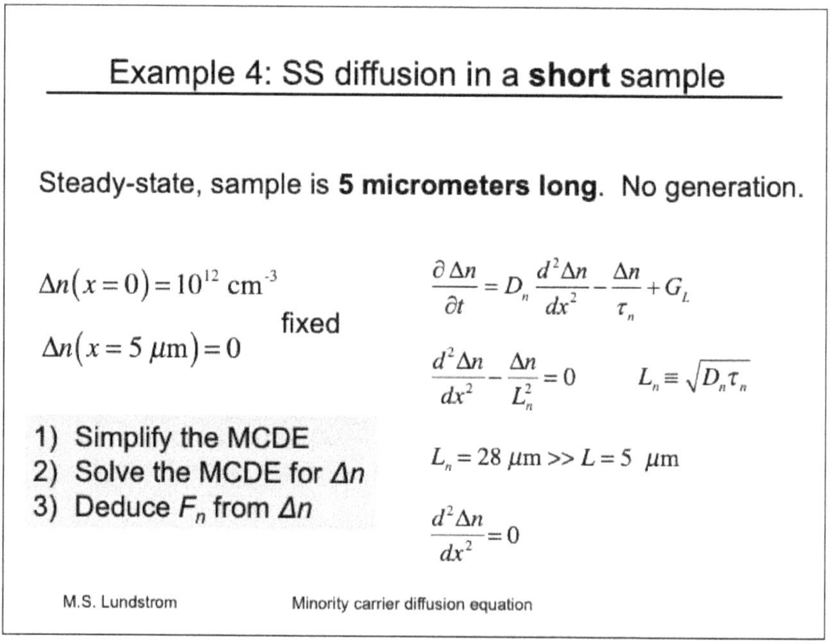

26.6 Ex. 4: Steady-state diffusion in a short semiconductor

For Example 4 the length of the semiconductor is much shorter: $L = 5\ \mu\text{m} \ll L_n = 28\ \mu\text{m}$, which qualitatively changes the solution.

Although the semiconductor is short compared to the diffusion length, the boundary conditions are the same. As shown above on the right, we begin with the same differential equation, the MCDE, but because $L_n \gg L$, carriers will diffuse across with little chance of recombining. The recombination term in the MCDE can be dropped, and we get the even simpler equation shown above as the last equation on the right.

The solution to this differential equation is not an exponential; in fact, you can see that a linear function satisfies that equation:

$$\Delta n(x) = Ax + B\,.$$

(Just plug $Ax + B$ into the differential equation, and you will see that it satisfies the equation.) The two constants, A and B, are determined by the two boundary conditions. First, set $x = 0$ to see that $B = \Delta n(x = 0)$, one of the boundary conditions. Next, set $x = L$, to see that $\Delta n(x = L) = 0 = AL + B$, which we can solve for A. Instead of an exponentially decaying solution for $\Delta n(x)$, we get a linear solution, $\Delta n(x) = \Delta n(0)\,(1 - x/L)$.

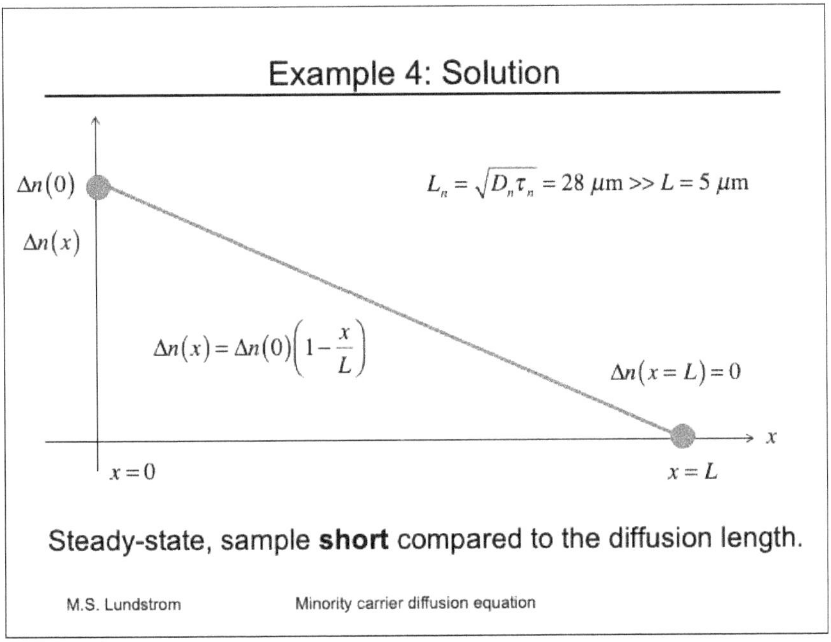

Shown above is $\Delta n(x)$ vs. x for Example 4. The minority carrier concentration goes linearly from its value at the left to its value at the right. **We find linear solutions to the MCDE when the semiconducting region is short compared to a minority carrier diffusion length.** Physically, what that says is that if electrons enter the semiconductor at $x = 0$, they diffuse across with essentially no chance of recombining. If we evaluate the flux of electrons diffusing, $F = -D_n(d\Delta n(x)/dx)$, it is the same everywhere. At any location, what diffuses in diffuses out, no electrons are lost due to recombination.

Be careful not to confuse this example with ballistic transport. The length of the sample is assumed to be much greater than the mean-free-path but much shorter than the diffusion length. Minority carriers diffuse across the semiconductor.

Example 5

Steady-state, sample is **30 micrometers long**. No generation.

$\Delta n(x=0) = 10^{12}\ \text{cm}^{-3}$

$\Delta n(x = 30\ \mu\text{m}) = 0$

fixed

$$\frac{\partial \Delta n}{\partial t} = D_n \frac{d^2 \Delta n}{dx^2} - \frac{\Delta n}{\tau_n} + G_L$$

$$0 = D_n \frac{d^2 \Delta n}{dx^2} - \frac{\Delta n}{\tau_n} + 0$$

1) Simplify the MCDE
2) Solve the MCDE from Δn
3) Deduce F_n from Δn

$$\frac{d^2 \Delta n}{dx^2} - \frac{\Delta n}{L_n^2} = 0 \qquad L_n \equiv \sqrt{D_n \tau_n}$$

$$L_n = 28\ \mu\text{m} \qquad L = 30\ \mu\text{m}$$

M.S. Lundstrom Minority carrier diffusion equation

26.7 Ex. 5: Steady-state diffusion when $L \approx L_n$

Let's consider a more general example. Example 3 was a long semiconductor in steady-state, and Example 4 was a short semiconductor in steady-state. For Example 5, the length is about one diffusion length, $L = 30\ \mu\text{m}$, which is approximately equal to L_n.

We begin with the same MCDE, eliminate the time derivative because we're in steady-state, and then set $G_L = 0$ because there is no optical generation. In this case, however, we have to retain the recombination term because recombination is not negligible.

The simplified MCDE for this problem is shown above as the third equation on the right. The solutions to this differential equation are $\exp(+x/L_n)$ and $\exp(-x/L_n)$, and we must retain both of these terms:

$$\Delta n(x) = Ae^{-x/L_n} + Be^{+x/L_n}\,.$$

To evaluate the constants, A and B, we use the two boundary conditions.

$$\Delta n(0) = A + B = 10^{12}$$

$$\Delta n(L) = Ae^{-x/L_n} + Be^{+L/L_n} = 0\,.$$

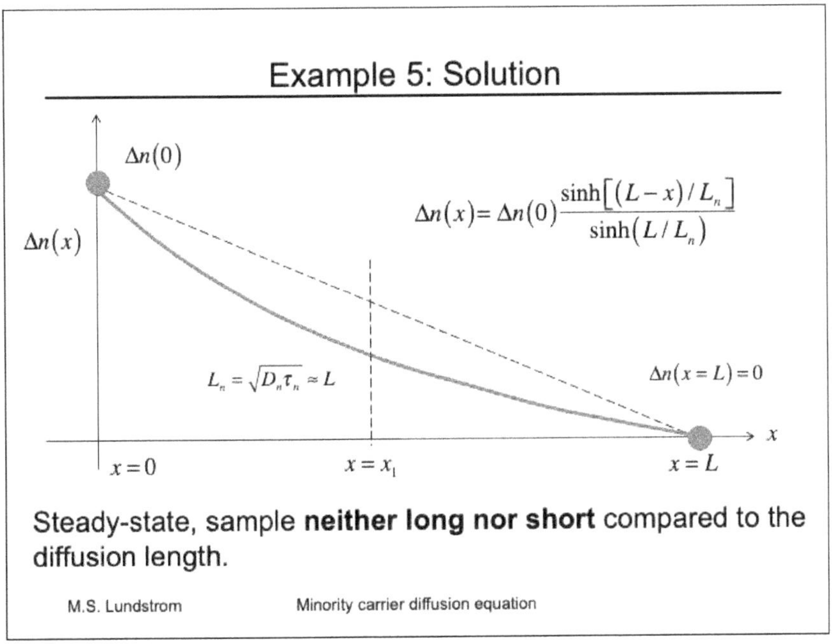

When we solve for A and B and express the result in terms of hyperbolic sines, we find the solution shown above. The graph shown above compares the solution to the linear solution of the previous example (the dashed line). At any point, x_1, we see that the gradient to the left of x_1 is larger in magnitude than the gradient to the right of x_1. More carriers are diffusing in than are diffusing out because some are recombining.

If we were to make the diffusion length shorter and shorter, the solution would become a decaying exponential (i.e. as the sample becomes long compared to the diffusion length). If we were to make the diffusion length longer and longer, the solution would become a straight line (i.e. as the sample becomes short compared to the diffusion length).

Semiconductor engineers frequently make the "long sample" or "short sample" approximation, because then it's easy to write down the solutions for those cases. The intermediate case takes a bit more work to solve, but the procedure is the same, and the final solution makes clear physical sense.

Example 6:

An infinitely long sample is uniformly illuminated with light for a long time. The optical generation rate $G_L = 1 \times 10^{20}\ \text{cm}^{-3}\ \text{sec}^{-1}$. The minority carrier lifetime is 1 microsecond. The surface at $x = 0$ is highly defective, with a high density of R-G centers, so that $\Delta n(0) = 0$.

Find the s.s. excess minority carrier concentration vs. position.

$$D_n \frac{d^2 \Delta n}{dx^2} - \frac{\Delta n}{\tau_n} + G_L = 0$$

Can we guess the solution?

M.S. Lundstrom Minority carrier diffusion equation

26.8 Ex. 6: Steady-state diffusion with surface recombination

For Example 6, the sample is infinitely long, uniformly illuminated with light, and in steady-state like Example 1. We'll assume the same optical generation rate as for Example 1 and a minority carrier lifetime of one microsecond, just as it was in all of the examples, but in this case, at $x = 0$, there is a surface, and we'll assume that the surface is highly defective. A highly defective surface will introduce defects in the forbidden gap that will facilitate the recombination of minority carriers. We'll assume that the surface is so defective that it forces the excess minority carrier concentration to be zero at $x = 0$. Any minority carrier that reaches the surface immediately recombines. The boundary condition is $\Delta n(x = 0) = 0$.

We'll get the same steady state equation that we've been dealing with in the previous two examples, but now with an optical generation term. This is an inhomogeneous differential equation with a forcing function, the optical generation term. We could solve this differential equation, but we can guess the solution using the intuition we've developed from earlier examples. Let's see how to do that.

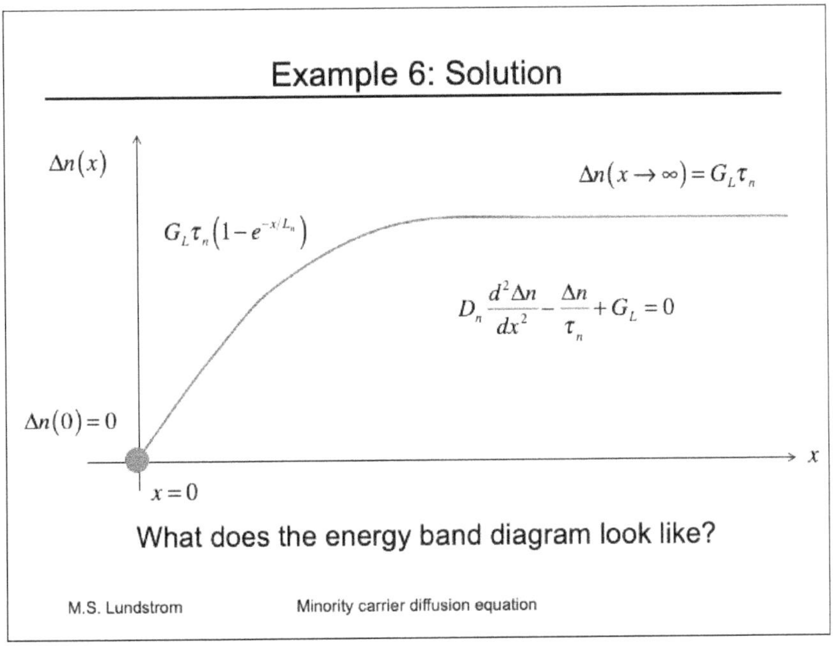

The slide above summarizes the problem and plots the expected solution. The defective surface forces the excess minority carrier concentration to be zero at the surface, that is the first boundary condition. Deep in the bulk, it looks like an infinite semiconductor with a uniform generation rate. We saw in Example 1 that for a uniformly illuminated infinite bulk semiconductor, $\Delta n(x) = G_L \tau_n$. How deep into the semiconductor do we need to go to find this solution? Well, it would be a few diffusion lengths deep because as we have seen, the characteristic length scale in these problems is the minority carrier diffusion length.

With this knowledge, we can sketch the solution. The excess electron concentration is zero at the surface. Deep in the bulk, it is $G_L \tau_n$. The solution will go smoothly from zero to $(G_L \tau_n)$ over a distance on the scale of a diffusion length. We can even guess the mathematical solution; it must be $\Delta n(x) = G_L \tau_n [1 - \exp(-x/L_n)]$. If we formally solve the MCDE, this is the answer we will find.

How do we draw an energy band diagram with quasi-Fermi levels for this case?

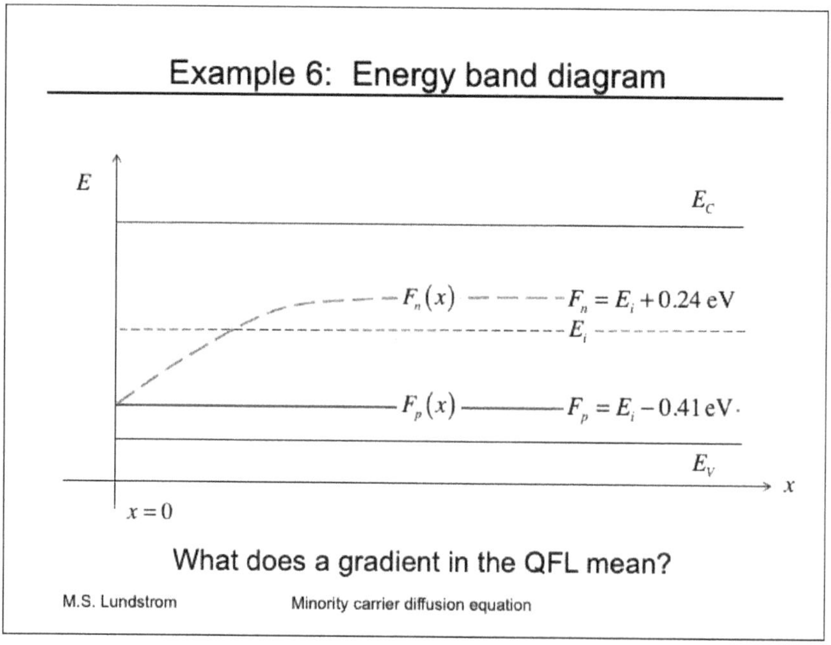

To draw the energy band diagram, remember that we have assumed that there is no electric field, so the bands are flat. This is a P-type semiconductor in low level injection, so F_p will be where E_F was in equilibrium, near the valence band. Deep in the bulk, the concentration of excess electrons is high, but at the surface, there are no excess carriers.

At the surface, where there are no excess carriers, $F_n = F_p$. Deep in the semiconductor, F_n must lie above the intrinsic level because the excess minority electron density is high. The QFL, F_n, will go smoothly from its value at the surface to its value in the bulk with the length scale being the minority carrier diffusion length. This reasoning allows us to sketch the QFL's as shown above. Note that from the solution for $\Delta n(x)$, we could find an equation for $F_n(x)$.

Looking at the energy band diagram above, we see a gradient in F_n near the surface, which means there's an electron current. This current is the flow of electrons to the surface. Electrons are generated in the bulk and diffuse to the surface where they recombine. There is also a gradient of F_p, but it is too small to see. The small gradient and the large hole concentration produces a flux of holes to the surface that is equal to the electron flux to the surface.

26.9 Summary

In this lecture, we have discussed how to solve the semiconductor equations under conditions for which we can simplify them so that analytical solutions are relatively easy to obtain. The particular simplification we discussed has to do with low-level injection of minority carriers in the absence of an electric field for which we can formulate a minority carrier diffusion equation (MCDE). The MCDE plays a key role in understanding the operation of minority carrier devices like solar cells or bipolar transistors [2].

While the MCDE can sometimes be complicated to solve, the qualitative form of the solutions are readily summarized. Transient solutions go exponentially with time with the characteristic time being the minority carrier lifetime. When the sample is long compared to a diffusion length, spatial solutions go exponentially with distance with the characteristic length being the diffusion length. When the sample is short compared to a diffusion length, spatial variations go linearly with distance.

This lecture completes Part 5 and the set of lecture notes. You should now:

- Be familiar with the three "semiconductor equations" that are the starting point for the analysis of most semiconductor devices.
- Understand how to qualitatively solve the semiconductor equations by drawing and reading energy band diagrams.
- Be able to explain what the electron and hole quasi-Fermi levels are and how they are used.
- Know when it is appropriate to simplify the semiconductor equations to the minority carrier diffusion equation (MCDE), how to do so, and how to solve the MCDE for important, simple cases.

26.10 References

For an excellent discussion of excess minority carriers in semiconductors including ambipolar diffusion and the inclusion of an electric field, see Chapter 10 in the classic semiconductor book by McKelvey.

[1] J.P. McKelvey, *Solid State and Semiconductor Physics*, Harper and Row, 1966.

For an introduction to minority carrier semiconductor devices such as bipolar transistors and solar cells, see:

[2] Robert F. Pierrt, *Semiconductor Device Fundamentals*, Addison-Wesley Publishing Co., 1996 (begin with Chapters 9, 10, and 11).

Epilogue

If we take the invention of the transistor at Bell Labs in 1947 as the starting point, semiconductor technology has been advancing and shaping the world we live in for more than 75 years. These lectures have been an introduction to semiconductor physics, to a way of thinking that has guided the development of the technology from the beginning. Since their invention in the middle of the 20th century, transistors, integrated circuits, and other semiconductor devices have transformed the world, but these lectures were not about devices or chips, instead, they were about the underlying physics that makes semiconductor technology possible. The treatment has been physical and intuitive — descriptive, not heavily mathematical. The lectures presented an electrical engineering perspective on semiconductors that others may find a useful introduction to the approach that has guided the development of this amazing technology.

My focus in these lectures has been on the "forest," not on the"trees." For those who use semiconductor devices, the lectures provide an understanding of the physics that underlies their operation. For semiconductor technology developers, the lectures provide a starting point for diving deeper into specific topics of condensed matter physics, statistical mechanics, thermodynamics, and materials science relevant to semiconductors. For those who have already taken advanced courses, the lectures provide a big picture view of how all the specifics come together. As one student told me, "now I understand what those advanced courses were all about."

I'm convinced that semiconductor technology will continue to advance for another 75 years. There will be many opportunities to address the world's grand challenges with new semiconductor technologies, and my hope is that the time you've spent with these lectures helps prepare you for that important work.

For those whose appetites have been whetted, a list of suggested references, some of them "semiconductor classics," follow.

The approach used in these lectures owes much to my long time colleague, the late Robert F. Pierret. Reference [1] below is on semiconductor physics; reference [2] also discusses devices.

[1] R.F. Pierret, *Advanced Semiconductor Fundamentals*, (2nd edition), Pearson Education, Inc., 2003.

[2] R.F. Pierret, *Semiconductor Device Fundamentals*, Addison Wesley Longman, 1996.

For an excellent, more recent treatment of semiconductor physics and devices, see:

[3] J.A. del Alamo, *Integrated Microelectronic Devices*, Pearson Education, Inc., 2018.

For a deeper treatment of semiconductor physics and devices, see the following:

[4] Sandip Tiwari, *Semiconductor Physics: Principles, Theory, and Nanoscale*, Oxford Electroscience Series, Vol. 3, 2020.

[5] Sandip Tiwari, *Nanoscale Device Physics*, Oxford Electroscience Series, Vol. 4, 2017.

[6] Debdeep Jena, *Quantum Physics of Semiconductor Materials and Devices*, Oxford University Press, 2022.

[7] Jasprit Singh, *Physics of Semiconductors and Their Heterostructures*, McGraw-Hill, Inc., 1993.

[8] B.K. Ridley, *Quantum Processes in Semiconductors*, 4th Ed., Clarendon, Press, Oxford 1999.

[9] John H. Davies, *The Physics of Low-Dimensional Semiconductors*, Cambridge University Press, Press, 1998.

Carrier transport, the flow of electrons and holes flow in semiconductors, is discussed in:

[10] S. Datta, *Lessons from NanoElectronics: A. Basic Concepts*, (2nd Ed.), World Scientific, 2017.

[11] Mark Lundstrom *Fundamentals of Carrier Transport*, 2nd Ed., Cambridge University Press, 2000.

Several excellent, classic semiconductor books are still, well worth consulting.

[12] J.S. Blakemore, *Semiconductor Statistics*, Dover Publications, Inc., 1962, 1987.

[13] A.C. Smith, J. Janek, and R. Adler, *Electronic Conduction in Solids*, McGraw-Hill, 1965.

[14] J.P. McKelvey, *Solid State and Semiconductor Physics*, Harper and Row, 1966.

[15] A.S. Grove, *Physics and Technology of Semiconductor Devices*, John Wiley and Sons, 1967.

[16] Charles M. Wolfe, Nick Holonyak, Jr., and Gregory E. Stillman, *Physical Properties of Semiconductors*, Prentice Hall, 1989., 2003.

Index

absorbing contact, 120, 237
absorption coefficient, 317
acceptor, 65
  deep and shallow, 68
  ionized, 66, 67
Airy function, 110
Auger recombination, 295
  and impact ionization, 321
avalanche multiplication, 320
avalanche photomultiplier, 320

B-coefficient, 291, 294
ballistic
  devices, 282
  limit, 239
  transport, 239
band bending, 345
  and inversion, 348
band gap, 10, 13, 43
  direct, 45, 53, 56
  indirect, 45, 55, 56
bandstructure plot vs. energy band diagram, 54
bandstructure plots, 52, 53, 56, 136
  GaAs, 143
  graphene, 144
  heavy hole, 142
  light hole, 142
  silicon, 139, 142
  split-off band, 142
  upper valleys, 143
black body radiation, 81
Bloch function, 134
Bohr model, 5, 83
Boltzmann transport equation, 236, 373
bonds
  covalent, 8
  dangling, 44, 148
breakdown electric field, 46
Brillouin zone, 135, 149, 150
built-in
  electric field, 286, 357
  voltage, 353

carrier concentration
  calculation in energy space, 182, 184
  calculation in k-space, 180
  equilibrium, 185, 188, 190, 203
  excess, 290
  extrinsic, 64
  extrinsic region, 215
  freeze-out region, 216, 225
  intrinsic, 64
  intrinsic region, 216, 218
  nondegenerate, 191
  relation to doping, 209, 213
  relation to Fermi level, 198
  relation to intrinsic level, 201, 202
  vs. temperature, 71, 212
channels, 236, 240

chemical potential, 169
computer aided design, 337
conductance, 267
conduction band, 11, 136
conductivity, 248, 250, 265, 270
constant energy surfaces
  ellipsoidal, 141
  spherical, 140
continuity equation, 331
  electrons, 332
  holes, 332, 333
Coulomb
  force, 5
  potential, 133
crystal momentum, 51, 54, 136, 145
crystal potential, 49, 132
current equation
  bulk semiconductor, 246, 248, 250
  diffusion, 273, 278
  drift, 251, 258, 265
  drift-diffusion, 281, 283, 287
  electrons, 341
  holes, 341
  hydrodynamic analogy, 232
  isothermal, 235
  Landauer approach, 236
  nano device, 234, 242
  relation to quasi-Fermi level, 249
  small bias, 245
  total current, 287, 341

de Broglie
  relation, 85, 91
  wavelength, 108
degeneracy factor, donors and acceptors, 222, 223
degenerate semiconductor, 175
density of states, 147
  1D, 151, 153
  1D vs. 2D vs. 3D, 161
  2D, 151, 156, 157
  3D, 151, 158
  effective mass, 159
  graphene, 162
  k-space, 148, 151, 181
  per subband, 155, 157
  vs. energy, 152
detailed balance, 338
diffusion coefficient, 46, 272
  microscopic definition, 275, 278, 284
diffusion current
  and general current equation, 278–280
  electrons, 273
  holes, 273
diffusion length, 277
diffusive
  limit, 239
  transport, 252
dispersion, 98, 136
displacement current, 287, 341
displacement vector, D, 334
distribution of electrons in the conduction band, 182, 183
divergence of a vector, 332
donor, 61
  binding energy, 62
  deep and shallow, 68
  ionized, 61, 222
doping
  amphoteric, 69, 70
  N-type, 57, 60
  P-type, 57, 65
  substitutional, 57, 60, 65
drift current, 251
  and general current equation, 268–270
  electrons, 258
  holes, 258
drift velocity, 253, 255–257
drift-diffusion equation, 281, 283, 287
  and general current equation, 283, 287

E(k), 52, 135, 145
effective density of states, 203
  conduction band, 185, 188, 190
  silicon, 191
  valence band, 189, 190
effective mass, 46, 48, 49, 56, 135
  density of states, 159

density of states, silicon, 191
longitudinal, 141
transverse, 141
Einstein relation, 279, 280, 284, 287
electrochemical potential, 169
electron
1D, 115, 116
2D, 109, 113, 116
wavevector, 90
force due to electric field, 51
momentum, 90, 132
rest mass, 48
temperature, 263
trap, 298
traveling wave, 95, 131
wave packet, 96
wavelength, 51, 90
electron concentration
equilibrium, 185, 188, 190
equilibrium, nondegenerate, 188, 190
energy
quantized, 104
energy and momentum, 50, 135
energy band diagram, 12, 343, 349
and carrier concentrations, 347, 349
and electric field, 347, 349
and electrostatic potential, 347, 349
and space charge density, 349
nonequilibrium, 358, 359
NP junction in equilibrium, 351, 353
NP junction under bias, 359
energy bands
ellipsoidal, 141
Kane bands, 138
nonparabolic, 138
parabolic, 137, 145
spherical, 140
energy levels, 5
epitaxy, 111
equilibrium vs. nonequilibrium, 329, 364
extrinsic region, 212, 213, 215

Fermi energy, 169
Fermi function, 167, 169, 170, 177, 179
at T = 0 K, 171
effect of temperature, 171
Fermi level, 169, 177
relation to carrier concentrations, 193
vs. temperature, 227
Fermi window, 241
small bias, 243, 244
window function, 244
Fermi-Dirac integral, 179, 186
order 1/2, 185
relation to exponential, 186, 187
Fick's law, 271, 272
field ionization, 309, 318
forbidden gap, 9
free carriers, 47
freeze-out region, 215, 220

Gauss's law
differential form, 335
integral form, 334, 335
generation
thermal, 305
grain boundaries, 25
group velocity, 98, 99

high field transport, 263
hole concentration
equilibrium, 189, 190
equilibrium, nondegenerate, 189, 190
hole trap, 298
holes, 11
hot carrier transport, 263
hot carriers, 14
transport, 263
hydrogen atom, 5

impact ionization, 309, 318, 319
and Auger recombination, 321
insulator, 15
intervalley transfer, 264

intrinsic carrier concentration, 10, 13
  GaAs, 43
  Ge, 43
  Si, 43
intrinsic Fermi level, 175, 199
intrinsic level, 199
  numerical value for silicon, 200
intrinsic region, 215, 217
intrinsic semiconductor, 13
ionized impurity scattering, 259

Landauer approach, 236, 282
lattice
  basis, 19
  body centered cubic, 20
  Bravais, 20
  constant for Si, 21
  crystal, 18
  diamond, 7, 21–23
  face centered cubic, 20
  nearest neighbor spacing,Si, 24
  simple cubic, 20
  unit cell, 19
  Zinc blende, 69
lattice scattering, 260
linear transport, 263
low-level injection, 292, 366

Matthiessen's rule, 262
mean-free-path, 238, 274
metal, 15
metal-insulator transition, 226
Miller indices, 29
  angle between two planes, 37
  direction normal to a plane, 35, 39
  equivalent directions, 39
  equivalent planes, 32, 39
  specific direction, 34, 39
  specific plane, 30, 39
minority carrier diffusion equation, 375, 376
  low-level injection, 378
  solution procedure, 377
  ss diffusion in a finite sample, 390
  ss diffusion in a long sample, 385
  ss diffusion in a short sample, 388
  ss diffusion with surface recombination, 392
  steady-state, uniform generation, 380
  transient decay to equilibrium, 382
minority carrier diffusion length, 379, 385
minority carrier lifetime, 307
  Auger, 296, 300
  defect-assisted, 299, 300
  radiative, 293, 300
  with multiple processes, 301, 307
mobility, 46, 259
  electron, 256, 257
  hole, 257
  ionized impurity scattering, 259
  lattice scattering, 260
  Matthiessen's rule, 262
  microscopic definition, 284
  relation to conductivity, 270
  vs. temperature, 261
modes, 240
momentum relaxation time, 256
  relation to mobility, 256

nanowire, 92
native oxide, 46
near-equilibrium, 233, 252, 287
  transport, 263
Newton's laws, 48, 49, 80
nondegenerate semiconductor, 172, 174
  N-type, 172
  P-type, 173
np product, 193, 203, 306
  numerical value for silicon, 197
  out of equilibrium, 329, 364, 369
  relation to band gap, 194, 195
  relation to effective density of states, 194, 195
  relation to temperature, 194, 195

occupation function, 373
optical absorption, 309
  direct gap semiconductor, 314
  direct vs. indirect gap, 316
  indirect gap semiconductor, 315

optical generation rate, 317
orbital, 5

particle in a box, 102
  finite barrier, 112
passivation, 44
permittivity of free space, 334
phase velocity, 98
photoelectric effect, 82, 310
photon, 14
Plank's constant, 81
Poisson equation, 336
Poisson-Boltzmann equation, 340
principal quantum number, 5

quantized current, 282
quantum confinement, 101
quantum mechanical reflection, 117, 128, 129
quantum mechanical tunneling, 117
quantum mechanics, 79, 99
quantum well, 109
quasi-equilibrium, 373
quasi-Fermi level, 247, 249, 329, 363
  and low-level injection, 367
  assumptions, 373
quasi-Fermi potential, 248

R-G center, 298, 305
reciprocal space, 134
  Brillouin zone, 135
recombination, 11, 45, 289
  and generation, 304
  Auger, 290, 295
  band-to-band radiative, 290, 291
  defect-assisted, 290, 297
  in a direct gap semiconductor, 302
  in an indirect gap semiconductor, 303
relative dielectric constant, 334
resistance, 267
resistivity, 265
  vs. doping density, 266

satellite valleys, 264
semi-metal, 144
semiclassical particle, 145
semiconductor, 15
  1D, 92
  2D, 93
  3D, 94
  polycrystalline, 25
  amorphous, 17, 26
  crystalline, 17
  degenerate, 175
  equations, 327, 330, 336, 337, 341
    equilibrium, 338
  group IV, 42
  II-VI, 42
  III-V, 42
  nondegenerate, 172–174
  polycrystalline, 17
  quasi-neutral, 357
semiconductor equations, 327, 330, 336, 337, 341
  and Maxwell's equations, 337
  equilibrium, 338
  unknowns, 328, 330
silicon dioxide, 44
solar spectrum, 311
space charge neutrality, 206
SRH recombination, 290
states
  de-localized, 9
  donor, 63
  electron, 5
  extended, 47, 63
  localized, 9, 63
  surface, 148
statistical force, 277
subbands, 113–116
surface states, 148

thermal energy, 10
thermal velocity, 252
thermalization, 312
thermoelectric effect, 235
transit time, 254
transmission, 236, 237
  relation to mean-free-path, 238, 239
  through a barrier, 124
  tunneling, 123

transport
  ballistic, 239
  diffusive, 252
  hot carrier, 263
  Landauer approach, 236, 282
  linear, 263
tunneling, 119
  dependence on height of barrier, 125, 129
  dependence on mass of the particle, 125, 129
  dependence on thickness of barrier, 125, 129
  gate oxide of a MOSFET, 126
  resonant, 127

uncertainty relations, 97, 99, 137
units
  eV, 4
  SI, 4, 74

valence band, 11, 136
valence electrons, 5, 6
valley degeneracy, 141
velocity
  drift, 253, 255–257
  group, 98, 99, 132, 137
  phase, 89, 98
velocity saturation, 264
vertical transitions, 314

wave equation, 79, 86
  time independent, 87, 99, 101, 131
wave function, 99
  Bloch wave, 134
  interpretation, 87, 99, 101
  particle in a box, 105
  solutions, 88, 95
  time independent, 87
  traveling wave, 131
wave function penetration, 112
wave packet, 85, 96, 137
wave vector, 90
wave-particle duality, 84
wavelength, 90
  de Broglie, 108
wavevector, 51
window function, 244
work function, 82, 310

Zener tunneling, 318

www.ingramcontent.com/pod-product-compliance
Lightning Source LLC
LaVergne TN
LVHW020754231125
825833LV00002B/129
* 9 7 8 9 8 1 9 8 1 8 3 6 5 *